UNEARTHED

UNEARTHED

THE ECONOMIC ROOTS OF
OUR ENVIRONMENTAL CRISIS

Kenneth M. Sayre

University of Notre Dame Press • Notre Dame, Indiana

Library of Congress Cataloging-in-Publication Data

Sayre, Kenneth M., 1928–
Unearthed : the economic roots of our environmental crisis /
Kenneth M. Sayre.
        p.   cm.
Includes bibliographical references (p.   ) and index.
ISBN-13: 978-0-268-04136-6 (pbk.)
ISBN-10: 0-268-04136-9 (pbk.)
1. Ecological engineering. 2. Energy dissipation—Environmental aspects.
3. Entropy—Environmental aspects. 4. Economic development—
Environmental aspects. 5. Energy conservation—Philosophy.
6. Energy consumption—Philosophy.   I. Title.
GE350.S39   2010
363.7—dc22

                                                    2010024338

to Jean and John,
    sister and brother

Contents

PART 3. Ethics and Economics

Preface

0.1 In a Nutshell

One common meaning of the term "unearth" is *to disclose* or *to reveal.*
The present study is intended, among other purposes, to reveal the basic
source of our environmental crisis. As the subtitle suggests, the crisis is
primarily of economic origin. Another common meaning is *to uproot* or
to eradicate. A further purpose of the study is to show how, if its roots
can be eradicated, the crisis can be contained and made less urgent. This
will require fundamental changes in our economic system.

A less common sense is *to be deprived of earthly existence.* This study
joins a growing number of other studies in concluding that the crisis, if
not contained, will lead to the demise of human society as we know it.
Homo sapiens might continue to roam the earth, but without the social
institutions that make us fully human.

From the perspective of the biosphere at large, human society has
proved to be a mixed blessing. Given the resources of cooperative inquiry, humanity is the only species capable of understanding the earth it
inhabits. And this understanding, applied humbly and wisely, has great
potential for the good of other creatures.

But humanity is also the most destructive species on earth. The economic practices that have spawned our current crisis are due in large

part to human contrivance. These practices not only have devastated vast extents of the earth's surface but have brought many species to the point of extinction.

Regrettable as it is, humanity is even capable of self-destruction. One doomsday scenario is destruction by nuclear warfare. A small portion of the world's stock of nuclear weapons would be enough to destroy the whole human race.

A less dramatic but more insidious scenario is the gradual destruction, by human action, of the ecological fabric on which society depends for its very existence. The present study examines this second scenario, with particular focus on how we fell into this predicament and on what we can do to extricate ourselves from it.

0.2 The Basic Problem

At first recognized only by a few far-sighted ecologists, the dangers of environmental degeneration have come into the limelight recently under the aspect of global warming. Climate change, however, is only part of the problem. And the extensive publicity given global warming has obscured other sorts of damage that human industry has imposed upon the biosphere. One such is depletion of the stratospheric ozone layer, allowing dangerous levels of ultraviolet radiation to reach the earth's surface. Although this problem was seriously addressed by the Montreal Protocol of 1987, holes in the ozone layer are still expanding in certain parts of the globe.

Another problem we hear of occasionally is the horrific release of toxins by human industry. These range from heavy metals accumulating in food chains, through synthetic poisons used in agriculture, to lethal wastes retained on the premises of nuclear power plants. Such toxins impose serious threats to the health of human beings, to say nothing of other creatures in the biosphere.

Problems of toxic waste, ozone depletion, and global warming are critical in their own right and fully justify any effort we can muster to combat them. What has yet to be generally recognized, however, is that these are only symptoms of a more fundamental problem—separate tips of the iceberg, so to speak. Even if global warming and the rest were solved individually, this deeper problem would remain unresolved.

Expressed as simply as possible, the fundamental problem is that human economic activity is generating far more degraded energy than the biosphere can cope with without becoming dysfunctional. If this impairment progresses much further, the biosphere will lose its ability to support human society as we know it.

The main reason this more basic problem has gone largely unnoticed may be that sciences dealing with the environment lack the vocabulary to express it. Climatology studies global warming, meteorology deals with ozone depletion, and toxicology treats the effects of poisons on living organisms. But even if the respective theories of these disciplines were somehow merged, they still could not articulate the underlying problem.

Given existing divisions among scientific disciplines, ecology seems closest to being up to the task. To describe the problem in its full complexity, however, ecology would have to join with thermodynamics. It would also have to enter territory often claimed by economics, insofar as economic production and consumption are the very activities with which the problem originates.

Let us restate the problem from an economic perspective. Both per capita and species-wide, human beings consume far greater amounts of energy than other creatures. This excess is due in large part to the energy consumed by industrialized economies, both in the production of goods and in their modes of consumption (e.g., use of automobiles). When this energy is used up, it does not disappear, but rather is transferred to the biosphere in degraded form.

As matters stand, the biosphere cannot cope with this massive influx of degraded energy and has responded by submitting to various forms of degradation itself (e.g., ozone depletion and global warming). The upshot is that the biosphere becomes progressively unable to support the activities primarily responsible for this excessive load of used-up energy. As human industry discharges ever increasing amounts of degraded energy, the biosphere's ability to support that industry is progressively diminished.

This constitutes the suicidal threat alluded to previously. Industrial activity is part and parcel of human society as we know it. As industrialized economies dump increasing amounts of degraded energy into the biosphere, their very existence becomes increasingly at risk. At the beginning of the twenty-first century, this process may already be too far advanced for industrialized society to disengage from its suicidal path. Unless we find other ways of living together that require less energy, human society as we know it seems doomed to extinction.

The present study undertakes to describe this predicament in sufficient detail to make it unmistakable to a receptive reader. Needless to say, not all potential readers will be receptive. To avoid being assailed by epithets like "luddite," "naysayer," and "prophet of doom," I have tried to lay out lines of argument sufficiently detailed that even skeptical readers will feel obligated to take them seriously. The general line of argument is summarized in the remainder of this preface.

0.3 Summary of Part 1

The central concept of this study is that of *entropy*, which originated in thermodynamics over a century ago and has since been taken over by many other disciplines.[1] As might be expected, use of the concept varies from discipline to discipline. Uses familiar in some contexts will seem out of place in others.

Chapters 1 and 2 aim at a "normalized" definition of entropy and related concepts, which is to say a definition accessible by all of the various disciplines involved in the study. In one basic form, entropy is energy that has lost its capacity for work. A common example is the low-grade heat generated by the metabolisms of living organisms. Another form of entropy is disorder, illustrated by a collapsed building or a dead tree. The connection between these two forms of entropy lies in the mutual convertibility of their high-grade counterparts. High-grade energy can be used to produce orderly structure (as in the construction of machines and buildings), and orderly structure can generate usable energy (as dammed-up water can be used to produce electricity).

Thermodynamics is founded on two fundamental laws of nature. The First Law of Thermodynamics is that the total amount of energy in a closed system (e.g., the total universe) remains constant. The Second Law is that the amount of entropy (degraded energy and disorder) in a closed system tends to increase with time. Taken together the two laws entail that energy in usable forms is constantly being expended and that entropy (useless energy) at the same time is constantly increasing.

These laws deal with measurable quantities. Expended energy is measured in standard units like calories and joules. Disorder in turn is measurable in quantities defined by information theory, which has been shown closely related to thermodynamics. The quantities in question are explained in the appendix to chapter 2. The main purpose of this appendix is to demonstrate that degrees of order in fact can be measured quantitatively. Readers not concerned with this issue can skip the appendix without losing track of the ongoing argument.

With one major exception, all physical processes are marked by an expenditure of usable energy and a corresponding increase in entropy. The major exception is the life process itself. Chapters 3 and 4 deal with relevant details. An essential feature of a living organism is its ability to extract useful energy and order from its environment, in exchange for the resulting entropy that the environment must somehow get rid of. From the viewpoint of thermodynamics, life is basically a process of gaining energy and sloughing off the resulting entropy.

A consequence is that life is possible only in environments (1) capable of providing energy in forms and amounts needed to run the metabolisms of resident organisms and (2) capable at the same time of carrying off the entropy resulting from this metabolic activity. The more complex the organisms involved, the more entropy they produce for their ecosystems to deal with. As far as we know, human beings are the most complex organisms of all.

At this stage in human development the environment supporting human life extends to the limits of the biosphere itself. This means that the entire biosphere is involved in the interchange of energy and entropy underlying human activity. It also means that human life as we know it can continue only as long as the biosphere is capable of dealing with the entropy that human activity inevitably generates.

For most of its time on Earth, the human race has relied primarily on energy sources that can be renewed as quickly as they are depleted. Most notable among these are various forms of biomass (such as cereals and wood) that are replenished by photosynthesis. Given the adaptive processes involved in a healthy ecosystem, the biosphere generally has been able to rid itself of all the entropy produced by its constituent organisms. This is accomplished ultimately by radiation of low-grade heat into space. In effect, the biosphere's ability to discharge used-up energy back into space has remained in balance with its ability to make fresh energy available for use by the life-forms residing within it.

With our growing reliance on fossil fuels over the past few centuries, however, this balance has been radically disrupted. Humankind has begun producing far more entropy than the biosphere is able to get rid of. The problem with fossil fuels is not just that (being nonrenewable) they will eventually be used up; more ominous in the long run is the fact that use of fossil fuels leaves behind residues that clog up the biosphere. The result is not unlike what happens when the sewer backs up in an apartment building. Unless the flow of materials in and out is returned to equilibrium, the building (in our case, the biosphere) becomes unfit for human habitation.

The next two chapters examine the severity of the problem and discuss various forms of entropy that have been accumulating in the biosphere since the Industrial Revolution. Chapter 5 points out some of the more obvious forms in which entropy has become impacted within the biosphere. Most widely noted in recent years, to be sure, is the build-up of low-grade heat known as global warming. Equally pervasive, and no less ominous, are forms of disorder constituting disruptions in the water cycle, the accumulation of toxic substances in our landfills, and the loss of biodiversity in our food chains. To escape strangulation by the growing glut of entropy being dumped into our environment, we have to do more than cut down on the emission of CO_2 (a major cause of global warming). In whatever form, we have to stop dumping more entropy into the biosphere than it is capable of handling.

Chapter 6 presents data showing that per capita energy consumption increased in a roughly linear fashion between 10,000 BCE and the time of the Roman Empire, and then increased more rapidly up to the time of the Industrial Revolution. From that time to the present, per capita energy consumption has been growing exponentially. World population itself has been increasing in a comparable fashion. The result of these two trends acting together is that, whereas total human energy consumption initially doubled every four or five millennia, its average doubling rate during the twentieth century was once every twenty-five years. This amounts to a veritable explosion of human energy consumption.

A corollary of this growth pattern is another veritable explosion—in amounts of entropy that human activity has passed off into the biosphere for its disposal. But there has been no corresponding increase in the ability of the biosphere to get rid of this entropy. After scarcely three hundred years of heavy reliance on fossil fuel, we have fouled our nest to what may be a point of no return.

This is our environmental crisis in a nutshell. We are consuming energy at rates far exceeding the ability of the biosphere to dispose of the resulting entropy. This puts human society as we know it in jeopardy. The burning question at this point is what can be done about it.

0.4 Summary of Part 2

Our problem stems from the fact that the biosphere is suffering from excessive consumption of energy by human society. To confront the problem realistically, we need to pin down the forms of human activity primarily responsible for this excessive consumption. This brings up the subject of economics.[2]

Chapter 7 explores the link between energy consumption and economic production. Data available from many sources show that these two factors are directly related. A relevant measure in this regard is the gross product an economy can generate with a given quantity of energy input. In the early 1960s, for example, the U.S. economy produced about a dollar's worth of goods for each 60,000 BTUs of energy consumed. Among developing (not fully industrialized) countries, Ghana required only 20,000 BTUs for a dollar's production. These figures are part of a set of data showing that developed countries are less productive in terms of energy efficiency. Although the amounts of goods they produce are unusually much higher than in developing countries, this is accomplished by consuming prodigious amounts of energy.

A consequence of this coupling of economic production to energy consumption is that volume of entropy discharged increases with volume of economic goods produced. Given that large portions of this entropy currently are retained within the biosphere, this means that increasing economic production results in progressively greater ecological damage. A neglected economic principle (one not found in standard textbooks) is that, under current conditions, a given quantity of economic production typically results in a corresponding quantity of environmental degradation.

Given that economic production is the main source of the problem, what should we try to do about it? Any effective remedy will have to reduce the burden of entropy imposed by human enterprise upon the environment. Three possible strategies are (1) repairing environmental damage on a piecemeal basis by innovative technology, (2) replacing fossil fuel with clean energy, and (3) reducing economic production to levels the biosphere can tolerate. Strategies 1 and 2 are discussed in chapters 8 and 9 respectively. The remaining strategy is examined in chapters 10, 11, and 12.

An illustration of strategy 1 is the installation of smokestack scrubbers to cut down emissions of sulfur dioxide, a major cause of acid rain. Inasmuch as many environmental problems stem from industrial technology in the first place, it seems not unreasonable to think they can be solved by technological means. While this approach should be pursued whenever feasible, some environmental problems are too pervasive to admit remedies of this sort. One such problem is the massive disruption of the planet's water cycle, subjecting increasing numbers of people to conditions of drought and hunger. Although desalination technology (transforming seawater to potable form) can provide relief in isolated cases, shrinkage of fresh water supplies continues unabated in most parts of the globe.

Chapter 9 discusses the alternative (strategy 2) of substituting clean energy for fossil fuel. Although several forms of clean energy are currently available, the most highly publicized forms are solar and wind power. Using clean energy should be encouraged whenever possible; however, this strategy falls short of providing a panacea. One reason is that there are many uses (e.g., for aviation) in which fossil fuel is not replaceable by clean energy. Another is that use of clean energy produces low-grade heat no less than use of fossil fuel, so that under some conditions clean energy itself can contribute to global warming. Many environmental problems would remain even if clean energy replaced fossil fuels in all feasible applications.

The remaining strategy goes to the heart of the problem. For the past 250 years or so, industrial economies have been consuming energy at rapidly increasing rates. Entropy resulting from that expenditure of energy is the ultimate source of our present crisis. Strategy 3 calls for a radical reduction in the amounts of energy we commit to economic production.

This strategy obviously runs counter to the conventional economic doctrine that continued growth is necessary for a healthy economy. If this doctrine is to be trusted, we are confronted by a choice between a healthy economy and a healthy biosphere. Coming to terms with this doctrine is necessary for any workable solution to our environmental predicament. The last three chapters of part 2 are occupied with a detailed examination of this commitment to economic growth.

Chapter 10 traces the history of the concept of economic growth from the land-based economies of the medieval period, through the mercantile economies in place prior to the Industrial Revolution, up to the classical and neoclassical economies of the past few centuries. The latter era has been marked by an emphasis on the role of the consumer class and on the emergence of what came to be known as the free-market system. The chapter concludes with a discussion of neoclassical growth models, leading to the conclusion that neither economic history nor standard growth theory demonstrates that growth is essential to a healthy economy.

Although mainstream economists tend to take the desirability of growth as self-evident, various arguments are sometimes put forward in its favor. Chapter 11 examines some of the more notable among them.

These include arguments relating economic growth to size of population, arguments relating growth to quality of life, and a key argument dealing with the so-called trickle-down effect. On balance, none of these arguments is sufficiently compelling to show that the alleged social benefits of growth outweigh its costs in terms of environmental degradation.

The main topic of chapter 12 is a relatively new discipline known as ecological economics (EE). A distinctive feature of this discipline is that it takes issue with the mainstream, arguing against quantitative economic growth in favor of what it sees as qualitative development. Ecological economists typically are sympathetic with the claim that entropy resulting from excessive economic activity is causing severe damage to the biosphere. This encourages a careful look at EE to see what it can contribute to the resolution of our environmental predicament.

Chapter 12 begins by distinguishing between environmental and ecological economics, the former being an offshoot of the mainline variety. Next it traces antecedents of EE in the economic literature of the late 1960s and early 1970s. It then discusses the emergence of EE itself in the late 1980s and attempts to articulate the values ecological economists believe to be embodied in qualitative development. Included are values of justice, equity, and human dignity, along with values of environmental quality.

Advocates of EE sometimes talk of a "paradigm shift" in which their discipline would replace neoclassical economics in the formulation of economic policy. The chapter ends with reasons for thinking that such a shift will not take place. One reason is that EE is too eclectic to come up with a unifying paradigm capable of replacing the mainstream model of the competitive market. Another is that EE lacks the descriptive power of mainstream economics, being less concerned with how markets *actually* operate than with delivering advice (however sensible) on how they *should* operate.

The upshot of part 2 for the ongoing argument is that mainstream dedication to economic growth does not override the imperative that society at large must cut back on its economic production. Strategy 3 is the only path that leads back to environmental health. Part 3 addresses the difficult question of how society might make headway along this path.

0.5 Summary of Part 3

Part 3 is organized around the following considerations. Free-market economies continue to flourish at the expense of the environment.[3] Such economies flourish only in societies where particular kinds of social values are prevalent. Thus one effective way of halting environmental degradation is to neutralize the particular values that support free-market economies.

Paramount among such values is the high esteem we assign to wealth. Conditioned by the value society ascribes to wealth, people grow up thinking of wealth as desirable in itself. Chapter 13 argues that desire for wealth is the driving force behind modern economies generally, and in particular behind economies with a free-market orientation.

The primary beneficiaries of free-market economies are private parties in a position to influence market activity. These include corporate executives, financiers, and investors, along with public officials acting in behalf of private interests. It is desire for wealth on the part of such beneficiaries that provides the impetus toward growth in free-market economies. A similar account could be given of managed (e.g., communist) economies, but with a different list of beneficiaries.

To describe how this works, something needs to be said about the flow of goods in market economies. Generally speaking, goods flow from supplier to consumer, while money flows in the opposite direction. Goods are drawn in the former direction by consumer demand, which likewise sets up a flow of money back to the supplier. To the extent that these flows can be controlled by interested parties, control is exercised by manipulating consumer demand.

Consumer demand can be manipulated by various forms of marketing. Forms of marketing distinguished in chapter 13 include product presentation, preference management, and neuromarketing, representing increasingly invasive forms of control over the consumer. Money derived from consumption will be divided among capital investment, taxes, further marketing, and profit, allotted in a manner to maximize the latter. Desire for additional wealth thus leads to additional consumption, entailing additional production and ultimately increased damage to the biosphere.

This dynamic has led to gross inequities in the distribution of wealth, both within particular societies and among societies. For example, the average income of the wealthiest 1 percent in the United States at the end of the twentieth century was more than 400,000 times greater than that of one-fifth of the world's total population. The excessive economic activity that is destroying the biosphere is benefiting only a comparatively small group of privileged individuals.

A common reaction to disparities like this is to denounce them as morally unjust. Chapter 14 examines inequities of this sort through the lenses of environmental ethics. Although it seems antecedently unlikely that our environmental crisis can be resolved by ethical theory alone, the insights afforded by environmental ethics might nonetheless prove helpful.

Chapter 14 begins by identifying three moral quandaries with an environmental bearing: (1) whether it is morally acceptable for a few people to maintain lifestyles unavailable to the rest while causing enormous damage to our common biosphere, (2) whether these few are morally accountable for their wasteful use of resources that might be vital to future generations, and (3) whether our extensive destruction of other species can be morally justified. The chapter then singles out two ethical perspectives particularly germane to these issues and proceeds to evaluate their potential contributions.

The first perspective treated is that of utilitarianism, cofounded by the nineteenth-century economist John Stuart Mill. This choice is dictated not only by the economic origin of the quandaries, but also by the similarities between the so-called utilitarian calculus and the cost-benefit analyses employed by mainstream economists. Despite these affinities, however, utilitarianism turns out to be conceptually unprepared to deal with these particular quandaries or with their underlying environmental causes.

Next is the perspective of Aldo Leopold's *A Sand County Almanac*, which has been a dominant force in environmental ethics for the past several decades. Although his so-called Land Ethic is not a full-fledged ethical theory like utilitarianism, it has useful things to say about the quandaries in question. Most useful, perhaps, is that contemporary society has no moral guidelines in place pertaining to the interaction between humanity and the rest of the biosphere. In Leopold's estimation, ecological health depends on a "durable scale of values" curtailing human activities that are environmentally destructive.

Leopold's perspective reinforces the unifying thesis of chapter 13, to the effect that our environmental crisis can be traced back to the desire for wealth dominating free-market activities. The value contemporary society places on wealth has no place in a "scale of values" conducive to environmental health. Among other ecologically damaging values indicated in chapter 13 are those involved in the manipulation of consumer demand. Were it not for a large number of people who value possessions, for example, it would be hard to stimulate consumer demand by standard techniques of marketing.

The perception emerging at this point is that the root cause of our present crisis is not one or another sort of economic activity as such, but rather the pervasive social values that make free-market activity profitable. To the extent that this is so, the remedy appears to lie in somehow replacing society's current set of ecologically damaging values with others more conducive to environmental health. Here we enter into unexplored territory. How can we tell what values are ecologically damaging? And what can be done to replace these damaging values by others that are ecologically salutary?

These questions set the agenda for the remainder of the study. The first task of chapter 15 is to explain what it is for a given set of values to be operative (in force, current) in a given social context. A key point of the explanation is that a value is established as operative not by ethical reasoning but rather by coming to play certain roles in actual social behavior. Value replacement is a matter of certain values supplanting others in their social roles.

The other task of this chapter is to distinguish various roles that social values might occupy and to describe what is involved in occupying those roles. Of primary concern are (1) *approbatory* values that serve to sanction certain behaviors as socially acceptable, (2) *commendatory* values that encourage certain social behaviors, and (3) *normative* values that mark certain behaviors as right or wrong. An example of the first category is gratification. A society with this value in force will consider behavior aimed at gratification to be generally acceptable. Examples of the second and third categories are convenience and tolerance, respectively. Contemporary society encourages convenience in our choice of transportation. And it rules out intolerance as wrong and hence unacceptable. Of particular note is the status of wealth as both an approbatory and a commendatory value. As matters stand, pursuit of wealth is both accepted and socially encouraged, whereas under other circumstances it might be regarded as a social transgression.

Chapter 16 attempts to identify other values current in industrial society that have contributed significantly to our environmental crisis. One such is the approbatory value of gratification mentioned previously, which enables consumers to gorge themselves on tasty hamburgers without qualms of conscience. Among the adverse effects of this value is the fact that as much as a million acres of rain forest have been destroyed to produce the meat consumed in fast-food restaurants. Another adverse value is that of comfort, which has led to acceptance of air-conditioning as a "basic necessity." The use of ozone-depleting coolants in air-conditioning equipment has caused extensive damage to both terrestrial and aquatic food-chains. Still another is the value of convenience, also mentioned previously, which stands behind society's preference for private transportation. All said and done, the private automobile is probably responsible for more ecological damage than any other institution of contemporary society.

Chapter 17 responds with a survey of social values that show promise of promoting ecological health. The list of values under this heading is provisional in that the values in question remain largely untried in contemporary social contexts. Although their probable effects on the environment remain conjectural, there is ample reason to expect that these effects would be largely beneficial.

Given the urgent need to *restrain* our profligate use of environmental resources, most of these values will be either commendatory or normative (rather than approbatory). An example is the value of simplicity, which would encourage people to walk when possible rather than driving an automobile. Another is that of moderation, which would encourage people to look beyond taste in their choice of diet. As far as consumerism is concerned, a key example is the countervailing value of contentment. People content with what they have will not be motivated by the prospect of acquisition and thus will tend not to respond to techniques of preference management.

It should be clear by this stage of the study that our environmental crisis will not abate as long as social values like those of chapter 16 remain in force. Although the list of salutary values in chapter 17 is tentative, it should also be clear that if alternative values of this sort were in force instead, the biosphere would benefit appreciably. The final chapter of the study addresses the substantial question of how value change of this sort might be brought about. What might be done of a practical nature to change the values that shape society's interaction with the rest of nature?

Chapter 18 puts forward suggestions under three general headings. First is the motivational equivalent of fighting fire with fire. As things work presently, marketing is used to manipulate consumer preferences in ways that increase corporate profit. If appropriately funded, marketing could also be used to inculcate values of an environmentally friendly sort. Instead of selling beer and cigarettes, that is to say, ads could be tailored to encourage people to prefer walking over driving large vehicles.

Second is the heading of individual action. Once people become aware of the environmental consequences of their personal activities (become aware of their "environmental footprint," so to speak), they will think of (a) things they do presently that should no longer be done (e.g., paying attention to manipulative advertising), (b) things they do currently that should be done differently (e.g., cooling homes by natural ventilation rather than air-conditioning), and (c) things not currently done that they should begin doing (e.g., restricting food purchases to local produce). Personal values reinforced by such activities have a way of spreading to friends and neighbors.

Third is the heading of cooperative action with like-minded individuals. One possibility is to form alliances that withhold patronage from organizations whose operations are ecologically damaging while supporting those that conduct business on an environmentally sound basis. The more people interact in such activities, the more prominent become the values motivating them within society at large.

Having progressed this far in the summary, some readers may be convinced that the crisis is urgent and want to go directly to chapter 18 for thoughts on what to do about it. If so, well and good. Nonetheless, the conclusion that the survival of human society as we know it depends upon renouncing the values of consumerism is sufficiently counterintuitive to require sustained argument for its justification. For readers not convinced antecedently, it is best to pick up the argument from its beginning.

PART 1

Entropy and Ecology

1

Two Laws of Thermodynamics

1.1 Things Run Down

Schoolchildren are sometimes shown movies of growing plants, speeded up to make leaves and blossoms unfold in a few brief moments. Imagine such a movie skipping forward past the growing period to the point where leaves and pedals begin their quick descent to the ground. Our present concern is illustrated by this latter part of the movie.

Take the case of a mature gingko tree whose leaves have yellowed and are about to fall. A notable thing about gingko trees is that they tend to lose the bulk of their leaves in a brief golden shower. Having brought a camcorder to the scene at the opportune moment, we capture the motion of the leaves as they fall to the ground. By speeding up the display we can watch the tree shed its leaves in what appears to be just a matter of seconds.

Let us consider the distribution of leaves before and after their fall. During the summer months the leaves were distributed in an orderly manner along the tree's branches. After their fall they are spread randomly around the base of its trunk. In filming their transit from the branches to the ground, we have recorded a progression from order to relative disorder.

Once this progression has been recorded, of course, we can view it in reverse by running the display backwards. A backward viewing of the falling leaves would show them streaming upward to rejoin their branches. Seen in rapid sequence, this backward flow of leaves would be dramatically opposed to the way things work in the ordinary world. In the ordinary world, order generally gives way to disorder. Change in the other direction runs distinctly contrary to the observable course of nature.

Next imagine that we are viewing a movie of an erupting volcano. First we see rock from the mountain top blasted high in the air and lava beginning to surge over the lip of the crater. Then we watch falling rocks destroying acres of forest and flowing lava igniting everything that stands in its path. Steam appears when the lava encounters water, and smoke fills the air above the shattered peak.

On this occasion we have observed a series of transformations brought about by a massive discharge of energy. The force of the erupting magma imparts kinetic energy to the rocks being lofted outward from the peak. At the height of their trajectory this energy receives a gravitational boost before their energy is dissipated in the destruction of the trees. Similarly, the high-grade thermal energy of the lava sends up columns of glowing particles and smoke until finally giving way to low-grade heat. What we see, from the perspective of energy transformation, is a process of energy being dissipated as these effects are accomplished.

Our display of these events could also be reversed in the manner of the movie of the falling leaves. One point of doing this would be to dramatize the temporal sequence of the events involved. Consider the contrast between the sequence initially recorded and that presented when the display is reversed. Among other things, the reverse display would show lava moving energetically up the face of the mountain and rocks springing upward to form a new peak. When the sequences forward and backward are compared, we have no trouble telling which corresponds to the ways of nature. Energy normally is expended as time advances; any reversal of this process appears unnatural.

The transformations we have been talking about thus far are correlated with the progression of time from past to future. On one hand, the forward progression of time is marked by a progressive slide from order to disorder. Other examples of order degrading into disorder are weeds taking over gardens, machines breaking down, and dust accumulating on bedroom floors. On the other hand, temporal progression is marked as well by a progressive expenditure of energy. Other examples of energy depletion, albeit less dramatic than an exploding volcano, are flashlight batteries running down, heated rooms becoming cold, and runners becoming fatigued in the course of a race.

As we shall see presently, indeed, the degradation of order and the degradation of energy are equivalent processes.[1] This equivalence follows from the fact that high-grade order and high-grade energy are mutually convertible. To prepare ourselves for understanding why this is so, we must first become familiar in general terms with what happens when energy becomes degraded. Let us begin with a brief look at the scientific underpinnings of such commonplace events as batteries running down and rooms in winter losing heat.

1.2 The First and Second Laws of Thermodynamics

Illustrations of progressive energy degradation like these are largely anecdotal. As such, they have little scientific value. Scientific investigation of such commonplace phenomena began approximately 150 years ago with the articulation of two fundamental laws of thermodynamics.

The First Law of Thermodynamics states that the amount of energy in a closed system[2] remains constant through time. Since the universe by definition is a closed system, a consequence is that the universe contains a fixed amount of energy. While energy within a specific locale (e.g., on Earth) might change both in quantity and quality, the total amount of energy remains the same overall.

The Second Law of Thermodynamics deals with changes in quality. The most important distinction to be made regarding quality is between energy capable of producing work[3] and energy lacking this capacity. What work amounts to here, in its most general sense, is physical alteration that occurs on other than a random basis. Illustrations include lifting weights (by an athlete), synthesizing molecules (by an organism's metabolism), and increasing a body's heat content (by solar radiation). According to the Second Law, the amount of energy capable of producing work (free energy) in a closed system tends always to decrease with time.

An alternative formulation of the Second Law is in terms of degraded energy. Energy becomes degraded as its capacity for producing work is lost. A quantity of energy within a given system might lose this capacity either by being wasted or by being expended as the system actually accomplishes work. This loss typically occurs through a series of changing system states during which the work potential of the system becomes increasingly degraded (recall the example of a flashlight battery gradually losing its charge).

Another way of putting the Second Law, accordingly, is that the energy in a closed system, while remaining constant in quantity (the First Law), tends to lose its ability to produce work with the passage of time. An equivalent statement is that the amount of degraded energy (i.e., lost work potential) in a closed system tends to increase with time.

Apart from carefully engineered approximations in the laboratory, the universe at large may be the only closed system in actual existence. The fact that the First and Second Laws are formulated in terms of closed systems, however, does not preclude their application to systems that are to some extent open. Otherwise thermodynamics would have few practical applications.

Application to open systems is assured by an important consequence of the two laws taken together. The First Law says that the total amount of energy in the universe remains constant. The Second Law says that energy in the universe tends to be degraded in use. The consequence is that degraded energy remains part of the universe. Energy once used does not just go away but continues to exist in degraded form.

This consequence holds for energy expenditures generally, obviously including those occurring in open systems. There is no thermodynamic requirement that energy degraded by use in an open system remain within the system where it was used. The requirement is that energy degraded by use in an open system remains somewhere in existence, whether or not in the system where the degradation occurred.

This consequence plays a crucial role in the discussion that follows. At various points in discussing the ecosystem, for example, we will be concerned with the effects of degraded energy that remains within the system. And in discussing problems arising from excessive use of energy by industrial technology, we shall see why these problems stem from the inability of the biosphere to rid itself of all the energy degraded within it.

1.3 Entropy

The term "entropy" (from the Greek *entrepō,* meaning "to alter") was coined (circa 1865) by Rudolf Clausius, the originator of thermodynamics, in connection with his work on problems of heat exchange. In line with the general principle that heat passes spontaneously only from hotter to colder bodies, Clausius conjectured that the transmission of heat in the opposite direction (e.g., when bodies are heated by friction) requires some sort of work. The Second Law emerged with his observation that this work can be accomplished only at the expense of some irreversible alteration in the surrounding environment. The alteration produced by work is an increase in what he called "entropy." Clausius's expression of the Second Law was the simple statement that the entropy of the universe tends always to increase.

What Clausius observed, in effect, is that the natural flow of heat from hotter to colder bodies can be reversed only by the expenditure of energy (e.g., rubbing cold metals together to make them warm). The term "entropy" designated the change undergone by the source of this energy. Put in terms introduced previously, this change amounts to a degradation of the energy involved. Claudius's statement that entropy in the universe tends always to increase thus converges with our expression of the Second Law in the previous section, to the effect that the amount of degraded energy tends to increase with time.

Whereas Clausius's original use of the term "entropy" applied specifically to contexts involving the exchange of heat,[4] its use soon became standard in other contexts as well. It was not long before it had become an important part of the conceptual apparatus of both physics and chemistry. By the mid-twentieth century, various biological and social sciences had also adopted the term, as had the burgeoning discipline of information theory.

As a result of this considerable diversity in use, the term "entropy" has been defined in several different ways.[5] It may be assumed that for the most part these definitions are mutually compatible. For present purposes, the term will be used only in ways that have been explicitly introduced as the discussion progresses. In this chapter so far, the term has been introduced as a designation for lost work potential, or (in the sense previously specified) for degraded energy. Its use as a designation for degraded structure (disorder) is explained in the following chapter.

1.4 How Energy Degrades

Energy is analogous in some ways to monetary value. The value of an ounce of gold can be converted into currency, which then can be used to purchase a valuable commodity. But there is an important disanalogy as well. Whereas under favorable circumstances the commodity (say a blue-chip stock) can be exchanged back for currency, which can be used to buy gold in turn, not all forms of energy are mutually interchangeable. Solar energy, for example, can produce electricity, but electricity, regardless of the amounts involved, cannot be reconverted into solar energy.

Other examples should help make this point clear. Electricity can be used to pump water uphill, and water running downhill through turbines can produce electricity. Thus electrical and mechanical energy are mutually convertible. Electrical and kinetic energy likewise are mutually convertible, as shown by electric fans and wind-driven generators.

But most processes of energy transformation in everyday experience involve forms of energy that are convertible in one direction only. The chemical energy produced in plants by photosynthesis cannot be transformed back into solar energy. The rotational and gravitational energies (on the part of the earth and the moon, respectively) involved in the production of tidal energy cannot be generated out of tidal energy in turn. And the thermal energy put out by a common space heater cannot be recovered to energize further cycles of space heating.

Energy for the most part is degraded in use, meaning that it cannot be reconverted to its previous form. This is the manner of energy degradation featured in the Second Law.

Even in transformations between mutually convertible forms of energy, some energy is always degraded to forms not reconvertible to the original. In conversions between electrical and mechanical, for instance, some energy is always degraded to the form of low-grade heat. Thus all transformations producing work involve some manner of energy degradation, which is to say that energy expenditures producing work are never 100 percent efficient. Processes that are reversible without loss of usable energy (if in fact there are any) by definition are not productive of work.

1.5 Degrees of Degradation

Forms of energy can be ranked with respect to convertibility. At the top will be forms convertible into every other form. If there is only one such form (perhaps the energy of the Big Bang thought to have originated the universe), it alone will have top ranking. If there are more than one at the top, each will be convertible into the others as well. Candidates for top ranking include gravitational and orbital energy, which (when not producing work) do not invariably degrade with time.[6]

At the very bottom of the ranking fall forms of energy incapable of being converted into any other form at all. In current thinking, one such form is the cosmic background radiation into which (following the Second Law) all energy ultimately will be converted. Inasmuch as work typically involves conversion to different forms of energy, this lowest form is incapable of doing work.

In between are forms of energy that can be converted into forms with lower (or equal) rankings, but not into forms above them on the scale. Fairly high within this intermediate range will appear the internal heat of stars, which (in the case of the sun) is convertible into solar radiation but not vice versa. Lower will be forms of energy into which solar radiation is convertible, such as electrical, mechanical, and kinetic, but which are not convertible to solar radiation in turn. Lower yet will be waste heat of terrestrial origin, which is emitted from the earth in the form of black-body radiation.[7]

Abstract as it may be in general outline, this ranking establishes a complex network of paths along which energy can be expended in doing the world's work. Apart from a few that can be traveled in either direction (e.g., that between electrical and kinetic energy), these paths are mostly unidirectional.

One might think of it this way. The lines of energy flow by which the world's work is accomplished lead inexorably "downward," with an excursion now and then in a "horizontal" direction. This downward trend is a consequence of the Second Law of Thermodynamics: the amount of energy available for work inevitably diminishes with time.

1.6 A Graphic Model

A simple model might help us pull these concepts together. The model is based on a series of bar graphs ordered along a horizontal baseline, as seen in figure 1.1. Each bar graph represents a particular (here unspecified) form of energy (solar, electrical, mechanical, etc.), and the bar graphs are ordered left to right according to the convertibility rankings discussed above. For example, inasmuch as solar energy is convertible into electrical energy but not vice versa, a bar graph representing the former would appear to the left of one representing the latter.

Figure 1.1. Energy Rankings 1

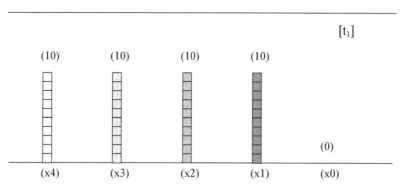

In keeping with the requirement that work capacity is always lost in transformations among energy forms, the bar graphs are broken down into segments so conceived that more usable energy is contained in segments of bar graphs to the left than in comparable segments to the right. These segments will be called "usable-energy packets." Due to the abstract character of the model, no empirically significant values are assigned to comparable segments. Difference in usable-energy content is represented instead by "multipliers"—(x4), (x3), etc.—specified below the bar graphs. The sense of the (x4) multiplier, for instance, is that each segment (packet) in its column contains 4 units of usable energy. These units are for comparison only and have no specific value in terms of standard measures like watts and joules.[8]

Each bar graph in this figure is divided into 10 packets, as indicated by the number above it. The leftmost column thus contains (10 x 4 =) 40 units of usable energy. Taking all four columns into account, we see that the bar graph as it stands has (40 + 30 + 20 + 10 =) 100 units of work capacity overall. The place to the right marked "(x0)" is reserved for a bar representing energy with no work potential to be added in the following figures.

As it stands, figure 1.1 is static, showing the state of the system at its initial moment only (time t_1 at the upper right). Progression in time is represented by an ordered series of bar graphs, each step in the series showing a change in at least one column. Figure 1.2 represents a possible second stage (t_2) in a series beginning with figure 1.1. In comparison

Figure 1.2. Energy Rankings 2

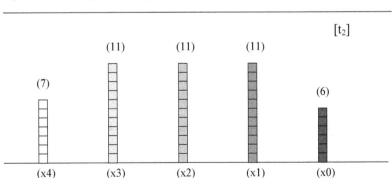

with figure 1.1, this figure shows a decrease of 3 packets in the (x4) column (12 units of usable energy), an increase of 1 packet in each of the (x3), (x2), and (x1) columns, and a new column in the (x0) position measuring 6 increments high. This latter column is heavily shaded to indicate that it contains no usable energy, in accord with the significance of the (x0) multiplier explained previously.

These changes are to be interpreted as follows. Twelve (3 x 4) units of usable energy have been expended from the supply of the (x4) column. Of these, 3 units—one (x3) packet—have been converted to the (x3) column, 2 units to the (x2) column, and 1 unit to the (x1) column. Each of these 3 columns, accordingly, is 1 packet higher. This accounts for 6 of the 12 units removed from the (x4) column. The remaining 6 have lost all potential for useful work and hence show up in the (x0) column. All of the 12 units removed from the leftmost column have been degraded, but while 6 still retain work capacity, the remaining 6 are incapable of further work. At the stage represented by figure 1.2, the system of graphs in question still has 100 units of energy overall, of which (7x4 + 11x3 + 11x2 + 11x1 =) 94 remain available for work.

To continue the demonstration, consider that during the next stage of operation (t₃) additional work is done involving the conversion of 1 packet of (x3) energy into a single packet of (x1) energy and 2 units of (x0) energy, while 1 packet of (x2) energy is "wasted" (no work accomplished) by conversion into (x0) energy directly. As a result of these latest

Figure 1.3. Energy Rankings 3

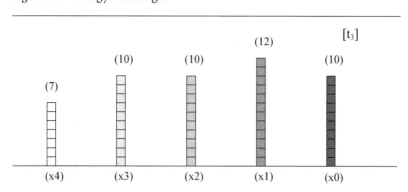

transactions, both the (x3) and the (x2) columns have been diminished 1 packet from figure 1.2, the (x1) column has gained another packet, and the (x0) column has been increased an additional 4 increments. At this stage, the system represents 90 units of usable energy (7x4 + 10x3 + 10x2 + 12x1) and 10 units of energy without work capacity. As with stages t_1 and t_2, however, 100 units of energy remain within the system overall.

1.7 The Model Applied to Open Systems

Given that its total amount of energy remains constant, the model thus far corresponds to a thermodynamically closed physical system. As operation of a corresponding physical system progresses, more and more energy will end up in the state represented by the (x0) column. Its final stage of operation would be reached when all energy initially within the system has reached this final state, indicating that the system is incapable of further work.

The model can be altered to illustrate an open system with two additional provisions. First, any given column can increase without a corresponding decrease in any column to its left. Such increases would represent additional energy being brought into the corresponding physical system. When the increase is in one of the columns with positive multipliers, the imported energy contains capacity for work. When an increase is in the (x0) column, however, it represents energy whose work capacity has previously been exhausted. The significance of this latter case is that useless energy has been "dumped" into the system from some outside source.

The second provision is that any column can decrease without a corresponding increase in any column to its right. When the decrease occurs in one of the columns with a positive multiplier, this indicates that usable energy is being exported to another system. When the decrease occurs in the (x0) column, this indicates that the corresponding physical system is getting rid of useless energy.

Addition of these provisions does not change the basic requirement of the model that energy conversions occur exclusively in the rightward direction. This requirement signifies the fact that energy conversions in physical systems always involve some degree of degradation, regardless of whether the system in which they occur is open or closed. Nor do these provisions affect the basic physical fact that used-up energy does not simply go away. While (x0) units may disappear from the model, the corresponding depleted energy is still present elsewhere in the physical system's environment.

1.8 The "Heat Death" of the Universe

Energy that has lost all its work potential constitutes low-grade heat. In its terrestrial form, low-grade heat is exemplified by the body temperature of a living animal or the warmth of an operating engine. In its cosmic form, it is exemplified by the black-body radiation by which terrestrial heat leaves the surface of the earth.

High-grade thermal energy, on the other hand, is still capable of doing work. Energy present in boiling water, for example, is capable of driving steam engines, and intense heat generated by electricity can be used to melt metals. But heat emitted from the earth by black-body radiation is too far degraded to retain capacity for further work.

The so-called "heat death" of the universe is the state at which all energy in the universe has been degraded to a form no longer capable of doing work. One way of conceptualizing this state is to think of it as a condition in which all energy in the universe has been reduced to a form thermodynamically equivalent to the energy emitted from the earth's surface into space.[9] In this state, all temperature differences capable of work have been exhausted and the universe at large has become thermally inert.

So the "heat death" of the universe is not a state at which the universe has become "too hot" to survive. It is a state at which the universe has become "dead" in the sense of containing no heat capable of doing work. At this state all energy has been degraded to useless entropy, and the universe has reached the end intimated by the Second Law.

1.9 Entropy Retained on Earth

The earth itself, of course, is an open system. It receives a constant stream of high-grade energy from the sun and emits a corresponding stream of low-grade energy back into space by black-body radiation. Since the first appearance of life on earth (roughly three and a half billion years ago) up to the present era, the amount of depleted energy leaving the earth has roughly matched that of the high-grade energy entering from the sun. During the past two or three centuries, however, changes in the earth's atmosphere have impeded the normal flow of low-grade energy back into space. This has led to substantial amounts of depleted energy being retained within the atmosphere, a phenomenon currently known as global warming.

Global warming is the result of entropy being prevented from leaving the earth's surface via its normal channels of black-body radiation. For reasons soon to be examined in detail, abnormal amounts of entropy are accumulating on and about the earth's surface in the form of degraded structure as well. To understand how degraded structure (disorder) ties in with degraded energy, we need to take a careful look at the relation between order and energy in their undegraded forms.

2

Entropy and Disorder

2.1 The Relation between Energy and Order: A Classic Example

The concept of entropy was introduced by Clausius in connection with his formulation of the Second Law of Thermodynamics (previously noted in section 1.3). In its original use it referred to used-up energy, and this use remains prevalent in various sciences. By the middle of the twentieth century, a broader understanding had developed associating entropy with randomness and disorder.[1] Although seemingly disparate upon first consideration, these two conceptions of entropy are closely interconnected.[2] To see why, we need to consider the relation between order and usable energy.

A standard illustration of this relation is a container of gas divided into two interconnected compartments. As long as gas molecules pass freely between compartments, their average kinetic energy (energy of motion) will be the same in both. Since heat is directly proportional to average kinetic energy, the two compartments will also be at the same heat level.

If the molecules are distributed so that one compartment contains a significantly larger proportion of fast-moving molecules than the other, however, there will be an appreciable difference in heat level between the two compartments. If the difference is sufficiently great (say, enough to boil water), the system will contain heat energy capable of doing work. Energy in this form could be used to produce electricity (steam turbines), for example, or to power machinery (steam locomotives) for transporting goods.

The work potential of the gas chamber in this illustration can be described in two equivalent fashions. On one hand, the temperature difference between the two compartments constitutes such-and-such an amount of usable thermal energy. On the other, the degree of order present in the distribution of high- and low-energy molecules between the two compartments is sufficiently great to produce such-and-such an amount of work as the order is dissipated (i.e., as the average kinetic energies of the molecules in the two compartments return to parity). For the molecules to be distributed with a certain degree of order is tantamount to the chamber's containing a certain amount of usable thermal energy.

2.2 Order and Energy: Other Common Examples

Similar illustrations are provided by other energy sources. Electrical energy is available from a socket only when it is connected to a power source with opposing polarities (+ and –). The negative pole of a power source is the terminal characterized by a concentration of negatively charged atoms, while the atoms of the positive pole are predominantly positive. When the circuit is closed (e.g., by switching on an appliance), the positive pole attracts electrons, and the negative pole repels them (by Coulomb's Law). The resulting movement of electrons constitutes a flow of electrical current.

The opposing concentrations of charged atoms at the terminals of a power source constitute a departure from purely random arrangements. Inasmuch as atoms arranged randomly with respect to polarity are incapable of inducing an electrical current, the flow of power between terminals is enabled by this departure from randomness. Usable energy shows up again as a correlate of order, and disorder as an absence of usable energy.

Another illustrative example comes with the fuel cells currently being developed as a replacement for fossil fuels. In a typical fuel cell arrangement, hydrogen entering the cell is broken down into electrons and protons. The electrons then are diverted into an external circuit as a flow of electricity, while the protons remain within the cell to be combined with returning electrons to reconstitute hydrogen in the presence of oxygen. The main outputs of the process are usable electricity (from the flow of electrons), water (from the combination of hydrogen and oxygen), and low-grade heat (which escapes into the atmosphere).

In this process, the atomic structure of the two gases is manipulated to produce usable energy in the form of electricity. The order induced on the atomic level is the source of energy made available for everyday use.

Radiational energy also fits into this picture. Solar radiation is highly ordered insofar as it is (1) directional and (2) characterized by a relatively high number of oscillations per unit distance.[3] Because of its highly ordered wave structure, solar radiation in the midfrequency range serves as a medium conveying solar energy directly to chlorophyll-bearing plants, where it is converted into biomass, providing food for other organisms. The structure of the highly directional incoming radiation, as it were, is transformed into energy for use by plant-eating organisms. The low-grade heat energy that results, on the other hand, exists at wavelengths lacking sufficient structure for further work. When this useless energy is emitted from the surface of the earth, it spreads randomly (nondirectionally) through surrounding space.

Structure is an orderly arrangement of parts within a system. What examples like these show, accordingly, is that usable energy and order go hand in hand. The flip side of this relation is that degradation of energy is tantamount to degradation of order. Expended energy and disorder are correlative forms of entropy.

2.3 Gradations of Structure Correlated with Gradations of Energy

Transformations among energy forms proceed according to rankings with respect to convertibility (see section 1.5). It follows that the structures implicated in energy transformations are subject to ranking as well. In general, transformation from a higher to a lower form of energy is accompanied by change from a higher to a lower degree of structure.

By way of illustration, consider the conversion of (1) solar energy to (2) electricity (by photovoltaic receptors) and then to (3) high-grade heat (in a stove burner), which in turn produces (4) kinetic energy (in boiling water) and (5) low-grade heat (in the surrounding air). Correlated with each energy source (1–4) in this series, there is a characteristic structure that accounts for its capacity for accomplishing useful work. Usable energy in case 1 is conveyed by a highly ordered configuration of wave oscillations, in case 2 by an orderly movement of electrons through the conductor, and in case 3 by a nonrandom pattern of molecular activity within the stove burner. Usable energy is still available with the structural turbulence of the water in 4, although this does not figure in the illustration. When the low-grade thermal energy of 5 has been dissipated in the air, however, it no longer contains potential for accomplishing work.

At each stage of this process, structure is expended in making energy available for work. The wave structure of the solar radiation, for instance, is spent producing electrical energy, and the orderly flow of electrons through the conductor is spent in producing high-grade heat.

At each stage, moreover, energy made available by the expenditure of structure is used in establishing further structure at a subsequent stage. Solar energy is used to induce the orderly flow of electrons in the conductor, electrical energy is used to induce nonrandom agitation of molecules in the burner, and so forth. The net effect is that, at each stage of the series, structure is expended in making energy available for work, which then is used to generate structure at a later stage.

What happens in the conversions from (1) solar to (2) electrical to (3) thermal to (4) kinetic energy, accordingly, is a series of transformations in which one kind of structure is exchanged for another. The stage-wise progression of energy forms can also be viewed as a stage-wise progression involving different kinds of structure—the wave structure of the solar radiation, the arrangement of electrons in the conductor, and so on.

Moreover, just as there is a ranking of progressive degradation in the series of energy transformations 1 through 5, there is a sense as well in which the associated structures exhibit decreasing gradations of order. Intuitively, the wave structure of solar radiation represents a greater departure from randomness than the molecular activity of the stove burner, and the latter in turn a greater degree of nonrandomness than the molecules in the hot air above the stove.

Intuitions aside, we need a characterization of order that makes comparisons like this possible on an objective basis. The groundwork for such a characterization is laid out in the following two sections, which examine the reciprocal relation between randomness and order.

2.4 Degrees of Order

Order and disorder are comparative states, which means that both can be present in varying degrees. Things may be well ordered in one comparison and relatively disordered in another. A deck of cards segregated by color only (spades and clubs coming first, say, with hearts and diamonds following), for example, exhibits more order than a deck that has been thoroughly shuffled. But the former is relatively disordered in comparison with a brand-new deck in which each suit is arranged internally by rank.

In most contexts, nonetheless, there will be arrangements in which disorder is maximal and others in which order reaches a peak. Maximum disorder occurs in a deck of cards when the sequence within the deck is entirely random (a state approximated by repeated shuffling), whereas maximum order is present when each card is located both by rank (i.e., "taking order," 10 over 9, jack over 10, etc.) and by strength of suit (spades over hearts, etc.).

Other commonplace examples are easy to find. A row of spice containers in a kitchen cabinet is maximally ordered when arranged alphabetically by names of contents (e.g., anise, basil, coriander, dill) and maximally disordered when their arrangement is random. A set of socket-wrench heads is completely ordered when each is placed in its container according to size and completely disordered when scattered haphazardly across the garage floor. A set of professional journals is shelved in perfect order when arranged uniformly by sequence of publication dates—and so forth.

Speaking generally, we may say that maximum disorder is a state of completely random distribution and that maximum order is a maximal departure from an entirely random state. What we need next is a working grasp of comparative degrees of order that fall between these two extremes.

2.5 Factors Determining Degrees of Order

Given this understanding of maximal disorder as a state of complete randomness, it is natural to think of degrees of order as degrees of departure from a completely random state. When departure from randomness is complete, of course, the degree of order is maximal.

To clarify the sense of which departure from randomness admits degrees, we need a working definition of randomness. As a first approximation, randomness is equivalent to statistical independence. Two events are statistically independent if the occurrence of one does not affect the probability of the other's occurrence. A set of events thus is completely random when the occurrence of one has no bearing on the occurrence of any other.

Departure from randomness occurs when the occurrence of certain events begins to influence the likelihood that certain other events will occur as well. The more extensive this influence among events within the set, the more extensive their departure from a completely random state. The greater this departure, in turn, the more the events are statistically *inter*dependent. The degree of order of a given set of events is directly correlated with the degree of interdependence among the events themselves.

By way of illustration, let us return to the example of the playing cards. To simplify matters, we may stipulate that the arrangement of cards in a series is entirely random if the identity of any given card is independent of its place in the series. This holds both for the arrangement of the deck itself and for the sequence resulting when cards are dealt off the top of the deck.

In the case of a newly opened pack (arranged in maximal order at the factory), a series of cards dealt off the top will exhibit complete regularity as the sequence unfolds. The sequence accordingly is completely nonrandom. The identity of each card is maximally interdependent with the identities of adjacent cards in the unfolding sequence.

In the case of a sequence dealt from a thoroughly shuffled deck, by contrast, there will be no appreciable interaction among successive members. The purpose of shuffling is to arrange the cards randomly, which is intended to ensure that each card's identity is independent of its place in the series. Randomness in arrangement of the deck goes hand in hand with a card's independence from its neighbors in the sequence dealt.

Between the maximal order of a new deck and the randomness induced by shuffling, there will of course be many intermediate degrees of order. Generally speaking, we may say that degree of order varies directly with degree of interdependence among members of the sequence. But degree of interdependence varies inversely with degree of randomness in their arrangement. One variable affecting degree of order, accordingly, is degree of randomness. As the latter increases, the former decreases.

Another factor affecting an arrangement's degree of order is the number of items it includes. To get a feel for this, compare a series of six cards dealt from a newly opened deck with a series that continues until all fifty-two cards (excluding jokers) have been dealt. Under assumptions laid out previously, numbers of both series are entirely regular in sequence. Nonetheless it appears natural to think that the series of fifty-two is more highly ordered than the series of six. An entire deck in proper sequence represents a greater departure from randomness than does a smaller series also in proper sequence.

Think of it in terms of a mathematical analogy. There are ($3 \times 2 \times 1 =$) 6 ways in which the first three cardinal numbers can be arranged, ($4 \times 3 \times 2 \times 1 =$) 24 for the first four, ($5 \times 4 \times 3 \times 2 \times 1 =$) 120 for the first five, and so forth. This means that the probability of a correct ordinal sequence (1, 2, 3) for the first three is 1/6, that for the first four is 1/24, and that for the first five 1/120. But the lower the probability of a given occurrence, the less likely that it would happen on a strictly random basis. The occurrence of five numbers in correct ordinal sequence (1, 2, 3, 4, 5) thus is a greater departure from randomness than in the case of three or four numbers.

Similarly, an arrangement of fifty-two cards all in proper order is a greater departure from randomness than a proper arrangement of a smaller number. Given the relationship between randomness and order, it follows that the regular arrangement of all fifty-two exhibits a higher degree of order than the arrangement of only six.

These considerations enable a working definition of orderliness (degree of order). An arrangement's degree of order is tied to its incidence of nonrandom occurrences. In upshot, an arrangement's degree of order (1) varies directly with the number of its featured members and (2) varies inversely with the degree of randomness (independence) among these members.

Technical Addendum[4]

Our concern with order and disorder for the remainder of this study will have to do mainly with operating systems, in contrast with decks of cards and number sequences. As defined in chapter 1, an operating system is an open system of physical variables interacting through time. Of primary concern in what follows will be systems constituting biological organisms (chapter 3) and ecosystems in which various species of organisms interact (chapter 4).

In the case of ecosystems and of individual organisms alike, order is created and maintained by a steady flow of high-grade energy into the system. This energy is put to work in building biomass (e.g., plants in ecosystems, tissue in organisms) and in running the metabolisms of the organisms involved. In both kinds of system, likewise, entropy resulting from this work is discharged back into the surrounding environment. As with operating systems generally, some of this entropy will take the form of low-grade heat. Some also will take the form of degraded structure, such as dead plant mass and animal excreta.

Entropy in the form of low-grade heat (degraded energy) typically can be quantified in measures applicable to the higher-grade forms of energy consumed in producing it. A standard measure of entropy in chemistry, for instance, is joules per Kelvin. As matters stand, however, there is no comparable measure of entropy in the form of degraded structure or disorder. The reason for this lack is not that degrees of structure are inherently unmeasurable. At the very least, a given degree of structure can be quantified with reference to the amount of energy required to raise it to a higher degree.

The reason there is no standard measure of structure per se, rather, must be that no need has been perceived to define such a measure independently of the amount of energy necessary to produce it in the first place. Need now arises with the present project of comparing degraded (expended) energy and degraded structure (disorder) as separate but interacting forms of entropy. Meeting this need does not require finding a measure of structure that is applicable in fieldwork or in the laboratory. This is a separate project.[5] For present purposes, we must be content with a demonstration showing how different degrees of order can be quantitatively distinguished.

A mathematical (thus quantitative) measure of order in operating systems is explained in the appendix to this chapter. Its purpose, as already indicated, is to show that entropy in the form of degraded structure or disorder is subject to quantitative measurement no less than entropy in the form of degraded energy. Readers not concerned with this matter may pass over the appendix without losing track of the continuing discussion.

2.6 Entropy and Randomness

Discussion of entropy in chapter 1 was confined to expended energy, which is energy no longer capable of work. Work is done when physical occurrences are brought about by other physical occurrences rather than occurring randomly (section 1.2). A standard example of energy incapable of further work is the low-grade heat expended by metabolic activity (e.g., the body heat of living animals).

Another conception of entropy was introduced into the discussion at the beginning of the present chapter. This second conception equates entropy with disorder. In photosynthesis, solar energy is expended in creating biomass that provides chemical energy to plant-eating organisms. The highly ordered wave structure of sunlight is converted into chemical structure, which then is further degraded into the waste products of metabolic activity. This process of increasing degradation culminates in the nondirectional wave structure of black-body radiation by which fully expended energy from the sun is eventually returned to space.

The conceptions of entropy as expended energy and as structural disorder can be further integrated in terms of a third conception equating entropy with randomness. This conception was implied in our discussion of orderliness in the preceding section but needs to be more explicitly articulated. On one hand, random events are events whose occurrence exhibits no particular order. The entropy present in disorder thus is equivalent to that present in random occurrences.

On the other hand, random occurrences are incapable of doing work. This follows from the definition of work as a physical alteration that occurs on a nonrandom basis (see section 1.2). Random events do not bring about nonrandom occurrences. Also by definition, usable energy is energy with work potential, which means that occurrences incapable of work are devoid of usable energy. The joint consequence is that random occurrences ipso facto are devoid of work potential. Randomness and absence of energy capable of work are equivalent forms of entropy.

In connection with operating systems particularly, increased disorder is equivalent to increased randomness, and increased randomness is equivalent to increased amount of energy unavailable for work. The upshot, thermodynamically speaking, is that (1) expended energy, (2) randomness, and (3) disorder are equivalent forms of entropy. Different forms may come into play in different contexts. Our main concern in the present study will be with forms 1 and 3.

2.7 Negentropy

Let us summarize and simplify. One way of conceiving entropy equates it with expended energy. This conception occupied center stage in chapter 1. The opposite of entropy thus conceived is usable energy, which is to say energy capable of being used for work.

Another way of conceiving entropy equates it with random occurrences, which by definition are incapable of producing work. This sense was laid out in the preceding section. The opposite of entropy thus understood is departure from randomness which, if sufficiently extensive, amounts to potential for doing work.

Yet another conception equates entropy with disorder. Entropy in this sense has been the primary focus of the present chapter. The opposite of entropy in this sense is orderly structure. Orderly structure constitutes a departure from randomness, which in suitable forms provides energy capable of doing work (see section 2.3 above).

Along with each conception of entropy goes a specific way of understanding its opposite. The opposite of form 1 is energy capable of work. The opposite of form 2 is nonrandom occurrences, and that of form 3 is orderly structure. Under certain circumstances, as we have seen, both nonrandom occurrences and orderly structure provide potential for doing work.

Scientists concerned with these matters have found it convenient to have a single term covering the opposite of entropy in these several senses. Schrödinger chose the expression "negative entropy."[6] Others have opted for the truncated version "negentropy."[7] Thus understood, the term "negentropy" can be used to refer to usable energy, nonrandomness, and order or structure indifferently.

2.8 Other Formulations of the Second Law

In chapter 1, the Second Law was formulated as saying that the amount of energy in the universe capable of work tends to decrease with time. An obviously equivalent formulation is that the amount of expended (degraded) energy in the universe tends always to increase. The First Law states that the amount of energy in the universe (usable or not) remains constant through time. A joint consequence of these two laws, however expressed, is that the energy expended in doing work does not simply go away but rather remains in a form no longer capable of doing work.

At this point in the discussion, other formulations of the Second Law are also available. An alternative formulation in terms of randomness is that the incidence of nonrandom occurrences tends always to decrease.

Put in terms of order and disorder, the Second Law states that the disorder present in the universe tends always to increase while the incidence of order constantly decreases with time. In these terms, the so-called heat death of the universe (see section 1.7) is a state of complete disorder. A consequence is that all happenings in the universe at this conjectured point occur on a strictly random basis.

Since expended energy, randomness, and disorder are all forms of entropy, a more comprehensive formulation of the Second Law is to the effect that the amount of entropy in the universe tends to increase with time. An equivalent formulation is that the amount of negentropy in the universe tends always to decrease. This latter formulation will play a central role in chapter 3, when we begin to consider the implications of the Second Law for living creatures.

APPENDIX
An Information-Theoretic Measure of Order in Operating Systems

A.1 The Task of the Appendix

An operating system, as characterized previously, is an open system of physical variables interacting through time. What distinguishes operating systems from static physical systems (e.g., stone bridges) and nonphysical systems (e.g., systems of geometry) is the temporal change among their interacting features. Put differently, an operating system is an arrangement of physical entities in configurations that undergo regular patterns of temporal change.

According to the definition in section 2.5, a system's degree of order (1) varies *directly* with the number of interacting features and (2) varies *inversely* with the degree of independence (randomness) among those features as they change with time. What counts as a relevant feature in a given system will vary from case to case. While all operating systems have parts of one sort or another, what typically changes with time are not the parts but their interrelationships. For present purposes, we focus on the configurations in which parts of the system are arranged rather than on the parts themselves. The interacting features we will be concerned with are the configurations of parts as they undergo temporal change.

A simple example is a children's teeter-totter, consisting of a plank and a fulcrum over which it pivots (the parts). Let us say that these parts are configured in one of three ways during successive stages of normal operation: (a) both ends level with the fulcrum, (b) one end higher, and (c) the other end higher. We think of the apparatus operating in normal fashion when *b* and *c* alternatively are followed by *a* (*b, a, c, a, b, a, c, . . .*). Irregularities occur when one end approaches the level of the fulcrum but then falls again before rising above it (as when one user is significantly heavier than the other).

For a somewhat more complex example, consider an ordinary (non-digital) alarm clock. While the number of parts will vary with design, the basic purpose of the device (let us say) is to show a different configuration of hands on the face for each passing second of a twelve-hour period and then to make a noise when a predetermined configuration is reached. This requires a total of at least (12 x 60 x 60 =) 43,200 different configurations through which the system must pass, plus a few more when the alarm is engaged.

According to the preceding definition, both devices exhibit order in their operation to the extent that their successive configurations are interdependent. This amounts to each configuration's yielding to the next in regular sequence. The definition also provides that the clock's degree of order is much higher than that of the teeter-totter. This is because there are many more configurations relevant to the clock's proper operation. The clock's operation becomes disorderly when its configurations fail to succeed one another in proper sequence (as when the second hand starts to oscillate because the battery has run down).

The present task is to develop a measure enabling us to assign numbers to the degrees of order by which such systems might be characterized and hence to compare them on a quantitative basis. This measure should apply in principle to operating systems generally (not just mechanical devices), including those like ecosystems with biological components. Among resources we have available for the task is the mathematical theory of communication.

A.2 Mathematical Communication Theory

Communication theory began as a mathematical study of the efficient transmission of messages through communication channels. Its first systematic formulation came in 1948 with Claude Shannon's paper "A Mathematical Theory of Communication."[1] Since then, it has been employed extensively in the design of telephone systems, computer networks, and data links between space vehicles and their control stations. Less extensively, it has also been applied in certain branches of biology, economics, and psychology, and in humane disciplines like philosophy.[2]

Another title for this study is "information theory." The term "information" here is used in a precisely defined sense. This sense has nothing to do with gaining knowledge or intelligence or with ways in which a person might be well informed. It is not information *about* something or information that might be passed from person to person. In one way or another, information of an ordinary sort always involves *reference* (e.g., of symbols to things), which is a semantic concept entirely foreign to information theory.

Information in the technical sense boils down to changes in an event's probability of occurrence. By way of illustration, consider the flip of an unbiased coin. Before the flip, the probability of a head's coming up is 50 percent. When a head actually shows up, its probability has increased to 100 percent. The difference between the probabilities before and after is the information associated with the head's actual occurrence.

For various technical reasons pertaining to the widespread use of digital (binary) encoding devices, information is usually measured in bits (*bi*nary un*its*). One bit results when the probability of a given event is doubled, as in the illustration of the head's probability of occurrence changing from 50 percent to 100 percent. If the antecedent probability of an event is 25 percent (consider the spin of a four-sided dreidel), its actual occurrence yields two bits of information, inasmuch as two doublings are required to reach 100 percent from 25 percent.

Information can also be thought of as removed uncertainty. The more uncertain an event antecedently, the more uncertainty is removed by its occurrence. Since an event 25 percent probable is twice as uncertain as one with antecedent probability of 50 percent, occurrence of the former yields twice the information yielded by occurrence of the latter.

In its most general form, information can also be conceived simply as increased probability. This enables the association of information with events that never actually occur. If the probability of an event e_1 is changed from 25 percent to 50 percent by the occurrence of a distinct event e_2, then one bit of information regarding e_1 is produced by e_2's occurrence. This effect is the basis of transactions across information channels.

Since the probability of most events prior to occurrence cannot be specified in terms of progressive halvings of 100 percent, a more versatile function is needed for assigning numbers to quantities of information. The function adopted for this purpose by early information theorists is $log\ 1/P(e)$—that is, the logarithm of the reciprocal of the probability of event e conveying the information. Although logarithms to other bases have been used on occasion, use of base 2 went hand in hand with choice of the bit as the basic unit of information.

Technical Addendum

Inasmuch as the logarithm of $1/n$ is the negation of the logarithm of n, an equivalent expression is $-log\ P(e)$. Given the values –1 and –2 of the logarithms (base 2) of 0.5 and 0.25 respectively, occurrence of an event e conveys 1 bit of information if $P(e)$ is 50 percent and 2 bits if $P(e)$ is 25 percent, as before. Similarly, approximately 1.6 bits are conveyed if $P(e)$ is 33 percent (since $-log\ 0.33$ is 1.6), 0.6 bits if $P(e)$ is 67 percent, and so forth.

A.3 Information Channels

Communication theory is concerned primarily with the transmission of information across information channels. Considered in abstraction from particular physical embodiments (like an old-fashioned telegraph system), an information channel consists of a set of input symbols (e.g., dots, dashes, and spaces), a set of output symbols (also dots, dashes, spaces), and for all *pairs* of input and output symbols a set of conditional probabilities giving the likelihood that a specific output symbol will be received when a specific input symbol is transmitted.

In simple and well-designed physical communication systems, the probability of a faithful indication at the output of the symbol entered at the input should be fairly high. In a properly functioning telegraph system, for example, the probability of receiving a dash at the output when a dash is entered at the input is near 100 percent. But many communication systems are less reliable than this, and even telegraph systems are subject to static and other malfunctions. The conditional probabilities characteristic of a given information channel take uncertainties of this sort into account.

Technical Addendum

More exactly stated, an information channel consists of a set of input symbols A ($= \{a_i\}$, $i = 1, 2, \ldots, r$), a set of output symbols B ($= \{b_j\}$, $j = 1, 2, \ldots, s$), and a matrix of conditional probabilities $P(b_j/a_i)$ giving the probabilities for all i and j that b_j will be received at the output when a_i is entered at the input. In the case of the simple information channel diagrammed in figure A.1, $P(b_1/a_1) = 1/2 = P(b_2/a_1) = P(b_1/a_2) = P(b_3/a_2) = P(b_2/a_3) = P(b_3/a_3)$, while $P(b_3/a_1)$, $P(b_2/a_2)$, and $P(b_1/a_3)$ all equal zero.

Figure A.1. Simple Information Channel

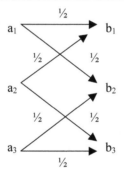

The conditional probability matrix characterizing this channel accordingly is shown in figure A.2.

Figure A.2. Simple Information Channel (Probability Matrix)

	b_1	b_2	b_3
a_1	½	½	0
a_2	½	0	½
a_3	0	½	½

A.4 The Average Information of a Channel's Source

The set of input symbols associated with a given information channel is sometimes known as the channel's "source." A channel's overall capacity to communicate information depends directly upon the amount of information that can be entered via its source. The amount varies both (1) with the number of distinct symbol-events the source provides and (2) with the probabilities of occurrence associated with those events.

Regarding factor 1, it is obvious that more uncertainty is removed by the occurrence of one out of ten equally uncertain events (each 10 percent probable before occurrence) than by the occurrence of one out of two equally uncertain events (each 50 percent probable). In general, the more distinct symbol-events it makes available, the more information a source can introduce into its channel.

Factor 2 comes into play when symbol-events at the source are not all equally probable. For the purpose of comparison, imagine a source providing three equiprobable symbols-events (each approximately 33 percent probable). Each of these events provides the same amount of information upon occurrence (about 1.6 bits). Since each occurs equally often, moreover, the source provides the same amount of information on the average as is provided by any individual occurrence (1.6 bits).

Now imagine a three-symbol source, one symbol-event of which is twice as probable as the other two (one 50 percent probable, the other two 25 percent each). Whereas occurrence of the more probable event results in just one bit of information, occurrence of the others results in two bits each. But the average information of the source is not just the average of these three amounts (⅓ x (1 + 2 + 2)). The reason is that the event providing the least amount of information occurs more frequently than the others and hence has proportionately more influence on the source's average information.

In this latter case, the source's average information is the sum of the information provided by the three symbol-events, each weighted by its individual probability of occurrence (1.5 bits, compared with about 1.6 when the three events are equally probable). Comparable provisions apply whenever the symbol-events of a source diverge from equiprobability.

In general, the closer symbol-events at the source approach equi-probability, the more information the source makes available for transmission through the channel. This quantity will be referred to as the "average information" of the source.[3]

Technical Addendum

The formal definition of "average information" is:

$$H(A) = \Sigma_A \, P(a^i) \, log \, 1/P(a^i)$$

where "Σ_A" indicates the summation of the quantity following it for all values a_i within A. What this equation says, in effect, is that the average amount of information issuing from source A equals the amount of information represented by the occurrence of a symbol-event a_i multiplied by its probability of occurrence, summated over the entire membership of A.

When there are just two symbol-events, a_1 and a_2, available at a source with probabilities 2/3 and 1/3 respectively, individual occurrences of the two convey approximately 0.6 (= $-log$ ⅔) and 1.6 (= $-log$ ⅓) bits of information respectively. In this case, $H(A) = (2/3)(0.6) + (1/3)(1.6) = $ (approximately) 0.9 bits of information.

Similar calculations show that when $P(a_1) = 3P(a_2)$ in a two-member source, then $H(A) = 0.8$; when $P(a_1) = 4P(a_2)$, then $H(A) = 0.7$; and so forth. When a_1 and a_2 are equally likely to occur, however, $H(A) = 1.0$. This illustrates the principle that the average information of an information source is maximal when its symbol-events are equally probable, and it also shows that one bit is the maximum amount of information that can be introduced into a channel by a two-member source.

A.5 Channel Equivocation

The general purpose of an information channel is to make information available at its output pertaining to the occurrence of particular input events. A channel may be more or less reliable in that regard. A given channel is 100 percent reliable if every event at its output gives an accurate and unambiguous indication of the corresponding event at its input. If the identity of events at its input is sometimes left ambiguous by symbol-occurrences at its output, however, the channel is said to be characterized by a certain amount of equivocation. A channel without equivocation is sometimes called a noise-free channel, suggesting that no "noise" occurs to diminish the channel's reliability as a transmitter of information.

Technical Addendum

Channel equivocation is the average amount of uncertainty at the output (B) regarding the identity of symbol-events at the input (A) after the reception of corresponding symbol-events at the output. Overall channel equivocation ($H(A/B)$) is a composite quantity resulting from (but not the same as) the equivocations associated with each output event b_j ($H(A/b_j)$). This latter is determined according to the formula:

$$H(A/b_j) = \Sigma_A \; P(a_i/b_j) \; log \; 1/P(a_i/b_j)$$

for all members a_i of the input set A.

It will be noted that the conditional probabilities in this formula are of input events (a_i) given output events (b_j)—thus, $P(a_i/b_j)$. This is opposite the sense of conditionalities ($P(b_j/a_i)$) defining a channel's probability characteristics (see figure 3.2, for example). Probabilities $P(a_i/b_j)$ can be obtained from probabilities $P(b_j/a_i)$ according to what is known as Bayes's Law:

$$P(a_i/b_j) = P(b_j/a_i)P(a_i)/P(b_j)$$

where $P(b_j)$ (the probability of b_j at the output) is the summation for all cases a_i of $P(b_j/a_i)P(a_i)$.

By definition, in a channel without equivocation there will be a unique input event a_i associated with any given output event b_j, such that $P(a_i/b_j)$ is unity and $-log \; P(a_i/b_j)$ is zero. When these values are summated for all members of A, of course, $H(A/b_j)$ also turns out to be zero.

For this quantity to equal zero means that when b_j occurs at the channel's output, it does so without equivocation.

By way of contrast, we may observe that there are two events (a_1 and a_2) at the input of the channel in figure A.1 that lead to b_1 at the output. For simplicity, assume that the three input events in this channel are equiprobable. It follows from the conditional probabilities of the channel matrix (figure 3.2) and Bayes's Law that both $P(a_1/b_1)$ and $P(a_2/b_1)$ equal one-half. Since $log\,(1/½)\,(=-log\,½)\,(=log\,2)=1$, both $P(a_1/b_1)\,log\,1/P(a_1/b_1)$ and $P(a_2/b_1)\,log\,1/P(a_2/b_1)$ equal one-half. These two pairings together thus contribute $(½ + ½ =)$ one bit of equivocation. Inasmuch as $P(a_3/b_1)$ is zero, no more equivocation comes from this source. $H(A/b_1)$ for this channel, accordingly, is one bit of equivocation.

The equivocation of an informational channel overall $(H(A/B))$ is the summation with respect to B of all values of $H(A/b_j)$ each multiplied by the probability of occurrence of b_j as follows:

$$H(A/B) = \Sigma_B\,P(b_j)\,\Sigma_A\,P(a_i/b_j)\,log\,1/P(a_i/b_j)$$

For any a_i and b_j for which $P(a_i/b_j)$ is unity—that is, for which b_j is an entirely reliable indicator of a_i—the quantity $log\,1/P(a_i/b_j)$ is zero, and the probability $P(b_j)$ is not a factor in the summation. For any b_j such that there is an a_i for which $P(a_i/b_j)$ is less than unity, on the other hand, $P(a_i/b_j)\,log\,1/P(a_i/b_j)$ will have a positive value, and its contribution to $H(A/B)$ will be a factor of $P(b_j)$ as well. When and only when $P(A/b_j)$ is zero for all members of B will the overall equivocation $H(A/B)$ of the channel itself equal zero.

A.6 Mutual Information

An information channel with low equivocation is characterized by a representation at the output of approximately the same events as those presented at the input. Conversely, the higher a channel's equivocation, the less reliable are events at its output as representations of input occurrences. The capacity of a channel to convey information thus varies inversely with the equivocation of its input with respect to its output.

As already noted in section A.4, a channel's capacity to communicate information also varies directly with the average amount of information that can be entered at its input. The difference between the average information of its input and the equivocation of its input with respect to its output thus measures a channel's capacity as a reliable conveyor of information. This quantity is referred to as the "mutual information" of the channel.

Technical Addendum

Given the preceding definitions of $H(A)$ and $H(Å/B)$, the mutual information $(I(A;B))$ of information channel A-B is defined as follows:

$$I(A;B) = H(A) - H(A/B)$$

The term "mutual information" reflects the fact that $I(A;B)$ is the amount of information shared at output B with input A.

A.7 Markov Sources

A "Markov source" can be defined most directly in contrast with a zero-memory source. A zero-memory information source is one such that the symbol-occurrences coming from it are statistically independent. What this means is that the identity of an event emitted by the source has no bearing on the identities of events either preceding or following it.

A Markov source, by contrast, is one in which the symbol-occurrences emanating from it are statistically *inter*dependent. In a first-order Markov source, the probability of a given event's occurring at a given place in the sequence depends to some extent on the identity of the immediately preceding event. When a letter q appears in an English text, for example, it is highly probable that the next letter will be a u. In a second-order Markov source, a given probability of occurrence depends upon the identities of the two preceding events (as qu in English almost always is followed by a vowel). And so it goes for Markov sources of higher orders.

This concept can be adapted to our purposes by thinking of the statistical relations among successive events issuing from a Markov source as equivalent to statistical relations between input and output events in an information channel. What this amounts to is easily illustrated by a specific sequence of symbol-occurrences—say, $s_1, s_2, s_3, s_4, \ldots s_n$—issuing from a first-order Markov source. Given any pair of events in sequence (e.g., s_1 and s_2), the first (s_1) is conceived as occurring at the input of an information channel and the second (s_2) at the channel's output, and so on for other pairs in the sequence (s_2 and s_3, then s_3 and s_4, etc.).

The result is to convert the statistical interdependencies characteristic of a Markov source into statistical interdependencies characteristic of an information channel. As seen in the following section, this will enable us to conceive of successive stages of an operating system as constituting an information channel. We shall also see how the mutual information of that channel can serve as a measure of the corresponding system's degree of order.

Technical Addendum

Consider a sequence of event-occurrences: e_n, e_{n+1}, e_{n+2}, etc. Now consider a series of pairs taken from this sequence, so ordered that the second member of a given pair (e.g., the pair e_n and e_{n+1}) becomes the first member of the pair immediately following (e.g., e_{n+1} and e_{n+2}). In a first-order Markov source yielding the sequence of event-occurrences indicated above, the antecedent likelihood of the pair e_{n+1}, e_{n+2} following the pair e_n, e_{n+1} is $P(e_{n+2}/e_{n+1})$. In a second-order Markov source, this antecedent likelihood is $P(e_{n+2}/e_n, e_{n+1})$. And so on for Markov sources of higher order.

The conditional-probability matrix characteristic of a given information channel can be converted into a matrix characteristic of its Markov-source equivalent by equating output events of the former with appropriate input events. The matrix of figure A.2, for example, is converted into the matrix of an equivalent first-order Markov source by relabeling the columns to match the rows, as shown in figure A.3.

Another way of displaying these relationships is illustrated by the state diagram in figure A.4. Figures A.3 and A.4 represent a first-order

Figure A.3. First-Order Markov Source

	a_1	a_2	a_3
a_1	½	½	0
a_2	½	0	½
a_3	0	½	½

Figure A.4. First-Order Markov Source (Flow Diagram)

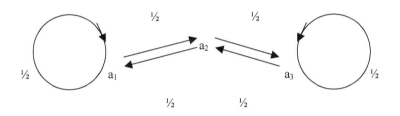

Markov source inasmuch as the probability of occurrence of a given event (heading the columns of figure A.3) is shown conditional upon the occurrence of the immediately preceding event only (indicated at the left of figure A.3).

A state diagram representing a second-order Markov source (with only two members, for simplicity) is shown in figure A.5.

In this example, $P(a_1/a_1, a_1) = P(a_2/a_2, a_2) = 0.7$, $P(a_2/a_1, a_1) = P(a_1/a_2, a_2) = 0.3$, and $P(a_1/a_1, a_2) = P(a_1/a_2, a_1) = P(a_2/a_1, a_1) = P(a_2/a_2, a_1) = 0.5$. In interpreting this figure, it should be noted that if a_1 follows either a_1, a_2 or a_2, a_2, then the next pair in sequence will be a_2, a_1. The designation "a_1, a_2," of course, indicates two events in that particular order, and so on for the other pairings.

Figure A.5. Second-Order Markov Source

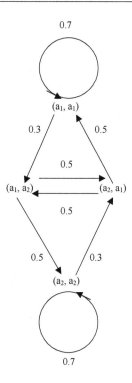

A.8 Operating Systems as Markov Sources

The definition of orderliness in section 2.5 specifies that an operating system's degree of order (1) varies directly with the number of interacting features and (2) varies inversely with the degree of independence among those features. Working equivalents of 1 and 2 can be expressed in terms of average information and equivocation respectively, leading to a definition of an operating system's degree of order in information-theoretic terms. Let us consider how.

The average information of an information source is the summation of the amounts of information represented by its individual members, multiplied in each case by probability of occurrence (section A.4). This quantity generally increases with number of members. When the source is a set of all configurations exhibited by an operating system during normal operation, these configurations count as members of the source. Accordingly, the average information of such a source generally increases with number of configurations.

Technical Addendum

In the case of sources with equiprobable members, their average information increases with number of members in an entirely regular manner. Qualification is required for sources whose members are not all equiprobable. Probabilities of occurrence of an n-membered source might be distributed among members in such a fashion that it has lower average information than a source with n-1 members. In a six-member source, for example, five might occur 10 percent of the time each and the other 50 percent of the time. The source's average information in this case is 2.16 bits, less than that of a five-membered source with equiprobable members (2.32 bits). Departures of this sort from a regular progression are unlikely to extend more than a step or two in the sequence. Thus a seven-membered source with one probability of 40 percent and the other six 10 percent has an average information of 2.52 bits, more than the 2.32 bits of an equiprobable five-membered source. Taking irregularities of this sort into account, it remains accurate to say that average information of information sources generally increases with number of members.

With regard to condition 1 in the definition of orderliness, we thus may say that an operating system's degree of order varies directly with the number of configurations it exhibits. This is to say that its degree of order varies directly with its average information as an information source.

We turn now to the second condition. An information channel A-B is reliable to the extent that events at output B provide unambiguous indications of corresponding events at input A. Ambiguity in that regard is referred to as equivocation. Channel equivocation is defined as the average amount of uncertainty left by the occurrence of events at B regarding the identity of corresponding events at A (section A.5).

To treat a first-order Markov source as an information channel is to treat each successive pair of events issuing from the source (events s_1, s_2, s_3, s_4, . . . s_n) as containing both an input and an output event (section A.7). In pair (s_1, s_2), for instance, s_1 is input and s_2 output; in pair (s_2, s_3), s_2 is input and s_3 output; and so forth. With higher order Markov sources, more preceding events are taken into account. In a second-order source, for example, the successive occurrences s_1, s_2, s_3 would contain s_3 as output and the pair s_1, s_2 as input. In any such case, equivocation of a Markov source is the average uncertainty regarding the identity of relevant antecedents after a given event has occurred.

When an operating system is treated as a Markov source, the successive configurations it exhibits through time are treated as a succession of events issuing from the source. The equivocation of an operating system so conceived is the average uncertainty left after the occurrence of a given configuration regarding the identity of relevant configurations preceding it. This uncertainty increases with increasing independence among configurations involved, and vice versa.

With regard to condition 2 of the previous definition of orderliness, accordingly, we may say that an operating system's degree of order varies inversely with its equivocation. The more independence on the average between successive configurations, the lower its degree of order, and vice versa. Conceived in this fashion, an operating system's equivocation measures the extent to which its successive configurations are independent.

As a final step, let us recall that a channel's mutual information is defined as the difference between the average information of its input and its equivocation (section A.6). This definition carries over to Markov sources generally and hence to operating systems considered as Markov sources. Inasmuch as an operating system's degree of order (1) varies directly with its average information as a Markov source and (2) varies inversely with its equivocation, the system's degree of order varies directly with its mutual information. In brief, the mutual information of an operating system treated as a Markov source serves as a measure of its degree of order.

A.9 Examples

In its normal operation, the children's teeter-totter considered in section A.1 functions as a Markov source issuing a series of repeating configurations (*b, a, c, a, b, a, c,* . . .). In this series, the probability of *a*'s occurrence at a given point is 50 percent, and those of *b*'s occurrence and of *c*'s occurrence 25 percent each. Operating normally (i.e., with complete regularity), this system loses no information to equivocation. Given these statistics, its mutual information is (1.5 – 0 =) 1.5 bits of information.

The other example considered in section A.1 is that of an alarm clock with both minute and second hands. In normal operation, the clock issues at least (12 x 60 x 60 =) 43,200 distinct configurations (with a few more when the alarm sounds). Its average information as a Markov source thus is between 15 and 16 bits of information (2^{15} = 32,768, 2^{16} = 65,536; see section A.2). Since no equivocation is present during normal operation, this is also the quantity of its mutual information.

These examples have been chosen for the ease with which their mutual information can be calculated. It would be considerably more complicated to determine the mutual information characterizing an automobile engine, and even more so in the case of a biological system. Our present purpose does not require actual calculations in such complex cases. The purpose of the appendix is merely to show how, in principle, such calculations might be made. To show this is to show that the orderliness of biological systems, in principle, admits quantitative measurement. Development of specific measures with practical application is a challenge this appendix does not address.

3

Life, Negentropy, and Biological Feedback

3.1 Earth before Life Appeared

Earth may have originated as a molten mass accumulated (over millions of years) from materials left over from the formation of the sun. In its early stages of cooling, if so, it would have radiated far more thermal energy outward than incoming solar radiation could account for. The trade-off between high-grade solar energy coming in and low-grade heat energy going out would have been tipped strongly in the latter direction.

At some stage along the way, however, the cooling process tapered off, a solid crust formed on the surface, and the surface temperature reached an approximate equilibrium. For this to happen, the amount of energy reaching the surface (internally by continued cooling and externally by solar radiation) would have to match the amount leaving the surface as low-grade heat. Once its temperature had reached a stage of approximate equilibrium, the earth's surface was ready for the appearance of life.

Other conditions, of course, had to be right for life to make an initial appearance. Once its surface had reached thermal equilibrium, the globe had to be spinning at a rate and angle relative to the sun that maintained periodic temperature variations within tolerable limits. Its magnetic field had to be aligned to deflect high-frequency solar particle radiation that would have been fatal to primitive organisms. And its gravitational pull had to be weak enough to allow the formation of a gaseous atmosphere but strong enough to keep molecules on which life depends (e.g., O_2, CO_2, H_2O) from drifting off into space. These are only a few such conditions that might be mentioned.

Our concern in this chapter, however, is limited to energy flows on and about the earth's surface. As we shall see, not only has the life process become a major factor determining the character of these flows, but it is now interacting with these flows in a manner that might prove self-destructive.

3.2 The Nature of Life

In one way or another, presumably, the question of the nature of life engages the interests of most reflective people. One does not have to be a philosopher to ask whether life has meaning, or a theologian to wonder about its ultimate source. Whatever one makes of them, such concerns do not preclude viewing life as fundamentally a biological process. This latter view will remain dominant through most of the present study.

Scientists commenting on the nature of life seem to agree on its highly improbable character. For example, Carl Sagan (astrophysicist) has remarked that the "improbability of contemporary organisms . . . is so great that these organisms could not possibly have arisen by purely random processes." Konrad Lorenz (ethologist) has spoken of life as "a steady state of enormous general improbability," and R. A. Fisher (statistician) has described biological evolution as "a mechanism for generating an exceedingly high degree of improbability."[1] Given the close association of improbability with order and usable energy (described in section 2.6), this suggests that life is essentially characterized by its orderliness and its use of energy.

There is less agreement among scientists regarding the nature of this phenomenon they find so improbable. Some think of life in terms of its physiological functions (reproduction, growth, response to stimuli), while others focus on its biochemical features (being based on proteins, metabolizing under the control of enzymes). Yet others stress genetic and evolutionary characteristics (genetic instructions passed from parent to offspring, species shaped through natural selection). While some of these features may be *necessary* and others *sufficient,* it remains debatable whether any are *both* necessary and sufficient for life as we know it.

There are enterprises for which a precise definition of life in terms of necessary and sufficient conditions is important. One such enterprise is inquiry into the moral status of certain medical procedures. If genetic duplication of cells in a fertilized ovum is sufficient for life, then a life is snuffed out by an early-term abortion. Again, if metabolic self-sufficiency is necessary for life, then removing a brain-dead person from life support is not letting a person die.

Another enterprise requiring a precise definition is our ongoing search for life on other planets. Is oxygen necessary for the metabolism of living organisms, or could life be present in an atmosphere (like that of Mars) consisting mostly of carbon dioxide? Again, are organic compounds like nucleic acids sufficient for life's occurrence wherever they are found, or would the discovery of such compounds on other planets leave open the question of whether life itself is also present?

Important as these enterprises may be in their own right, however, our present inquiry does not require a precise definition of life. As already noted, what our inquiry requires is an account of life that makes clear its relation to its surrounding environment. What we need in particular is an account of how a living organism exchanges energy with its environment and of how the environment is affected by this exchange. The present chapter is intended to provide an account of this sort.

3.3 Schrödinger's Characterization

In his now-classic description of metabolism, the physicist Erwin Schrödinger remarked that "the device by which an organism maintains itself stationary at a fairly high level of orderliness (= fairly low level of entropy) really consists in continually sucking orderliness from its environment."[2] To "suck up orderliness" is tantamount to receiving usable energy, which is equivalent in turn to acquiring negentropy (explained in section 2.7). Another remark of Schrödinger, in terms of the latter, is: "What an organism feeds upon is negative entropy. Or, to put it less paradoxically, the essential thing in metabolism is that the organism succeeds in freeing itself from all the entropy it cannot help producing while alive."[3] However one puts it, the point is simple: as a result of the interaction between organism and environment, the organism gains order, and the environment gains entropy.

Not only is a living organism able to "suck up orderliness from its environment," but it is able typically to acquire order (or energy) from components of its environment that exist at lower levels of order (or energy) than itself. For example, acorns and carrots are structurally less complex, and contain less energy for useful work, than the squirrels and rabbits they serve as food. A rabbit eating a carrot is roughly equivalent thermodynamically to a piece of warm toast receiving additional warmth from cold butter laid upon it (making the butter yet colder).

It is evident that the relation between rabbit and carrot is quite different from what actually happens when warm toast comes in contact with cold butter. In the latter case, heat energy flows from a higher level in the toast to a lower level in the butter. In the interchange between rabbit and carrot, however, energy is transferred from a lower level in the (cooler and structurally less complex) carrot to a higher level in the (warmer and more complex) rabbit. The same apparently "reverse" movement of energy occurs when a squirrel eats an acorn or when a hawk eats a squirrel. This raises questions about the relevance of the Second Law of Thermodynamics (as discussed in chapter 1) to processes involving the metabolic activity of living organisms.

3.4 An Apparent Conflict with the Second Law

At first glance, the metabolic process as Schrödinger described it appears contrary to the Second Law of Thermodynamics. According to this law, transformations of energy always tend to result in greater degrees of disorder and greater amounts of entropy. Yet in the energy transactions between organism and environment, the organism gains order and casts off entropy.

As commentators often note, however, the appearance of conflict with the Second Law arises only when we focus too narrowly on the organism itself. When the consequences for both organism and environment are taken into account, any hint of conflict disappears. The order gained by the organism is lost by the environment, which is what Schrödinger meant by the organism's "sucking orderliness from its environment." Another way of putting it is that the entropy lost by the organism is gained by the environment, which is tantamount to the organism's "freeing itself from all the entropy it cannot help producing while alive."

Far from conflicting with the Second Law, the metabolic activity of living organisms provides an instructive illustration. The amount of entropy passed off to its environment by a living organism invariably is greater than the amount of order it gains from its environment initially. The environmental cost of the organism's enhanced order is always a comparatively greater amount of disorder within the environment itself. When both organism and environment are considered together, accordingly, the upshot is that any interchange in which the organism gains structure or energy will result in an overall increase in entropy.

In effect, the life process serves as a "catalyst" speeding up the production of entropy in its surrounding environment. As a consequence of playing host to the organisms living within it, a given sector of the environment becomes disordered more quickly than it would if no life were present. This is what the Second Law leads us to expect in stating that entropy tends to increase as a result of any (irreversible[4]) natural process.

Another aspect of this interaction is that an environment tends to gain entropy more quickly as the organisms it supports increase in complexity. The reason is that increased complexity goes hand in hand with increased amounts of energy needed to support the metabolisms of the organisms in question. Since human beings have greater energy needs than squirrels and rabbits, for example, humanity inflicts more disorder than these other creatures upon its supporting environment. The fact that humanity at large discharges more entropy per capita into the biosphere than any other species figures prominently in the following discussion.

3.5 Feedback Connections between Organism and Environment

For an organism to gain energy from its surrounding environment, it must be appropriately coupled with specific energy sources. For example, a carrot can serve a rabbit as a source of energy only if the rabbit is near the carrot, can perceive it as such, and is capable of chewing and digesting it once the pieces are swallowed. Generally speaking, a given organism must be properly connected to its surroundings before it can gain the energy it needs for growth and sustenance. Conjoining the organism with its environment in the requisite fashion is the role of what biologists (following engineers) call "feedback."

Feedback occurs when the activity of an operating system[5] is influenced in turn by the effect this activity has upon its operating environment. While feedback occurs in many forms, all can be classified as either negative or positive.

Contrary to their common connotations, the term "positive" here does not indicate something particularly desirable (as with a "positive response" from a theater audience), and "negative" does not indicate something undesirable (as with a "negative review" by a music critic). In this context, both terms take on meaning with respect to deviation from a standard state—for example, a specific room temperature. Positive feedback increases deviation from the norm, as when heat from uncontrolled burning makes a fire burn yet hotter. Negative feedback, on the other hand, decreases ("negates") deviating activity, as when an air conditioner returns room temperature to a preset level.

Positive feedback is a source of instability that, if unchecked, can lead to destruction of an operating system. An example is a pothole in a heavily traveled road surface, which breaks up more rapidly as the hole gets bigger. Another is the explosion of a keg of gunpowder, in which the rate of oxidation (burning) increases proportionately to the heat produced by the oxidation already under way. The same dynamic is at work in a so-called population explosion (in which a population increases more rapidly the bigger it gets), except that the social process might continue through many generations, whereas gunpowder explodes within a matter of milliseconds. What happens in each of these cases is an increasingly rapid deviation from a standard condition (a smooth road, nonincendiary temperature, steady population level), exacerbated by changes introduced into the system by the ongoing process.

Standard examples of negative feedback are thermostatically controlled heating and cooling systems, which operate to maintain living spaces at stable temperatures. A familiar biological example is the process by which level of illumination is regulated on the retina. When light striking the retina exceeds an optimal level of intensity, the pupil of the eye contracts to reduce incoming light energy; when illumination decreases below optimal level, the pupil dilates to let more light enter.

Negative feedback is a process of stability and control and, as such, is essential to the life process itself. In the following section we look at various kinds of negative feedback operating within individual organisms. Then we consider negative feedback in the interactions among different organisms, which leads directly to the topic of ecosystems.

3.6 Biological Forms of Negative Feedback

There are two general types of negative feedback a living organism might depend on in maintaining a viable relation with its surrounding environment. One type works by making adjustments internal to the organism itself. This type is the basis of the biological process known as "homeostasis," and hence might be designated homeostatic itself. The other works by adjusting the relation between organism and environment. This second type has been referred to as "heterotelic" feedback.

As its name indicates, homeostasis is a process by which a system maintains itself (hence "homeo-") in a stable ("-stasis") operating condition. Stable operation typically requires holding certain key variables of the system (e.g., temperature in mammals) within a normal range of values. A key variable is protected by homeostatic feedback when its deviation from normality is countered by other changes within the system itself. Even when deviation is brought about by external causes, the system compensates by internal adjustments that return the values of its protected variables to a tolerable range.

Homeostasis is illustrated by the way some mammals control excessive body temperature by perspiration. Human body temperature, for example, must be held close to 98.6° Fahrenheit for normal metabolic activity. When ambient temperature is lower than body temperature, metabolic heat is discharged from the body surface by convection and radiation. But when ambient temperature exceeds body temperature, these same processes tend to transfer heat into the body instead. The body responds to this counterproductive tendency by emitting perspiration, which then cools the body's surface as it evaporates. Rates of perspiration vary both with ambient temperature and with humidity (under tropical conditions, the sweat glands of an adult human can emit over three liters of moisture an hour). Sweating thus is a homeostatic adjustment by which the body counteracts increases in temperature induced from without.

The other general type of negative feedback in biological systems is one in which the organism maintains stability under changing external circumstances by taking action that affects those circumstances directly. This type is called "heterotelic" because it helps bring about or maintain an optimal state of the system by adjusting its relation to other ("hetero-") factors. A nonbiological example is a target-seeking missile that responds to changes in its target's position by corresponding changes in its own direction, thus maintaining a state of active pursuit. Biological examples include the daisy that maximizes its input of solar energy by keeping its petals directed toward the sun (heliotropism) and the oak tree that assures itself a ready supply of water by sending its roots deep into the moist underground.

Technical Addendum

The process of perspiration has been mentioned as an example of homeostatic feedback. If we represent the protected states of a given organism (e.g., body temperature) by S, relevant states of its operating environment (e.g., ambient temperature) by O, and the effector mechanisms involved (e.g., sweat glands) by E, then homeostasis feedback can be represented schematically as in figure 3.1.

Figure 3.1. Homeostatic Feedback

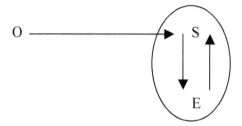

The point to be noted here is that the organic system (encircled) responds to external disturbances (the arrow from O to S) by engaging effectors (the arrow from S to E) that produce compensating changes within the organism itself. As indicated by the return arrow from E to S, these effectors are internally directed.

Heterotelic feedback, by contrast, is a process by which the system responds to external provocation by adjusting its relation to the environmental factors involved. Since heterotelic feedback works through its effect on the operating environment, it can be diagrammed as shown in figure 3.2 (reading S, O, and E the same as above).

Figure 3.2. Heterotelic Feedback

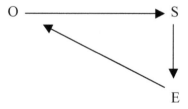

Because the activity of E is directed back to O, where the feedback is initiated, interactions of this sort ($O \to S \to E \to O$) are called "feedback loops." As we shall see presently, heterotelic feedback is the basis of more complex forms of feedback typical of organisms at the upper levels of the food chain.

Of the two types of negative feedback considered here, that underlying homeostasis seems to be biologically more fundamental. One reason is that any organism requires active homeostasis for continued vitality, whereas it is not uncommon for organisms to remain viable without active heterotelic involvement (e.g., when they are asleep). Another reason concerns the effect upon the organism when environmental perturbation is excessive. If an organism's only means of coping is by internal changes (homeostasis), external stress might produce results it cannot tolerate. (A deer cannot survive a forest fire merely by sweating.) But if it is also capable of altering its external circumstances when threatened, the organism can avoid effects that might otherwise prove fatal. (A deer capable of running can survive the fire.) Heterotelic behavior thus succeeds homeostasis in the course of ethological development.

3.7 Extensions of Heterotelic Feedback

Once heterotelic feedback is in place, it can provide the basis for other forms of feedback that are highly beneficial to the organisms involved. One such is "sentient" feedback, which enables perception at a distance of external objects. Heterotelic feedback by itself enables avoidance behavior, but the states it protects (S in figure 3.2) must be disturbed for this response to occur. If the effect of the operating environment (O) on S is particularly severe, the organism might lose its ability to escape further damage. (A rabbit can escape a hawk by taking cover under brambles, but this capacity is of little help if its first warning is the pain of talons in its back.) The survival value of heterotelic feedback is greatly enhanced if it is accompanied by an ability to receive notice of present danger without protected states S being put at risk. This is the role of sentient feedback.

Yet further protection of S can be achieved if the system is able to respond to threatening environmental states O before they occur and to take action preventing their actual occurrence. In point of fact, many organisms are capable of sentient feedback in which their behavior is governed by sensory response to *antecedents* of external states that would affect them adversely, in lieu of responding to these states only when actually present. This enhanced capacity may be designated "anticipatory" feedback.

On the part of nature at large, anticipatory feedback is supported by patterns of regular association among events that allow reliable expectation of a particular event before it occurs. On the part of behaving organisms, this facility is enabled sometimes by behavioral conditioning (Pavlov's dog was led to anticipate food by the sound of a bell) and sometimes by advanced cognitive faculties like memory and reason. As with sentient feedback generally, all anticipatory capacities involve the processing of information originating in the external environment.[6]

This brief discussion of sentient and anticipatory feedback has been confined to their usefulness in responding to potential dangers. It should be noted that there are parallel advantages to be had in pursuing opportunities from which the organism can benefit directly. A rabbit being chased will be served by these capacities, but so will the hound that is giving chase. An animal's role in the food web (see chapter 4) depends substantially upon the nature of the negative feedback capacities it employs.

Technical Addendum

Sentient feedback involves receiving notice of impending danger before it occurs and receiving this in a form not likely to harm the organism. These general features are illustrated in figure 3.3.

The significant difference here from figure 3.2 above is that S has been displaced by S' and connected to it by a delay factor (δ). The relation between S' and S is such that (1) for every circumstance O of the operating environment that affects a protected system state S there is another system state S' that is induced by O, (2) S' is not vital to the system's continued operation (and thus does not require the same level of

Figure 3.3. Sentient Feedback

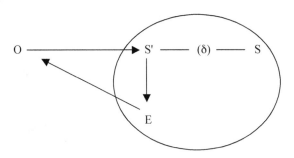

protection), and (3) change in *S'* is temporally prior to change in *S* should the latter occur. The symbol δ is intended to indicate the temporal delay of *S* relative to *S'*.

In actual biological systems, provisions 1, 2, and 3 may be achieved by the development of specialized receptors (*S'*) that respond more quickly to *O* than do the variables *S* protected by the system. As figure 3.3 makes apparent, these vital variables *S* are more fully protected than in simple heterotelic feedback by being removed from the primary feedback loop (*O* → *S'* → *E* → *O*). The increased survival value of such features would normally provide impetus for their development in the context of natural selection.

The enhanced protection afforded by anticipatory feedback, in turn, is illustrated by the features added to figure 3.4.

Figure 3.4. Anticipatory Feedback

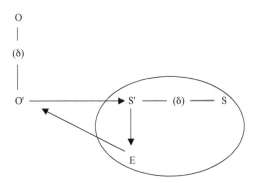

In this depiction, O' is an environmental circumstance that regularly precedes a threatening event O, and δ is a temporal delay (not necessarily the same) as before. Figure 3.4 shows how protected system state S is isolated from threatening event O by two distinct levels of "cushioning," in that both S and O are now removed from the primary feedback loop. The survival value of this anticipatory capacity should be obvious.

In section 3.6 it was noted that homeostatic feedback is a more fundamental process than the heterotelic version. An organism might have homeostatic without heterotelic capacities, but probably not vice versa. A similar relation exists between sentient and anticipatory feedback. Generally speaking, an organism with both heterotelic and homeostatic feedback is more complex than one with homeostatic alone, and among the former an organism with anticipatory feedback is more complex than one with sentient feedback only.

3.8 Ecosystems Based on Feedback Processes

By way of review, we may recall that most energy exchanges in the natural world involve a flow from higher to lower energy levels (e.g., hot toast to cold butter). A distinguishing feature of living organisms is that they are sustained by energy flows in the opposite direction (section 3.3). In Schrödinger's picturesque phrase, living creatures are capable of "sucking up" energy (and its equivalent, orderliness) from their environments. The capacities enabling living organisms to participate in this counterflow of energy are based on various forms of negative feedback (section 3.5).

Plants typically receive their energy from solar radiation, acting through the process of photosynthesis. This process is served by a plant's ability to orient its leaves and blossoms for maximum exposure to sunlight, a trait known as heliotropism. Heliotropism is a classic example of negative feedback (section 3.6). Inasmuch as herbivores feed on plants, and carnivores on herbivores, heliotropism is an essential aspect of the food web generally.

Although parasite larvae (e.g., those of the gypsy moth) feed on plants they already occupy, most herbivores (think of deer and rabbits) need to find suitable food sources before they can start feeding. This requires some form of sentient feedback (section 3.7; visual, tactual, olfactory, etc.). The same may be said of the carnivores that feed on the herbivores. As the squirrel relies on smell to find its buried acorn, so the hawk relies on vision to locate the squirrel. And both squirrel and hawk fall prey to the great horned owl, guided by both sight and hearing in its nocturnal predations.

Regardless of place in the food web, negative feedback is essential to the behavior patterns of prey and predator alike. Prey and predator play reciprocal roles that are determined by the feedback loops in which they participate. To be part of a food web ipso facto is to belong to a system of feedback loops that establish pathways of energy flow from organism to organism.

Imagine a system of this sort keyed to the needs of a particular group of predators—say, a mating pair of great horned owls. To keep it simple, let us assume that this pair exists on a diet of rodents and songbirds, supplemented by an occasional raccoon or skunk. During its lifetime of ten or so years, each owl will consume several thousands of these other creatures. And each of these latter likewise will have its own typical diet. Like skunks and raccoons, most rodents and songbirds are either carnivores or insectivores (insect eaters), and most will consume various plant products (roots, leaves, seeds, berries) as well.

To gain the energy they pass on to the owls, individuals of these other groups are occupied with eating during most of their waking moments. A field mouse will consume as much as a quarter pound of grain per month, along with myriads of berries, nuts, and insects. Songbirds consume roughly their weight in seeds and insects per week. Being omnivores, skunks and raccoons typically will eat whenever they find an opportunity. Given ample food supplies, individuals of these species can double their weight in a few weeks before hibernation.

Here are several stages in a network of energy transferal held together by distinctive forms of negative feedback. The owls have capacities of sentient feedback enabling them to locate and to capture their prey. Similar capacities guide the latter in avoidance behavior and enable them to locate their food in turn. The fortunate songbird can see an owl approaching and also can pick out edible berries on the basis of color. Sentient feedback enables the skunk not only to avoid its predators but also to seek out vegetation it can use as food. Other feedback loops come into play in the case of the insects (e.g., grasshoppers) that feed the raccoons and field mice and that feed themselves upon suitable stalks and leaves.

As far as the meat eaters, insect eaters, and plant eaters of this particular food web are concerned, the system can be thought of as organized in a treelike configuration. The pair of great horned owls is on top, followed by large numbers of rodents and songbirds, skunks and raccoons. On another level we find countless insects that complete the diets of the other species. Tying these levels together are various forms of sentient feedback, serving purposes of pursuit and avoidance alike.

The pattern continues on the level of the plants with which the upward flow of energy begins. At this fundamental stage, however, sentient feedback becomes less a factor and other forms of negative feedback take over. One form mentioned previously is the process of heliotropism, by which plants orient their leaves for maximum reception of energy from the sun. Other forms are the processes by which plants exchange oxygen for carbon dioxide and those by which matter from decomposing organisms is recycled in growing vegetation. These other forms of feedback will be examined in subsequent chapters.

Despite the ultra simplicity of this imaginary network (e.g., the great horned owl in fact preys on roughly 250 species), it serves as a model of what biologists call an ecosystem.[7] For present purposes, we may define an ecosystem as an operating system comprised of organisms and their sustaining environment in which every organism either supports or is supported by other organisms in the system. As this definition suggests, there is more to an ecosystem than its interactions between prey and predators, and more to these interactions than the feedback loops that sustain them. It remains the case, nonetheless, that ecosystems maintain their integrity through the operation of feedback loops among their constituent organisms. The flow of energy from plant to herbivore to carnivore to omnivore follows channels established by these feedback operations. In a manner of speaking, feedback interactions of this sort are the ligatures by which ecosystems are held together.

The same may be said of the biosphere at large. As may be recalled, this discussion of feedback and energy flows began with observations (section 3.1) about the effect of life on energy flows on and about the earth's surface generally. If we think of the biosphere as the totality of ecosystems on the face of the earth (more on this later), our initial interest boils down to a concern with energy flows pertaining to the biosphere at large. The final section of this chapter sets the stage for a more detailed discussion of this topic in chapter 4.

3.9 Energy Flows through the Biosphere

Apart from occasional excursions into space by astronauts and cosmonauts, life as we know it is confined to a thin envelope around the earth's surface. This envelope is known as the "biosphere." By definition, the biosphere includes all organisms living within that envelope. As early researchers were quick to realize, it also includes the many geological features of the surface (land masses, waterways, atmosphere) on which living organisms rely for existence.[8] The biosphere may be described as the totality of living creatures, together with essential nonliving resources provided by their surrounding environment.

One feature of the biosphere that early researchers generally overlooked is its discharge of "used up" energy in the form of low-grade heat. Although it has long been obvious that life depends on an input of solar energy, it has become apparent only recently that life is no less dependent upon the radiation of low-grade heat energy back into space.[9] By its very nature, metabolic activity within the biosphere produces heat. This heat is passed off into the surrounding environment, where it joins heat produced by nonbiological activity and eventually leaves the earth in the form of low-frequency (black-body) radiation.

These flows of radiational energy into, through, and out of the biosphere are represented in figure 3.5.

Figure 3.5. Energy Flows through the Biosphere

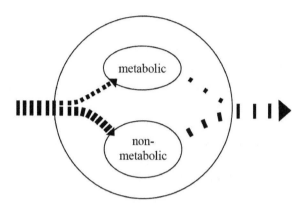

Here the circle represents the biosphere, the arrows with the densely packed line segments (to the left) represent incoming solar radiation,[10] and those with less densely packed segments (to the right) represent outgoing radiation in the form of low-grade heat. After entering the biosphere, solar energy is directed in part to metabolic activity and in (larger) part to nonmetabolic activity, both of which produce low-grade heat to be discharged back into space.

The difference in amplitude shown between metabolic and nonmetabolic flows is intended to indicate that relatively little of the solar radiation reaching the earth's surface becomes involved in biological activity. About one quarter of the sun's rays penetrating the atmosphere are in wavelengths useful for photosynthesis, and of these only about 1 percent contributes to biomass production.[11] Most solar radiation entering the biosphere falls on surfaces where it is converted directly into heat. Solar energy absorbed by desert surfaces during the day, for instance, is typically converted into heat and radiated back into space at night without being engaged in biological processes.

Even flows of energy with no biological involvement, however, pass through the biosphere before being released into space as infrared radiation. The reason is that all radiation arriving at and departing from the earth's surface passes through the atmosphere, and the earth's atmosphere is heavily involved in biological activity. This is one respect in which life has become a major force in determining the character of energy flows on and about the earth's surface, as observed at the beginning of this chapter.

The fact that earth's low-grade heat is discharged through its atmosphere leads directly to the phenomenon currently known as "global warming." Human industry contributes significantly to the accumulation of greenhouse gases in the atmosphere, which impede the radiation of low-grade heat into space. The buildup of waste heat that results is currently producing harmful effects on the biosphere at large. We return for a closer look at global warming in chapter 5.

Another feature of the biosphere to which early writers paid little attention is its distinctively hierarchical structure. Like the countless ecosystems contained within it, the biosphere consists of organisms that feed on other organisms, which feed on yet other organisms in turn. Our consideration of energy flows through the biosphere continues in chapter 4 with discussion of the hierarchical arrangement of the ecosystems involved in those flows.

4

Ecosystems and Top Consumers

4.1 The Trophic Organization of Ecosystems

In the previous chapter, an ecosystem was characterized provisionally as a hierarchy ordered by prey and predator, interconnected through a network of feedback loops. This characterization of ecosystems enabled us to make some preliminary observations about the flow of energy through biological systems, including the comprehensive ecosystem consisting of the biosphere at large. We turn now for a closer look at how ecosystems are organized.

Broadly conceived, an ecosystem is a system of living and non-living components that interact in filling the vital needs of the organisms involved (as discussed in section 3.8). Among their constituent organisms, ecosystems typically include *producers* (plants and protozoa capable of photosynthesis), *consumers* (organisms that eat other organisms), and *decomposers* (bacteria, worms, fungi, etc.) that break down organic materials cast off by other organisms. Within the class of consumers, a further distinction is made between *primary consumers* or *herbivores* (organisms that eat only plants) and *secondary consumers,* with the latter broken down into *carnivores* (meat eaters) and *omnivores* (animals that eat both plants and other animals).

Among the nonliving components of an ecosystem are the physical habitats (aquatic or terrestrial) in which its organisms interact, the supplies of minerals and other materials they rely on for growth, and various other features of the environment that affect their vital processes. There is nothing hierarchical about the arrangement of these physical components. An ecosystem owes its hierarchical structure entirely to the interactions among its constituent organisms.

Our previous characterization of this structure in terms of prey and predator applies mainly to consumers and does not take the roles of producers and decomposers into account. Let us extend our characterization in a way that brings the nutritional (trophic[1]) roles of producers, consumers, and decomposers together in a comprehensive picture. The hierarchical arrangement of these several trophic levels is commonly depicted in some form like in figure 4.1.

Figure 4.1. Trophic Structure of an Ecosystem

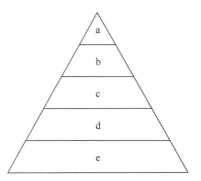

(a) omnivores (humans, bears)
(b) carnivores (lions, owls, snakes)
(c) herbivores (deer, rabbits, mice)
(d) producers (plants)
(e) decomposers (bacteria, beetles, worms, fungi)

The hierarchy represented in this figure is one of nutritional dependency. Organisms on each level are dependent on organisms below for nutrients essential to growth and metabolism. The obvious exception is the bottom level of decomposers, which take nutrition from organic matter left behind by organisms on the several levels above.

Clearly, figure 4.1 is simplified in many ways. For instance, it gives no indication that some plants are carnivores (pitcher plants feed on insects and sometimes small mammals) or that carnivores sometimes feed on omnivores (sharks eat people) or other carnivores (great horned owls eat snakes). Another simplification is that all organisms mentioned by way of example belong to terrestrial ecosystems. An aquatic illustration might have herring feeding on plankton, seals on herring, and polar bears on seals. It should also be noted that ecosystems have varying numbers of trophic levels, depending on the identity of their top-level consumers. An ecosystem with a population of golden eagles on top will have more levels overall than one with a colony of prairie dogs in that position.

An ecosystem supporting a population of wide-ranging omnivores (such as human beings), moreover, might contain many sequences of nutritional dependency, distributed through many trophic levels. The immediate lesson of figure 4.1 is simply that organisms on the top level of any given ecosystem depend on lower levels for their existence. When an ecosystem supporting a given top-level population collapses (consider the plight of the Easter Islanders[2]), the population it supports goes down with it.

4.2 Energy Flows in Ecosystems

Roughly 30 percent of the solar radiation reaching the earth's outer atmosphere is reflected back into space, and another 20 percent is absorbed by the atmosphere. The remaining 50 percent reaches the ground or ocean surface, where most of it is converted directly into heat. As already noted in connection with figure 3.5, about one-quarter of the solar radiation penetrating the atmosphere is in wavelengths useful for photosynthesis, of which only about 1 percent is actually engaged in that process. Our present concern is with this latter small portion.

From an energetic perspective, ecosystems may be viewed as a means of channeling solar energy through the metabolisms of their constituent organisms. Several stages are involved in this process, the most notable for present purposes being (1) the conversion of solar energy into biomass by photosynthesis, (2) the conversion of chemical energy in plants to more concentrated forms serving the energy needs of consumer organisms, and (3) the discharge of low-grade energy (entropy) expended in metabolic activity back into the surrounding environment.

Photosynthesis is a biochemical reaction, energized by sunlight, that converts carbon dioxide and water into carbohydrates and oxygen. A by-product of this transaction is the formation of phosphates with high-energy bonds (primarily adenosine triphosphate, or ATP) that store energy for eventual use by consumer organisms. Photosynthesis takes place in chlorophyll-bearing plants and algae (including phytoplankton), which serve as producers in their respective ecosystems (terrestrial or aquatic). This process is the primary source of nutrients on which other organisms in the biosphere ultimately depend.

Over half the energy fixed in photosynthesis is used in sustaining the metabolisms of producer organisms. Most of the rest is converted to forms usable by consumers. It has been estimated that between 10 percent and 20 percent of energy fixed by producers (less for aquatic than for terrestrial systems) is passed on to herbivores, about the same proportion of this is passed on to the first level of carnivores, and so on through successive trophic levels.[3] This means that in ecosystems with three or four levels of consumers (e.g., one supporting a population of grizzly bears) considerably less than 1 percent of the energy fixed on the producer level will reach the level of top consumer.

Despite this diminution in amount of energy passed from lower to upper trophic levels, there is a progressive concentration of energy made available to individual consumers. In the food chain from plankton to herring to salmon to polar bear, for instance, the salmon requires a great deal more energy than the herring, and the bear more yet than the salmon. This is one obvious reason why there are many more individual organisms at the bottom than at the top of a typical ecosystem.

The final stage in the passage of energy through an ecosystem is the discharge of the entropy that, in the words of Schrödinger (see chapter 3), an organism "cannot help producing while alive." All energy expended in the life process takes the form of entropy, and an organism must rid itself of this entropy on a continual basis just to stay alive. A consequence of top consumers consuming more energy than organisms below them (e.g., polar bears more than salmon) is that they produce more entropy to be discharged into the environment.

In a healthy ecosystem, some of the degraded energy produced by its constituent organisms will be transferred to neighboring ecosystems, and some will be directly emitted into space as low-grade heat. Within the economy of the biosphere overall, however, even low-grade energy that is diverted from one ecosystem to another is destined eventually to leave the planet as heat radiation (see figure 3.5).

The basic dynamics of energy flow in a typical ecosystem are represented in the figure 4.2, which retains the hierarchical structure of figure 4.1.

Figure 4.2. Energy Flow in an Ecosystem

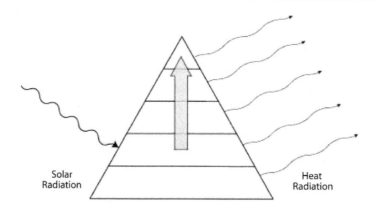

The "high-frequency" arrow on the left side represents incoming solar radiation, shown entering at the producer level (*d* of figure 4.1). The broad arrow in the middle represents the flow of usable energy upward from producers to consumers. To keep things simple, there is no indication of the relatively minor counterflow of energy from the upper levels that fuels the metabolisms of the decomposers (level *e*). The "low-frequency" arrows to the right represent low-grade heat. Heat leaves the ecosystem at all levels shown, inasmuch as all organisms produce metabolic heat.

Each type of arrow in this figure indicates a factor that is essential to the ecosystem's integrity. Ecosystems that lose their solar input for more than a few days stand in danger of total collapse. This effect is behind a current theory purporting to explain the sudden disappearance of dinosaurs from the fossil record about sixty-five million years ago.[4] According to this theory, a giant asteroid that slammed into the earth during this period propelled enough debris into the atmosphere to destroy most plant life around the planet. This spelled doom not only for giant herbivores like Triceratops and Stegosaurus but also for Tyrannosaurus Rex and other carnivores that preyed upon them.

The flow of usable energy from producers to consumers (the broad arrow in the middle) is the factor around which the ecosystem is organized. Anything that disrupts this upward flow will break the link between the upper and lower levels of the system. A common cause of such failure is an interruption of solar radiation needed for photosynthesis, as in the case of the vanishing dinosaurs. Another potential cause of disruption is the introduction of invasive species into the ecosystem. Purple loosestrife, for example, might crowd out other vegetation on which the herbivores of the system have come to rely. Although photosynthesis continues with the loosestrife, the plant is too coarse for most herbivores to eat and no longer produces biomass in forms usable by upper-level consumers.

No less essential to the vitality of an ecosystem is its ability to rid itself of the degraded energy (entropy) resulting from its interior biological activity. The "low-frequency" arrows to the right of figure 4.2 represent low-grade heat that must be cast off for this activity to continue. If for any reason this heat is retained within it, the normal operation of the system will be impeded. When this condition persists, the ecosystem's ability to supply its top consumers with needed energy will be increasingly impaired.

4.3 Transformations of Physical Structure

As entropy can take the form of either disorder or degraded energy (see chapter 2), so negentropy might be available as either order or high-grade energy. Although order in turn can take various forms, our present concern is with two specific sorts. One is order in the form of material structure (i.e., of physical objects). Examples of ecological significance are chemical elements (e.g., carbon, hydrogen, and nitrogen) and various compounds that remain intact as they move through the ecosystem. The other is order in the form of nonmaterial structure, having to do with functional interactions among the system's biological components. The present section deals with the ecosystem's management of material structure.

Whereas the flow of energy in an ecosystem is primarily from lower to higher levels, material structure moves in both upward and downward directions. Matter is passed up the food chain through successive trophic levels and then passed down in the form of excreta, molted feathers, decaying leaves, and so forth. This corresponds to the two-way movement of matter in individual organisms, inward by way of ingesting food and outward by way of discharging waste products.

In contrast with the passage of food through the digestive systems of individual organisms, however, interchange of materials within a given ecosystem tends to be a self-contained process. Material structure is generated by an ecosystem's chlorophyll-bearing components and passed upwards for incorporation in higher-level organisms. Once degraded by use at upper trophic levels, matter is then passed downward for decomposition into forms that can contribute to further photosynthesis.

The basis patterns of material transformation in ecosystems are depicted in figure 4.3.

Figure 4.3. Structural Transformations in an Ecosystem

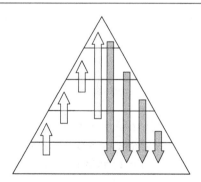

The upward arrows in this figure depict the passage of material structure from lower to higher trophic levels. Chemical nutrients move upward from decomposers to producers (levels *e* and *d,* respectively, of figure 4.1), and plant materials move onward to primary consumers. Also indicated in the figure is a movement of biomass from herbivores to carnivores. Although the upward arrow nearest the middle indicates the transfer only of plant material to the top-level omnivores, food channels should be assumed that lead to omnivores from all other trophic levels as well (people eat ants, jackrabbits, and snakes, as well as plants). The shaded arrows pointing downward, in turn, depict the passage of waste materials from all the other trophic levels down to the level of decomposers.

All arrows in figure 4.3 extend across trophic levels. Not indicated is another kind of structural transformation that is confined to the level of decomposers itself. The role of decomposers in the ecosystem is to convert waste materials from other levels into chemical elements that serve as building blocks from which new plant material is produced by photosynthesis. More will be said subsequently about the processes involved.

For now it is enough to note that the movement of material structure through the ecosystem actually constitutes a closed circle. As far as chemical elements are concerned, the same bits of matter might cycle through the system indefinitely.[5] Although ecosystems might exchange materials with other ecosystems during the course of normal operations, this is not always necessary to keep a given ecosystem up and running. What *is* required to keep an ecosystem in operation is a constant input of solar radiation (figure 4.2).

4.4 Functional Interactions among Biological Components

We turn now to the nonmaterial structure of ecosystems. The structure in question is nonmaterial insofar as it has to do with the functional interactions of the organisms involved. For a simple parallel, think of a basketball going through a hoop. This is an interaction between physical objects (the ball and the hoop), but the interaction itself is not another physical object. Similarly, when several organisms cooperate in one or another biological process, there is an interaction among physical entities. But the functional relation among them is not an additional physical entity.

There are functional interactions among organisms on any given trophic level. For example, a mating pair of golden eagles might cooperate in raising young, or they might fight off other birds who threaten their territory. There are also functional interactions among organisms on different levels. A certain stand of plants might feed a population of prairie dogs, which in turn serves as food source for a pack of coyotes. And the decomposers of an ecosystem are functionally related as a group to organisms producing the waste material they help to break down.

In section 3.8, ecosystems were characterized as systems of prey and predators held together by biological feedback interactions. But this description says nothing about how such systems themselves are organized. What the characterization leaves out is any indication of the functions served by these feedback interactions and of the way these functions relate to one another.

To remedy this lack, let us shift to a conception of an ecosystem as a network of functional interactions organized around its participating organisms. This is analogous to thinking of a road map as a system of lines (the roads) connected by dots (the towns) indicating how motorists can get from one place to another, as distinct from thinking of it as an arrangement of dots connected by lines that shows places one can reach as destinations. The conception of ecosystems in question focuses on the interactions among organisms (the lines) rather than on the organisms that participate in them (the dots). In brief, we are to think of an ecosystem as a network of interactions and the participating organisms as the nodes (connections) holding the network together.

The functional structure of an ecosystem may be defined as the overall configuration of functional interactions in which its constituent organisms participate. These configurations admit a variety of characteristic features. A given ecosystem will comprise interactions over a specific number of trophic levels, will include a certain number of populations among its primary consumers, will feature a top consumer population sustained by a number of functional interactions with lower-level consumers, and so forth. As a consequence of such features, a given ecosystem might be more or less adaptive, more or less robust, and more or less vulnerable to environmental disturbances.

Like their resident organisms, ecosystems are in constant flux. Stability on the part of ecosystems is the equivalent of health on the part of their constituent organisms. Briefly characterized, a healthy organism is one capable of maintaining its organic structure while responding to changes in its living environment. In like fashion, an ecosystem is stable to the extent that it is able to maintain its functional structure in the face of environmental change.

For an individual organism to maintain its structure does not require that its biological functions remain wholly unaltered. An infant mammal does not lose its health in shifting from milk to more solid food. Similarly, an ecosystem does not inevitably become unstable with changing populations on its lower trophic levels. An ecosystem supporting a great horned owl as top consumer would not necessarily lose its stability if chipmunks replace field mice in the owl's diet.

For present purposes, let us say that an ecosystem retains its structural stability as long as it provides a firm foundation of support for its top consumers.[6] An unstable ecosystem, on the other hand, puts the continued sustenance of its top consumers in jeopardy. In thinking about this topic, we should bear in mind that the top consumers of the biosphere at large are living members of the human race. In the following discussion of factors responsible for an ecosystem's stability, we are in effect considering factors that influence the ability of the biosphere overall to provide continued support of its human population.

4.5 The Contribution of Nutritional Diversity to Ecosystem Stability

To continue our discussion of figure 4.1, we may observe that the segments representing different trophic levels get narrower as they approach the top. With this in mind, we may think of the widths of these segments as corresponding to relative numbers of individual organisms functioning on the trophic levels in question. By way of illustration, a certain number of carnivores on level b will rely for nutritional input on a much larger number of herbivores on level *c*. A single great horned owl, for instance, consumes many thousands of smaller birds and animals over a lifetime of ten or so years. For another example, a standard-size bison will take in around thirty pounds of plant material (at several hundred plants per pound) in a single day. And literally countless bacteria and other decomposers are required to break down the waste products of the animal that consumes those plants.

On one hand, ecosystem stability requires ample populations on lower trophic levels to meet the needs of its top consumers. On the other, it also requires that these populations be distributed through a considerable variety of different species. What this latter requirement means in effect is that the functional stability of an ecosystem depends on the availability of alterative pathways through which nutrients can flow from lower to upper trophic levels.

This effect can be demonstrated by comparing the diagrams in figure 4.4.

Figure 4.4. Ecosystem Stability as a Function of Alternative Nutrient Pathways: Three Cases for Comparison

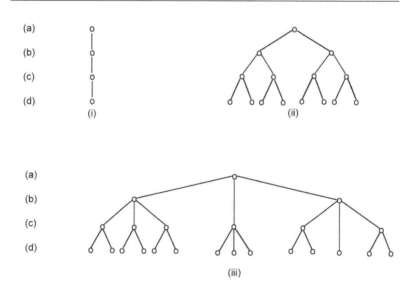

Each of the three diagrams in figure 4.4 represents the functional structure of a conceivable ecosystem. The diagrams are made up of circles, representing populations of organisms, and the lines between them, representing functional interactions among these populations. Trophic levels are labeled in the manner of figure 4.1. The functional structure symbolized by i, for instance, consists of an interaction between a plant population d and a population of herbivores c, an interaction between c and a population of carnivores b, and one between b and a population a of top consumers (other carnivores or omnivores, as in figure 4.1). For simplicity, decomposer populations are left out of the picture.

Diagram i corresponds to a functional structure that is too unstable to exist in nature. A conceivable instance, nonetheless, would be a single population of jackrabbits c feeding on an isolated stand of clover d, providing food in turn for a specific group of red-tailed hawks b, which are consumed by a single family of great horned owls a. If any one of the lower populations were severely depleted, the ecosystem would promptly collapse, and the owls at its top would die of starvation (unless their support shifted to another ecosystem). This imaginary ecosystem would be highly unstable because of lack of diversity in the nutritional sources available to its upper-level populations.

An ecosystem structured in the manner of diagram ii would provide alternative channels of nutrition at each trophic level, making it potentially more stable than an ecosystem patterned after diagram i. A conceivable instance might have a family of great horned owls as top consumer, feeding alternatively on a group of red-tailed hawks and a population of raccoons. The red-tailed hawks might have a diet of squirrels and rabbits, while the raccoons rely on songbirds and mice. Squirrels and rabbits might have a diet of fruits and grains in turn, as might the songbirds and the mice concerned. If one of these herbivore populations were to become severely diminished, the owls might still survive on nutrients channeled through alternative pathways.

We should note in passing that all the consumer species mentioned in connection with diagrams i and ii actually have diets far more varied than specified in these illustrations. Most herbivores (including mice and rabbits) eat more than fruits and grains. Red-tailed hawks feed on reptiles and smaller birds as well as on a wide variety of rodents. Being omnivorous, raccoons feed on crayfish, insects, and frogs, along with various fruits and vegetables—and so forth.

Diagram iii is unrealistic in the same respect and is included for purposes of comparison only. Whereas the top consumers in diagrams i and ii feed on carnivores exclusively, those in iii feed on carnivores (left), herbivores (middle), and omnivores (right). This contributes to the most important difference between diagram iii and the other diagrams, which is that the ecosystem it represents contains substantially more nutritional pathways serving its upper trophic level. If several of these pathways were eliminated, the remainder might still provide adequate nourishment for its top consumers. In contrast with the others, ecosystem iii rests on a sufficiently broad base not to be severely affected by a few disturbances on its lower levels.

4.6 Ecosystem Stability and Biodiversity

Maintaining a steady supply of nutrients is one among many services an ecosystem provides to its resident organisms. Any ecosystem of more than minimal complexity will probably involve symbiotic interactions (e.g., pollination of plants by insects), relationships of shelter (birds nesting in trees) and transport (seed spread in bird droppings), and so forth. Generally speaking, services like these will also be more reliable if provided by a variety of different species.

This brings us to the topic of biodiversity and its effect on ecosystem stability. Ecologists appear unanimous in viewing loss of biodiversity as a bad thing. It is regrettable aesthetically (the ivory-billed woodpecker must have been a magnificent sight). It is medicinally unpropitious (substances contained in extinct plants might have led to new medications). And it leaves fewer species for biologists to collect and study. Loss of these human benefits is certainly unfortunate.

A less anthropomorphic benefit was suggested by Eugene Odum in his book *Fundamentals of Ecology,* published almost sixty years ago.[7] Odum's proposal was that the homeostatic (self-stabilizing) capacities of ecosystems increase with the diversity of networks through which energy can flow to upper trophic levels. His line of reasoning, basically the line followed previously in section 4.5, is that if adjacent levels of an ecosystem are linked by several parallel pathways, then loss of energy flow through one pathway can be compensated by increased flow through others.

Odum's hypothesis has obvious intuitive appeal, and numerous attempts have been made to test it empirically. Energy flows through ecosystems proved difficult to quantify,[8] as did homeostasis on the part of ecosystems (the definition of homeostasis in section 3.6 does not carry over to ecosystems automatically, inasmuch as they are not biological organisms). This has made it difficult to establish a direct causal relation between number of species in an ecosystem and its overall stability.

On balance, however, it seems safe to say that enough empirical research has been done to show that ecosystem stability and biodiversity are positively correlated. Surveys of work in progress suggest that in some cases diversity is a source of ecosystem stability, while in other cases stability is an occasion for diversity instead.[9] The upshot in either case is that loss of stability and loss of biodiversity go hand in hand. Particularly in ecosystems supporting top consumers with extensive energy needs, severe loss of diversity can be a sign of potential collapse.

While thinking about this correlation, we should bear in mind that biodiversity is more than sheer number of species involved. An ecosystem's stability (or fragility) is an aspect of its functional structure. Considerations of biodiversity in an ecosystem are not addressed to numbers of species in isolation, but rather to the functional interactions in which resident species participate. Stability and biodiversity both have to do with an ecosystem's structure in the nonmaterial sense explained in section 4.4. Discussion of details in the following chapter (see section 5.9) will help us see that the nonmaterial status of an ecosystem's functional structure does not prevent loss of structure from counting as a form of entropy.

4.7 Ecosystems Related by Top Consumers

Figure 4.1 shows the hierarchical arrangement among an ecosystem's various trophic levels. The occupants of the upper level comprise what we have been calling the ecosystem's top consumers. An ecosystem's top consumers are those organisms whose nutritional needs are met within the system and who do not serve as prey for other constituent organisms. The top consumers of a given ecosystem need not be omnivores, as shown in the figure, but we are assuming that they are consumers as distinct from producers.

The hierarchy among trophic levels is internal to the ecosystem in question. There is a hierarchical arrangement of another sort (external) generally found among ecosystems themselves. With the exception of the biosphere at large, all ecosystems are included within more comprehensive ecosystems. And all save the simplest have other ecosystems included within them in turn. Our present task is to explicate this mode of inclusion.

For an ecosystem to be included in another ecosystem is not for one to be *part* of the other. Consider an ecosystem supporting a colony of prairie dogs as top consumers, which is included in another whose top consumers are a pair of golden eagles. The prairie dogs are parts of their supporting ecosystem, along with the plants on which they feed, the decomposers that break down their waste materials, and so forth. The prairie dogs are also parts of the more comprehensive ecosystem supporting the golden eagles. But the *ecosystem* supporting the prairie dogs is not itself a part of this more comprehensive ecosystem. An ecosystem is a hierarchy of organisms ordered by trophic levels, and ecosystems themselves are not organisms occupying trophic levels.

In general, ecosystem A is included within ecosystem B if the top consumers of A are also components of B, but the top consumers of B are not among the organisms comprising A. Thus the ecosystem with the prairie dogs as top consumers is included within the ecosystem featuring the golden eagles. Likewise, an ecosystem supporting a group of bison as top consumers might be included in another ecosystem supporting a tribe of indigenous people. But since the people are not part of the former ecosystem, there is no inclusion in the other direction (i.e., B is not included in A). The relation of inclusion is not reciprocal.

Inclusion nonetheless is a transitive relation. If ecosystem A is included in B, and B in C, then A is also included in ecosystem C. For example, the ecosystem featuring the bison is included in that featuring the native people. And inasmuch as the latter is included in the biosphere at large, the former is included in that most comprehensive ecosystem as well.

This definition allows a single ecosystem to be included within two more comprehensive ecosystems at the same time. By way of illustration, an ecosystem supporting a colony of prairie dogs could be included simultaneously in one supporting a pack of coyotes and one supporting a pair of golden eagles. By reverse token, a single ecosystem might include several less comprehensive ecosystems, no one of which is included in any other. The ecosystem supporting the golden eagles as top consumers might include distinct ecosystems supporting prairie dogs and jack rabbits, respectively, without any relation of inclusion between the latter two.

This is the manner in which other, less inclusive ecosystems are included in the biosphere at large. At this stage in geological history, the human race is the biosphere's dominant species. For all practical purposes, human beings function as the biosphere's top consumers. The biosphere includes all ecosystems meeting the nutritional needs of its human population, along with others meeting human needs that are no less essential. Among the latter are ecosystems maintaining a generally reliable supply of fresh air and clean water, others contributing materials for clothing and shelter, and yet others providing suitable contexts for social existence. Apart from their roles in the overall biosphere, however, these latter ecosystems may or may not be included within each other.

4.8 Ecosystems as Processors of Negentropy

As matters stand, consumers on all levels rely on other organisms for food. This is shown clearly in figure 4.1. As illustrated by producers like plants and algae, however, there are other ways of obtaining nourishment than eating living things. What figure 4.1 does not show clearly is that upper-level consumers are *unavoidably* dependent on other organisms for their livelihood.

All living creatures on the planet are ultimately reliant on solar radiation for energy to drive their metabolisms. Unlike plants and algae, however, consumer organisms are not equipped to receive metabolic energy from sunlight directly. The simple fact of the matter is that consumers such as jackrabbits, great horned owls, grizzly bears—and human beings—could not exist without other organisms to convert solar energy into forms their metabolisms can accommodate.

Let us recall (from section 3.3) Schrödinger's characterization of metabolism as a means of maintaining an organism's low level of entropy by "continually sucking orderliness from its environment." Since an organism's ecosystem is an immediate part of its environment, the upshot, again in Schrödinger's terms, is that "an organism feeds upon . . . negative entropy" provided by its ecosystem. The simple fact at hand is that the negentropy "sucked up" by consumer organisms (human beings included) must be altered from direct sunlight into forms capable of driving their metabolisms.

In effect, the negentropy that keeps these organisms alive must undergo preliminary processing before it can meet their metabolic needs. This preprocessing is the work of other organisms on lower trophic levels of their supporting ecosystems. Although performed externally to the upper-level organisms in question, the preprocessing provided by the ecosystem is an integral part of the vital processes by which they maintain their existence.

Maintaining an organism in existence requires both an external supply of energy to run its metabolism and external provisions for getting rid of the entropy produced by its vital processes. As Schrödinger puts it, the organism must have a way of "freeing itself from all the entropy it cannot help producing while alive." This amounts mainly to ways of dissipating the low-grade heat produced by its metabolism and to means of discharging the waste materials resulting from its processes of growth and regeneration.

The organism's ecosystem thus plays a vital role both in transforming solar energy into forms usable by its metabolism and in getting rid of the entropy it produces while using this energy. The processes enabling an organism to maintain its vitality are not confined to the space occupied by the organism itself. These processes include functions performed within its supporting ecosystem as well.

Let us focus these observations on the plight of the biosphere's top consumer specifically. We human beings are unavoidably dependent on the biosphere at large both for converting solar energy into forms our metabolisms can accommodate and for getting rid of the entropy produced as an essential part of maintaining our existence. If the biosphere ever reaches a point at which it is no longer capable of providing those services, human life as we know it will become extinct.

The stark reality of the matter is that this point might be nearer than most of us realize. As far as energy input is concerned, human ingenuity has been so successful in creating "processed" food that it may be hard to convince ourselves that we cannot "go it on our own" without the help of other species. What we must bear in mind in this regard is that there is no reasonable prospect whatever of our learning how to engineer food directly from solar energy.

As far as the disposition of entropy is concerned, there are compelling signs already that the biosphere is rapidly losing its capacity to get rid of the waste products created by human industry. Failure to reverse this development portends the end of human existence as we know it. The following chapter examines the details behind this threat.

5

Entropy Trapped within the Biosphere

5.1 Overview: Behind and Ahead

Humankind's relation to its supporting biosphere resembles that of an unborn child to its mother's womb. Both are dependent on their surrounding environments for sustenance as well as for the disposal of waste materials. And both would perish if these services were withdrawn. This is a crucial lesson of chapter 4.

There are disanalogies that are no less important. The human race obviously is not a living organism. It is not headed toward a time of partition after which it can exist without the biosphere's assistance. And the ministrations it receives from its sustaining environment are not reserved for humankind alone.

The most important difference for present purposes, however, is that a fetus is not capable of self-initiated action inflicting harm on its containing matrix. In stark contrast, humanity has initiated a profusion of deliberate projects that are inflicting severe damage on the biosphere that supports it. The task of the present chapter is to inventory the damage in question.

With this chapter, our study reaches a point of transition. Previous chapters have laid down the scientific basis of the study, beginning with the thermodynamic concept of entropy in chapters 1 and 2. Our concern in chapters 3 and 4 was to see how this concept applies to living organisms and to the ecosystems in which they participate. Our main resources in this latter endeavor were the biological sciences, especially that of ecology, in which the concept of entropy plays an increasingly prominent role.

The transition in the present chapter regards both focus and resources. Up to now, we have been dealing with processes that might occur almost any time since plant life took hold on earth about 444 million years ago. From here onward, our focus narrows to things that have happened within roughly the last 250 years—the time since the onset of the Industrial Revolution. The resources on which we must rely for guidance, accordingly, shift from those of science to those of history. In particular, we will be drawing from recent work in the history of energy use, the history of technology, and the burgeoning field of environmental history.

5.2 Thermal Pollution

In chapter 4, negentropy sustaining life in the biosphere was divided into three categories: usable energy, material structure, and nonmaterial functional structure. Corresponding to these categories, entropy can take the form of degraded (useless) energy, degraded material structure (e.g., waste material), and degraded nonmaterial structure (e.g., dysfunctional systems). Entropy in each of these forms is becoming increasingly prevalent within the biosphere. Our discussion of this tendency begins with entropy in the form of degraded energy.

When its work potential has been completely expended, energy takes the form of low-grade heat. In the normal course of events, low-grade heat produced within the biosphere is eliminated as black-body radiation.[1] A common example is the heat produced by sunlight on a desert surface, which is transmitted back into space at night. If this process is unimpeded, the desert can become quite chilly before the next sunrise.

If its normal discharge is impaired, however, low-grade heat can accumulate in amounts that have significant biological impact. To visualize this, one may think of buildings without air-conditioning that get progressively hotter during a summer heat wave, making it progressively more dangerous for people to remain inside. Heat trapped in this fashion is often referred to as thermal pollution.

One type of thermal pollution studied by ecologists is that caused in streams and lakes used by electricity generating stations. Water that cools condensers of power plants can undergo temperature gains of more than 15° Fahrenheit, which is enough to cause significant changes in aquatic life near its point of discharge.[2] Spawning and egg development of salmon and trout, for instance, occur in lower temperatures than those of catfish and carp, making it difficult for the more delectable game fish to survive in niches incorporating water discharged from power plants.

Another effect of increased water temperature is a decrease in solubility of oxygen, which is needed for the respiration of aquatic animals like fish and for the decomposition of biodegradable wastes. This reinforces the process of eutrophication, which occurs when excessive amounts of nutrients lead to the growth of large quantities of algae and aquatic plants that use up oxygen required by animal species.[3] A further adverse effect of increased water temperature, in conjunction with eutrophication, is to make the water involved less suitable for drinking. Proliferation of certain algae can produce compounds that are toxic to livestock, and the nitrates present in eutrophied lakes can also be toxic to human beings.[4]

Thermal pollution of this sort is one form of entropy with deleterious effects on both aquatic and terrestrial populations. Another sort of thermal pollution with adverse effects is known as global warming.

5.3 Global Warming: Underlying Causes

In its current use, the term "global warming" designates a buildup of low-grade heat in the atmosphere that cannot be discharged into space by black-body radiation. The main cause of this heat concentration is a breakdown of the mechanisms by which heat is normally transferred through the atmosphere. An analogy would be the mechanical failure of a pump emptying water out of a flooded basement. A contributing cause is increasing heat entering the atmosphere as a result of human activity (see section 9.5). The analogy here would be more water entering the basement as the pump breaks down. In the current case, both kinds of causes seem to be operating simultaneously.

The warming trend in question is taking place in the broad context of incessant temperature fluctuation over the earth's surface. These variations fall within a narrow range of temperatures that must be maintained for life on Earth to continue. Recent studies show periodic fluctuations in average temperature over the long run within a range of about 11° to 16° Celsius (52° to 61° Fahrenheit), with glaciations occurring roughly every 100,000 years.[5] The average global temperature presently is about 15° Celsius, which is a full degree below the high preceding the most recent ice age.

These observations show that there is nothing new about heat building up in the atmosphere. To the contrary, large amounts of atmospheric heat must be retained for the biosphere to remain within a viable temperature range. As with all major changes in environmental circumstances, past periods of high temperature have been accompanied as a matter of course by adaptive changes in many of the ecosystems affected.

From the human perspective, however, two aspects of the current warming trend make it particularly worrisome. One is that it seems to be developing too rapidly to allow time for compensating human adaptation. The other is that the rapidity of this development appears to be a result of human interference with natural climatic processes.

Here is a brief sketch of the processes involved. Global temperature is held within a viable range by a complex system of negative feedback interactions centered around water vapor in the atmosphere and radiation toward and away from the earth's surface. Although details remain elusive, climatologists have begun to understand these interactions in general outline.[6] Low-lying clouds (typically cumulus) reflect sunlight back into space, keeping it from heating up land masses and oceans. High-flying clouds (typically cirrus), on the other hand, absorb large amounts of outgoing heat, keeping it from leaving the atmosphere.

The concerned feedback processes are set in motion when a given sector of the earth's surface becomes abnormally hot. Under such conditions, the proportion of low-lying to high-flying clouds begins to increase, with more of the former to block incoming sunlight and fewer of the latter to impede outward radiation of heat. The net result is a decrease in surface temperature. When the earth's surface cools down, the reverse process takes over. The proportion of low-lying to high-flying clouds decreases, resulting in an increase in surface temperature beneath. Although many other climatic variables are involved, such as humidity, latitude and longitude, and wind currents, a feedback process of this sort seems largely responsible for maintaining the earth's surface temperature within a stable range.

This is the point at which so-called greenhouse gases come into the picture. These include methane (CH_4), nitrous oxide (N_2O), and most notably carbon dioxide (CO_2). Although CO_2 has been part of the earth's atmosphere since the beginning of plant life, its recent notoriety is due to the fact that it is emitted in large quantities by internal combustion engines and fossil-fueled power plants. As it turns out, the additional amounts of CO_2 entering the atmosphere as a result of human industry seem to be disrupting the natural feedback loops that maintain the biosphere at a stable temperature.

By itself, atmospheric CO_2 absorbs far less heat than the water vapor at the heart of the negative feedback process described above. By comparison, the amounts of heat retained by CO_2 are negligible over most parts of the infrared (heat) spectrum. But in one part of the spectrum (around fifteen micrometers), where the absorption effect of water vapor itself is minimal, CO_2 takes over as the main heat-retention factor.[7] This extra absorption as a result of the presence of unusually large amounts of CO_2 causes the air to become warmer, and the warmer air takes up more water vapor in turn. Increasingly high levels of CO_2 thus introduce a positive feedback effect that throws the temperature-regulating mechanisms of the atmosphere increasingly out of balance. Although it works its mischief indirectly, CO_2 released by human consumption of fossil fuel deserves its reputation as a major source of thermal pollution.

On one hand, industrial society's massive consumption of fossil fuel has reached the point of almost doubling the amount of humanly produced low-grade heat the biosphere was called on to dissipate during preindustrial times. Details behind this claim will be examined in the following chapter. On the other hand, substances released by industrial processes have upset the natural regulatory mechanisms that govern the return of degraded heat energy to space. In terms of the analogy at the beginning of this section, the basement is getting fuller, and the pump is breaking down.

5.4 Global Warming: Major Effects

In the months following Hurricane Katrina, which occurred in August 2005, the attention of the thinking public was focused on the threat of rising sea levels. The dominant image was the devastation of New Orleans, which was already below sea level (about six feet on average) before the storm. It is hard to miss the causal links between increasing air temperature, increasing ice melts in polar regions, and increasing levels of water lapping at coastal cities.

Climatologists naturally are interested in establishing correlations between rising surface temperatures and rising ocean levels. Historical records indicate a temperature rise of about 1° Celsius since the beginning of the Industrial Revolution, accompanied by a rise in sea level of four to nine inches (depending on data source) during the twentieth century alone.[8] Although detailed predictions regarding climate are necessarily uncertain (as well as politically provocative), researchers have tried to extrapolate from existing data with the help of computer models.

An overly optimistic prediction (assuming that greenhouse gas emission is halted immediately) projects a mere 0.5° Celsius temperature increase within the present century, accompanied by a rise in sea level of four additional inches. A worst-case scenario has global temperature rising 3.5° Celsius by 2100, causing a rise in sea level of over a foot.[9] Somewhere in between would be enough to inundate various low-lying islands (e.g., the Maldives) and to wipe out several coastal cities already below sea level (e.g., San Jose and Long Beach, as well as New Orleans).

The 2005 flooding of New Orleans, of course, was due not only to higher ocean elevation but also to the sheer intensity of the hurricane that breached its levees. As the public was informed afterwards, the levees had not been built to withstand the force that Katrina brought to bear. Another problem spawned by global warming is that the rising ocean temperatures that come with it appear to be causing tropical storms to increase in both frequency and intensity. Empirical studies by climatologists show that hurricanes in the northern hemisphere have almost doubled in power since 1950, and that category 5 and 4 hurricanes (the successive ratings of Katrina) themselves have become twice as frequent within the last thirty years.[10] Although the issue is still being debated, it appears likely that global warming is a significant factor behind these increases.

Another concern is the possible effect of global warming on the system of currents comprising the Gulf Stream that brings temperate weather to Europe. When the warm water of the Gulf Stream reaches the Norwegian Sea, some of its heat is transferred to the colder air. Being more dense, the now cooler water at the top of the circuit falls to the ocean floor, giving impetus to a return current that flows southward and replenishes its heat. Global warming brings higher air temperatures to the northern regions, which hinders the transfer of heat from the ocean. Higher temperatures also hasten glacial melts, making cool water at the top of the circuit less salty and (being less dense) less forceful in its effect upon south-flowing currents. If the continuity of the Gulf Stream were interrupted for these and similar reasons, it would stop conveying warm water up from the equator and could plunge Europe into a mini ice age.

An oceanic impact already in progress comes with the effect of water temperature on the photosynthetic algae at the bottom of most marine food webs.[11] This is illustrated by the case of the zooxanthellae that provide tropical coral reefs both coloration and nourishment. When water temperature becomes too high, these microorganisms are separated from their coral hosts, which subsequently bleach and begin to die. Since coral reefs serve as habitat for millions of species worldwide, their demise puts countless ocean ecosystems in jeopardy.

The vitality of coral reefs and the algae that feed them also depends heavily on the pH level (relative acidity and alkalinity) of the water they inhabit. Since both coral and photosynthetic organisms like phytoplankton contain calcium carbonate (an alkali), both are threatened by increasing acidity (lowering of pH levels). During recent decades, more and more CO_2 has passed from the atmosphere into ocean surface waters. This CO_2 causes the pH of the surface to decrease, with the effect that both coral reefs and phytoplankton are literally dissolving.

The atmosphere's increasing burden of CO_2 thus impairs the stability of marine ecosystems in two distinct ways. As a major cause of global warming, it contributes to the starvation of tropical coral reefs. And by being absorbed from the atmosphere, it is driving ocean acidity to levels that neither corals nor their sustaining algae can tolerate.

Global warming may also be responsible for altering the flow of equatorial air currents in ways that are hastening the spread of arid land areas (desertification). This flow begins with the rise of hot air at the equator into the upper atmosphere, where it cools and spreads both northward and southward. When it reaches a latitude where it is no longer sustained by upwelling drafts, this cool air drops toward the ground and flows back to the equator, where it is warmed once again and continues the cycle. Because air tends to gain and lose moisture as it warms and cools, circulation following this pattern results in heavy downpours around the equator and desiccating winds at the other end.

This drying effect is a primary cause of subtropical deserts. As a probable result of increasing temperatures at the subtropics, which keep cool air aloft longer, the circulation loops have been extending to latitudes increasingly removed from the equator. Empirical surveys have shown that subtropical climates have expanded about seventy miles in either direction within the last quarter century.[12] This portends a continuing increase in desert land around the globe. Mediterranean countries such as Greece and Spain are currently undergoing this process, and desertification in northern China has reached crisis proportions.[13]

Other threatening effects of global warming include massive food shortages, widespread scarcity of fresh water, and mass extinctions of biological species. We return to these topics as the discussion continues.

5.5 Degraded Material Structure: Solid Wastes

High-grade energy and high-grade structure are both forms of negentropy (as explained in chapter 2). When degraded, both become forms of entropy. Global warming results from an accumulation of low-grade heat energy within the earth's atmosphere. Having looked at various undesirable effects of global warming, we turn now to problems posed by the growing presence in the biosphere of degraded structure.

In chapter 4, structure was divided into material and nonmaterial (functional). Examples of high-grade material structure range from the very small (e.g., molecular compounds) to the middle-sized (e.g., dams and buildings), up to and beyond the geophysical (e.g., Earth's magnetic shield that deflects potentially harmful solar particle radiation). This section deals with degraded structure of the mid-range sort.

Environmental problems posed by solid wastes are typically approached in terms of volume. Influences such as population growth, proliferation of consumer products, and rising incomes in industrialized countries have resulted in astounding accumulations of discarded stuff. New York City alone produced about five million tons of solid waste in the year 2000, requiring a nine-mile-long convoy of trucks daily to haul it away.[14] Some of this waste ends up in landfills as far away as Virginia, and some is simply dumped into the ocean. The mouth of the Hudson River is distinguished by over a hundred square kilometers of ocean bottom incapable of supporting life.[15]

Solid wastes generated by U.S. municipalities at large around the turn of the millennium totaled over 230 million tons annually, close to 1 ton per person on average.[16] Considerably larger amounts of solid wastes are generated by industrial processes in this country, estimated to exceed 7.6 billion tons.[17] Of this, between 5 and 10 percent is toxic.

A compilation of statistics like these on a worldwide basis would be sobering, to say the least.[18] A more important dimension of the problem, however, has to do with the ultimate fate of the wastes we are throwing away. Biologically generated structure (biomass in its various forms) by nature is biodegradable. While some kinds take longer than others, biomass eventually is decomposed and its chemicals recycled in the continuing life process (as described in chapter 4). Crude oil itself is no exception, being decomposable by various microorganisms found in the soil.

The case is otherwise, however, with many products created from oil by human technology. As a rule, products made from plastics and from other polymers like nylon are not biodegradable. The problem with such materials is that their molecules are too large and too tightly bonded together to be broken down by decomposer organisms. Ways have been found of recycling some of these materials in other plastic products. Progress has also been made recently in breaking polystyrenes down by microbial action. But the most common way of breaking these polymers down currently is by incineration, which not only requires outside energy but also releases toxic gases into the atmosphere. When products made of plastic are used up and discarded into the biosphere, accordingly, most of them are likely to remain there indefinitely.

Plastics are being used increasingly in containers and packaging, as well as in the manufacture of a wide range of electronic and other consumer items. Many plastic products, moreover, tend to wear out quickly. And many are expressly designed for one-time use (e.g., disposable diapers). A result is that increasing percentages of wastes thrown out by human society come in forms that are not biodegradable. In 2002, roughly fifty-four million tons of plastic were produced in North America.[19] Within a few years, a vast majority of the plastic products resulting will have found their way into our oceans and landfills.

The magnitude of the problem is illustrated by a Texas-size mass of floating plastic debris accumulated by the rotation of the North Pacific Gyre off the coast of California and Mexico. Included in the mass are small pieces of plastic that resemble zooplankton and enter the ocean food chain when consumed by jellyfish. Tens of thousands of sea mammals die annually in the North Pacific alone from the "plastic poisoning" that results. Waterfowl from a majority of species examined worldwide for this impact were found to have indigestible plastic debris in their stomachs.[20] This debris obviously can be passed on to other animals that eat waterfowl without disemboweling them.

The unvarnished fact of the matter is that the enormous accumulations of plastic junk we are discarding into the environment will continue to grow until we stop using plastic products. This inundation of junk consists of entropic matter that the biosphere cannot discharge as low-grade heat. The normal functions of the biosphere will be increasingly impaired as it becomes increasingly contaminated with our plastic wastes.

5.6 Breakdown of the Ozone Layer

Plastic products like those caught in ocean gyres are degraded material solids that remain solid when broken up into tiny pieces. Material structure comes in liquid and gaseous forms as well. Among structures of a gaseous nature, none is more important to life on Earth than the layer of ozone concentrated in the stratosphere at an altitude of about fifteen to thirty kilometers. This crucial structure is being destroyed by our careless release of petrochemical derivatives.

Ozone (O_3) is a slightly bluish allotrope of oxygen caused by electrical discharge (among other sources) and hence present to some degree wherever electricity is used. At ground level, it constitutes a pollutant that not only makes breathing difficult but reduces plant photosynthesis by about 20 percent.[21] This ozone is obviously undesirable.

However, an accumulation of ozone in the stratosphere, built up by the action of sunlight on atmospheric oxygen over hundreds of millions of years, has until recently blocked out about 99 percent of incoming ultraviolet (UV) radiation. This ozone is not only desirable but literally essential for human life.

In its shortest wavelengths, UV radiation is lethal enough to be used in medicine as a means of sterilization. At mid-range frequencies, it causes skin cancer, macular degeneration, and cataracts in susceptible humans. Mid-range UV radiation can also damage the photosynthetic capacities of plants, harm the eggs and larvae of terrestrial fauna, and destroy plankton near the ocean surface. These plankton are the main source of biomass for marine ecosystems, and account for a large percentage of the biosphere's production of oxygen. So important is the protective role of stratospheric ozone that terrestrial life could not have begun to proliferate until that ozone layer was largely in place.

Ozone breaks down when exposed to halogens like chlorine and bromine.[22] During the ages while the ozone layer was forming, there was no significant presence of these chemicals in the upper atmosphere. Then in the early 1930s, industrial chemists hit on a class of petroleum products known as chlorofluorocarbons (CFCs), of which freon used in air-conditioning is probably the best known. CFCs were first used as refrigerants and as propellants in spray cans. Subsequently they were used in the manufacture of electronic circuit boards and of Styrofoam products. Since CFCs are relatively cheap to produce, little effort was made to conserve their use. It is estimated that by 1987 (the date of the Montreal Protocol, the first international agreement limiting their use), about 650 thousand tons of CFC gases had escaped into the upper atmosphere.

Partial destruction of the ozone layer over Antarctica was detected in the late 1970s. By the late 1980s, warnings were being issued in Australia to avoid unnecessary exposure to the sun. "Holes" in the layer over several northern countries were discovered at about the same time, and by the mid-1990s parts of the United States were affected as well. It has been estimated that between one and two million additional cases of skin cancer were caused by ozone depletion between 1975 and 2000, which entails between ten thousand and twenty thousand premature deaths. Despite earlier hopes that the ozone layer had begun to heal as a result of the Montreal Protocol, research results released in 2005 showed that the protective layer over the Arctic was the thinnest on record.[23]

A contributing factor undoubtedly is that, although some heavily industrialized countries (the United States, Russia, Japan, those of the European Union) sharply curtailed their use of CFCs after 1987, others (India and China) continued to increase their use. Even if production of these gases were to stop entirely, depletion of the ozone layer from past use will continue into the second half of the twenty-first century. No one knows how many centuries it will take for our ozone shield to regain the full protective capacity from which we benefited prior to the invention of CFCs.

5.7 Degraded Functional Structure: Disrupted Cycles of Replenishment

Chapter 4 distinguished between material and functional (nonmaterial) structure and between their respective forms of degradation. The preceding sections (5.5 and 5.6) have looked at degraded material structure of two particularly damaging sorts. We turn now to consider several ways in which the functional structure of the biosphere is being impaired. We start with the disruption of certain natural cycles that maintain a ready supply of ingredients essential to life.

One such cycle consists of the reciprocity between nitrification (the conversion of free nitrogen to soluble compounds) and denitrification (the conversion of compounds to free nitrogen), which scientists refer to as "the nitrogen cycle." The main products of nitrification are ammonium (NH_4) and nitrate (NO_3), which plants assimilate as nutrients. Denitrification occurs as nitrates from dead organic matter are broken down into free nitrogen (N_2) and nitrous oxide (N_2O) by the action of certain bacteria. Some of the resulting N_2O is further reduced to N_2, and some usually escapes into the atmosphere.

Although human activity plays a limited role in denitrification (e.g., through sewage treatment), its influence on the nitrification phase of the cycle is much more extensive. Its major contribution in this regard comes with the industrial fixation of nitrogen for use in fertilizers. As a baseline for comparison, about 140 million metric tons of nitrogen per annum is converted into compounds by natural biological processes.[24] Around the turn of the millennium, about 80 million *additional* tons entered the biosphere in the form of commercial fertilizers. Use of commercially produced fertilizers continues to grow and is projected to reach 134 million tons per annum within another two decades.

Despite salutary effects in alleviating world hunger, this massive infusion of nitrogen compounds into the biosphere has had environmental consequences that are largely negative. One is its substantial contribution to global warming. As a general tendency, the more nitrogen compounds are involved in biological processes, the more nitrous oxide results from denitrification. Nitrous oxide, it turns out, is a heat-trapping gas that absorbs outgoing infrared radiation not captured by other greenhouse gases (see section 5.3). By retaining this heat in the atmosphere, nitrous oxide contributes a significant percent of overall greenhouse warming.

Another adverse consequence of industrially fixed nitrogen is its role in the eutrophication of lakes and waterways. As noted in section 5.2, eutrophication occurs when excessive amounts of nutrients find their way into a body of water, accelerating the growth of algae and aquatic plants that consume oxygen needed for animal life. A major cause of this condition currently is the runoff of commercially produced fertilizers from land unprepared to retain them. This condition affects inland and coastal waterways alike. According to an opinion made public recently by a group of prominent biologists, the eutrophication of estuaries and coastal seas is "arguably the most serious human threat to the integrity of coastal ecosystems."[25]

The same group of scientists found a correlation between use of commercial fertilizers and loss of species diversity. New sources of nitrogen lead to the dominance of a few plant species prepared to take advantage of the additional nutrients, enabling them to crowd out their less receptive neighbors. In certain parts of Europe and North America, where use of manufactured fertilizers has been particularly intense, the increased dominance of a few nitrogen-responsive grasses has caused decreases in other plant species of up to 80 percent.

Another cycle that must remain in approximate balance for most ecosystems to function is the complex set of interactions by which carbon dioxide (CO_2) is exchanged for oxygen, and vice versa. As previously described in section 4.2, plants absorb CO_2 from the atmosphere (in land-based systems), combine it with water and various other chemicals by photosynthesis, and produce new plant mass and oxygen as a result. At the other end of the cycle, animals take oxygen from the air, use it in the metabolic reduction of food, and emit the resulting CO_2 back into the atmosphere. While considerable variation in relative levels of these two gases in the atmosphere seems tolerable in the short term, ecosystems involving both plants and animals would suffer from an overbalance in either direction over an extended period.

Several factors are at work today that tend to upset the balance between these two vital gases. One is the rapid growth of land committed to roads and buildings, which decreases the amount of plant life using CO_2 for photosynthesis. Another is the prodigious destruction of the world's vegetation by burning, which not only decreases the overall amount of plant life but also introduces large amounts of CO_2 into the air. As already noted, CO_2 is the greenhouse gas primarily responsible for global warming. The major source of excess CO_2 currently building up in the earth's atmosphere, however, appears to be our profligate use of fossil fuels. Here is one more way in which our addiction to fossil fuel is upsetting the biosphere: it is disrupting the balanced interchange between carbon dioxide and oxygen on which most plants and animals are vitally dependent.

No less important for a properly functioning biosphere is a steady balance of the water cycle. Of the approximately 1.5 billion cubic kilometers of H_2O on or near the earth's surface, about 97 percent exists as liquid in seas and oceans. Of the remaining 3 percent, about three-quarters is frozen in polar ice caps and mountain glaciers, and most of the rest exists in freshwater lakes and underground aquifers.[26] At any given moment, a certain amount of water vapor in the atmosphere condenses as precipitation, and a comparable amount evaporates back into the atmosphere. Average rates of precipitation and evaporation, however, differ widely from place to place. These differential rates play a major role in determining the distribution of various life-forms across the earth's surface. Let us consider human life in particular.

While human life today can be supported almost anywhere in the biosphere, it is concentrated in ecosystems with a broad basis of photosynthetic productivity (i.e., with ample plant life). Since plant life requires reliable supplies of fresh water (by rainfall or aquifer), this means that the human population by and large is dependent upon a water cycle that makes water available on a regular basis in quantities needed by local plant life. The problem in this regard is that water is becoming increasingly scarce in many parts of the world, to the extent that available sources are no longer adequate to support indigenous human populations. Because of such phenomena as recent droughts in central Africa, receding aquifers in North America, and worldwide pollution of lakes and rivers, large-scale imbalances are occurring in those parts of the planet's overall water cycle that affect human existence directly. Another source of imbalance, of course, is the increasing demand for water in various sectors of human industry, such as recreation (e.g., golf courses), mining, and the cooling of electricity generators.

These examples are representative of many such cycles that must remain in equilibrium for the biosphere to remain hospitable to human existence. Major disruptions of these cycles introduce significant disorder within the biosphere. And disorder, we have seen, is a form of entropy.

5.8 Industrialized Agriculture and the Green Revolution

Human disruption of natural cycles can set forces in motion that lead to ecosystem collapse. There are other circumstances of ecosystem destruction to which humans contribute more directly. A case in point is the production of crops by "factory farming."

The earth's surface includes roughly twenty-two billion acres capable of supporting vegetation.[27] Of this, about one-sixth is used for the production of crops, as distinct from the grazing of livestock (which occupies an additional one-third). By its very nature, agriculture involves the replacement of naturally evolved ecosystems with contrived systems over which humans exercise at least partial control.

In their efforts to gain more complete control, however, food producers in recent decades have turned farming into a manufacturing process largely reliant on sophisticated technology. To begin with, the operation is usually given over to a single crop (monoculture), planted with commercially produced and often genetically engineered seed. The process requires getting rid of competing plant species ("weeds"), which is accomplished by petrochemically produced weed killers (herbicides). It also requires the eradication of damaging insects, accomplished by dosing the plants and surrounding soil with other kinds of petrochemical poisons (insecticides).

Once these poisons are in the soil, moreover, they kill off microorganisms responsible for converting dead biomass into compounds capable of nourishing new plant life.[28] The infusion of petrochemicals continues with the application of artificial fertilizers, which are now the monocrop's only source of nourishment. In effect, the factory farm is a region removed from the order of nature and converted into a holding ground for the production of food by petrochemical technology.

This is the stuff of which the Green Revolution was made. The Green Revolution was a technological package exported from First World laboratories to the cultivated fields of the world at large. Plant geneticists isolated strains of staple crops like corn and rice that produced high yields in response to artificial fertilization, were relatively pest-resistant, and could be planted and harvested mechanically. The impressive achievements of the Green Revolution during the 1960s and 1970s have been copiously documented. For a while it was widely touted as the technological salvation of an increasingly populous and hungry world.

But it was not long before the flaws inherent in this technological program became apparent. One socioeconomic problem is that developing countries often do not have the funds to pay for the chemical products and machines required. Another is that the monocrops grown in these countries are usually sold abroad to improve cash flow rather than used to feed their own peoples. Yet another is that the production techniques in question encourage large-scale industrial agriculture at the expense of small farmers, who subsequently lose their acreage to large corporations.

From an ecological perspective, however, the most severe problem is that the chemicals and production techniques characteristic of the Green Revolution are destroying the ecosystems by which indigenous people had long been sustained. Wild plants and animals once used for food are being poisoned,[29] local water supplies are being polluted by chemical runoff, and productive wetlands are being drained to accommodate heavy machinery. The quandary faced by the original occupants is that the new way of producing food is too expensive, but that time-tested ways are no longer serviceable.

Looking ahead to the next section, we may note that agribusiness is a significant factor in contemporary loss of biodiversity as well. Whereas humans at one time or another have used perhaps 3,000 species of plants for food, only about 150 of these (1 in 20) are commercially produced today, and most of the rest have gone extinct. Nine basic foods now account for over three-fourths of world agriculture, with four of these together outweighing all other plants consumed.[30]

Corn provides a typical example. Less than one-tenth of corn varieties grown a century ago are still in production. Over 97 percent of total corn production is grown with artificial fertilizers and poisons, almost entirely by monoculture. The few remaining varieties are increasingly at risk of being replaced with genetically modified (GM) strains; between 2001 and 2004, the proportion of GM corn produced in the United States increased from 25 percent to 45 percent.[31] This bodes ill for corn as a food staple if biological threats emerge that genetic engineers are not able to manage.

5.9 Loss of Biodiversity and Its Human Consequences

The more species included in a given ecosystem, the greater its level of biodiversity. The biosphere, of course, is the most diverse ecosystem of all. Varying estimates place the number of species worldwide at between five million and fifty million, of which about one and a half million are actually known.[32]

Large numbers of species have become extinct since the biosphere's origin, and many others have taken their place. It has been estimated that species extinction during recent geological time occurred at a rate of about one per year. By the end of the twentieth century, however, around thirty thousand to fifty thousand extinctions were occurring each year, mostly in tropical rain forests. At this rate, it seems likely that species now are being lost faster than they are being replaced.[33]

Unlike massive extinctions in the past (recall the passing of the dinosaurs mentioned in section 4.2), the loss now under way is due in large part to human activity. Humanity contributes to the downfall of other species in at least three distinct ways. One is by hunting, either for food or for exploitation. Passenger pigeons, once among the most populous bird species in the United States, became extinct late in the nineteenth century as a result of consumer demand for their flesh and feathers. In like fashion, there is concern among zoologists that demand for ivory currently is driving elephants toward the point of extinction.

Second is the introduction of invasive species, whether by commerce or by individual travel. The case of purple loosestrife was mentioned in section 4.2. Another frequently discussed example is the introduction of Nile perch into Lake Victoria some fifty years ago, for purposes of commercial fishing. Considered one of the world's worst invasive predators, the Nile perch has caused the extermination of two hundred or more species once native to the lake.[34]

It is the third category of humanly induced extinction, however, that is ecologically most damaging. This comprises the widespread destruction of habitats themselves, as distinct from the destruction of particular occupants of particular ecological niches. A frequently cited illustration is the obliteration of tropical rain forests by logging and burning. Although estimates vary with source and interest, a typical assessment puts the loss at about sixty thousand square miles (roughly the area of New York state) of rain forest a year. With each year's loss of habitat goes the loss of an additional twenty-five thousand or more biological species.[35]

Other examples abound. To be sure, most of the forms of ecological damage discussed in this chapter carry loss of species among their consequences. Global warming is destroying coral species in the Caribbean (discussed in section 5.4). A combination of habitat change, global warming, UV radiation, pollution, and overfishing during the past fifty years has cut species diversity in parts of the ocean by up to 50 percent.[36] And we have already looked at the tendency of industrialized agriculture to reduce the world's store of edible plant species (see section 5.8).

Why does humanity's large-scale destruction of other species constitute a problem for humanity itself? Here are a few reasons focused on the rain forests specifically. Before GM crops, human diets derived entirely (sometimes by deliberate cross-breeding) from species found in the wild. At least four-fifths of the developed world's present diet originated in tropical rain forests.[37] Well over one thousand edible fruits are still present in the rain forests that have not yet found their way to First World markets. One reason destroying other species constitutes a problem is that loss of biodiversity threatens these additional food sources.

Before the rise of the pharmaceutical industry, for a second reason, most cultures relied on medicinal plants for healing. Currently, about one-quarter of manufactured medicines derive from rain forest ingre-dients, and less than 1 percent of tropical plants have been tested for possible medicinal uses. Species destruction thus is problematic in that loss of biodiversity cuts back on our sources of new medications.

Third, destruction of plant species and of species that consume them threatens the habitats of upper-level consumers, and massive de-struction of such species threatens human habitats in particular. Five centuries ago an estimated ten million native people lived in the Ama-zon rain forest. Today there are less than two hundred thousand, due in large part to massive rain forest destruction. Similar loss of habitat has occurred in the subtropics as a result of desertification. Yet another rea-son destruction of other species constitutes a human problem is that loss of biodiversity leads to a loss of human habitation.

There is a more general answer to the question, however, of which considerations like these serve as mere particular illustrations. This an-swer has been at hand since the close of the previous chapter. Human beings are the biosphere's top consumers. At this point in human his-tory, the human race depends upon the biosphere at large for its very ex-istence. Most fundamentally, we depend upon the biosphere both for converting solar energy into forms we can assimilate and for getting rid of the entropy we produce in consuming that energy (see section 4.8).

Within the past few hundred years, our needs in both respects have increased enormously. In the aftermath of the Industrial Revolution, humankind has become dependent on modes of production and con-sumption requiring unprecedented amounts of negentropy (energy and structure) to sustain, which at the same time are dependencies produc-ing unprecedented amounts of entropy (waste heat, degraded structure) for the biosphere to expel. In order to provide services of this sort on the level required, the biosphere must maintain its functional stability (as described in section 4.4).

Inasmuch as functional stability and biodiversity go hand in hand (see section 4.6), however, it follows that a substantial loss of biodiver-sity will substantially diminish the biosphere's ability to provide these services. In upshot, humankind's massive destruction of other species has the effect of undermining the ecological support required for human society's continued existence.

6

The Rising Tide of Human Energy Use

6.1 A New Turn in the History of Energy Use

To prepare for this chapter, let us recall the definition of thermal equilibrium given in chapter 3. The earth's surface is in thermal equilibrium when the amount of energy reaching it from the sun is evenly matched by the amount of low-grade heat leaving as black-body radiation. Recall also the supposition that a state of near equilibrium must have been present for life to make an initial appearance on Earth (see section 3.1).

As life proliferated, the earth's surface became part of what we now call the biosphere (see section 3.9). Energy arriving at the Earth's surface perforce is energy entering the biosphere. Similarly, energy leaving the earth's surface must exit through the biosphere as well. So we can speak of the biosphere itself being in thermal equilibrium. Since departure from equilibrium means either excessive heat or excessive cold, the biosphere must maintain approximate equilibrium for its resident organisms to remain viable.

During most of the era since life began, thermal equilibrium has been maintained by the prevalence of two energy-related conditions. One is that the amount of low-grade heat needing to be discharged is limited to that resulting from use of high-grade energy entering the biosphere during roughly the same time period. This condition is met as long as energy consumed within the biosphere is all renewable, in the sense of being naturally replenished at roughly the rate of consumption. When this is the case, the biosphere is not called upon to discharge additional entropy stemming from use of nonrenewable (e.g., fossil) energy sources.

The other condition is that all entropy produced within the biosphere is discharged by natural processes that are continuous with those producing it. This second condition is met as long as all by-products of energy consumption are biodegradable. Entropic by-products that are not biodegradable (e.g., plastic wastes) are retained within the biosphere, being irreducible to low-grade heat that can be discharged into space.

The first condition assures that no more expended energy leaves the biosphere at a given time than entered originally as solar energy within the same time frame. The second condition assures that all high-grade energy entering the biosphere exits in due course as low-grade heat. Working together, these conditions maintain an even balance between incoming and outgoing energy, which means that the biosphere is held in thermal equilibrium.

Effects of nonbiological events like volcanic eruptions and asteroid impacts aside, these conditions presumably held steady for billions of years up to the onset of industrialization.[1] Around the middle of the eighteenth century, however, human affairs took a radical turn destined to jar the biosphere out of thermal equilibrium. An indication that something radical was happening at this point is that human energy consumption began to rise precipitously. Steadily increasing amounts of energy came from fossil sources (coal, oil, gas), which means that the energy involved was not renewable in the sense explained above. As a consequence, the biosphere was called on to emit increasing quantities of low-grade heat in excess of that stemming from concurrent solar radiation.

Another aspect of human industry's increasing reliance on fossil energy was that it soon led to the development of products that are not biodegradable. The first plastic compounds appeared in the 1860s,[2] and the quantity of plastic artifacts has been growing ever since. Human activity thus became responsible for increasing quantities of material stuff that will remain within the biosphere indefinitely.

As industrialization spread around the globe, however, its most ominous aspect came to be the plethora of polluting by-products that spread in its wake. Among these are the several gases discussed in chapter 5 (carbon dioxide, methane, nitrous oxide) that impede the discharge of heat through the earth's atmosphere. A consequence is the buildup of heat within the biosphere currently known as global warming (as discussed in section 5.3).

On one hand, industrialization has radically increased humankind's contribution to the amount of heat awaiting discharge by blackbody radiation. On the other, it has made the atmosphere increasingly opaque to the passage of this low-grade heat. Our current predicament is that the more fossil energy we bring to bear in fueling human industry, the less able the biosphere becomes to discharge the resulting waste heat.

The predicament is aggravated by other forms of structural breakdown within the biosphere also as a result of energy-intensive industry. Among those examined in the previous chapter are the progressive destruction of the stratospheric ozone layer, the poisoning of crop land by industrialized agriculture, and the accelerating extinction of other species within the biosphere on which humanity depends for its existence. There is near unanimity among scientists currently studying such developments that human life as we know it is increasingly in jeopardy.

One way or another, these developments are contingent upon the massive swell in human energy consumption beginning with the Industrial Revolution. The task of the present chapter is to gain a comprehensive perspective on how this predicament came about.

6.2 Energy Consumption in Preagricultural Society

Our approach in this and subsequent sections will be to characterize each era under consideration with respect to its typical energy sources, its approximate per capita energy consumption, and the population served by the energy used. We begin in this section with the era of hunter-gatherers.

Human beings probably have not always been the biosphere's chief consumers of energy. As long as our ancestors fended for themselves individually (or in small groups), they probably lagged several species in per capita consumption—polar bears, elephants, walruses, whales, and perhaps various smaller animals that hunted in packs.

One reason is that both hunting and gathering are usually unreliable and often inefficient ways of obtaining food. Only a relatively small portion of the planet's vegetation comes in forms humans can digest (fruits, grains, nuts, tubers), and these required foraging over wide and frequently shifting areas. As far as hunting is concerned, an individual's success rate with ungulates such as deer and antelope probably averaged one or two a week.[3] Although small prey like monkeys might be encountered more routinely, probabilities of hitting them with spears and arrows must have been rather low.

Estimates like these suggest that individual hunters had good reason to band together in groups where chances of a shared meal were several times greater. By working together, hunters could kill large animals like mammoths in open combat or could herd them into traps where they were easily slaughtered. Since a small mammoth provided the energy equivalent of about fifty reindeer,[4] for example, each participant could walk away from a kill with more food energy than could be acquired in several days of solitary hunting.

Beyond shared fortunes in the hunt, another advantage of group participation was efficient division of labor. Designated adults could cut up the meat and carry it back to homebase (perhaps caring for infants simultaneously), allowing others to continue the hunt. Similar advantages were available in foraging groups, where some could prepare the food at hand while others went looking for further gleanings.

Assuming bands of three to six hunters, and an average family size of six per hunter, we can estimate an average community size of twenty to forty individuals.[5] Given the success of this arrangement over many millennia, we can also assume that people generally could obtain energy inputs sufficient to meet their basic requirements. This amount can be estimated as roughly equivalent to one billion joules (about 278 kilowatt hours) per person per year, the amount needed to sustain the poorest of the world's people today.[6] Average worldwide human population during this era has been pegged at four to five million.[7]

6.3 The Introduction of Agriculture

Agriculture began with the transition from food collection to food production. The upshot of this transition is that food (plant and animal) is no longer gathered from the wild but rather grown deliberately under human supervision. Although accounts vary in specifics, there is general agreement that the agricultural era began between six and nine thousand years ago.[8]

In term of efficiency, raising crops had many advantages over foraging in the wild. Farmers could concentrate on crops best suited to their needs and avoid having to travel long distances in search of food. Once suitable farmland had been located, its yield could be improved by various techniques of irrigation and fertilization. And since farming communities tended to be relatively permanent, it became feasible to store food for use during unproductive periods. Another advantage was the diversification of labor made possible within the community. Some members could specialize in heavy work like clearing forests, others in repetitive tasks like weeding and harvesting, and yet others in domestic jobs like cooking and weaving.

As farming practices became more productive and surpluses developed, small communities expanded into established towns. Surpluses were needed to feed a growing number of specialists—administrators, soldiers, craftspeople, and merchants—not engaged directly in food production. Insofar as the services provided by these specialists tended to make food production more efficient and reliable, moreover, greater surpluses became available, and expansion continued. Whereas agricultural villages contained only a few hundred people, the earliest known cities (e.g., Ur in Mesopotamia) probably were inhabited by several thousand.[9]

Several millennia passed, however, before the growth of cities had a major impact on world population. By 5000 BCE the total was still around five million people, having risen perhaps a million since the era dominated by hunters and gatherers.[10] Annual per capita energy consumption, on the other hand, increased several fold, to the neighborhood of six billion joules.[11]

6.4 Muscle-Driven Technology

Early agriculture was powered by human labor. A major step forward in productivity came with the introduction of draft animals around 3500 BCE. Oxen and water buffalo apparently were domesticated first, followed by horses about 3000 BCE.[12]

Domesticated animals were employed to draw conveyances before they found use in farmers' fields. Once adapted to this latter use, however, animals became increasingly indispensable in the production of food. They provided a ready supply of nutrient-rich fertilizer deposited conveniently just where it was needed. And they provided muscle power for farming techniques that were previously impracticable.

Prior to the domestication of oxen, cultivation of farmland was done with hoes. It has been estimated that a farmer working with a hoe needed upwards of a hundred hours to prepare a hectare of land for planting. Given an average-sized ox pulling a simple wooden plow, the farmer could accomplish the same task in about thirty hours.[13] Although an investment of labor and land is required to keep farm animals fed, efficiency gains of this magnitude made farming with animals far more productive than anything achievable with human muscle power alone.

In addition to their usefulness in fertilization and tilling, animals also made big differences in irrigation. Although small quantities of water can be elevated (from wells and lakes) by hand-operated devices like buckets and Archimedean screws, traditional agriculture tended to avoid land that could not be reliably watered by either rainfall or gravity (via ditches from rivers). As animal power became available, however, devices were invented that could lift much larger quantities of water over greater heights. One such device was an escalator-like arrangement of clay pots dipped into a water source that filled at the bottom and discharged at the top; another was a series of metal buckets on the rim of a large wheel turned by the circular motion of animals through a right-angle gear train.[14] Animal-powered irrigation both increased the productivity of established cropland and opened up new areas to cultivation.

Further improvements in productivity followed with more effective plow designs, more efficient yokes and harnesses, and the development of animal breeds better suited for regimented labor.[15] It will be noted that improvements of this sort are results by and large of human ingenuity. While improvements in the use of animal power were still being made into the modern era, the basic technology was largely in place by 1000 BCE (roughly the beginning of the Roman Republic).

By this stage in history world population had doubled several times over, approaching the neighborhood of fifty million.[16] Per capita energy consumption also had increased significantly, to an estimated thirteen billion joules per year.[17]

6.5 Technology, Animal Power, and Additional Forms of Energy

Using animal labor in the production of human foodstuffs was a major step in humankind's growing control of energy employed for human advantage. The fact that cattle can be used to plow fields, to haul loads, and to turn waterwheels, and then eventually to be killed for meat, was enough to establish animal power as a mainstay of civilization up to the modern period. In many parts of the world today, including Amish farms in North America, agriculture still relies on draft animals as a major source of energy.

Nonetheless, there are obvious limits to the number of animals that can be made to cooperate in a given human venture. Further increases in per capita energy consumption during the centuries leading up to the Industrial Revolution were due largely to improvements in technology. There were improvements in the equipment by which animal power was brought to bear, along with technological advances that opened the way to harnessing new sources of energy.

An illustrative case of the former was the evolution of instruments pulled by animals for breaking up ground. Ards (hoe-like devices rigged for pulling), which were in use worldwide by 1000 BCE, gradually gave way to moldboard plows that could dig deep furrows in heavy soil. Cast-iron moldboards were common in China by the third century CE. And cast-iron plows with wheels came into general use after the technology was introduced in Europe during the Middle Ages.[18]

Parallel improvements were made in techniques for harnessing draft animals to their implements. Head yokes for oxen gave way to more efficient neck yokes, while throat-and-girth harnesses were replaced by breast bands as the preferred means of equipping draft horses. Given that a single properly equipped draft animal can produce the work equivalent of three or more human laborers,[19] the gain in foodstuffs produced can be substantial when large teams of animals are enabled to work together. Developing technologies like these not only increased the amounts of food energy available per capita within farming communities but generated surpluses for trade and for the expansion of urban centers as well.

Although animal labor continued to be important into the modern era,[20] technological improvements in the use of wind and water power were more influential in preparing for the arrival of the Industrial Revolution. Wind had been used to drive seagoing vessels since antiquity. During the early Roman era, sails were basically square in shape and were fixed perpendicular to the ship's main axis, which made sailing upwind difficult. By the end of the medieval period, however, major sea powers worldwide had ships with triangular sails and stern-post rudders,[21] which, in combination with deeper hulls, made them efficient converters of wind energy into humanly controlled transportation.

On land, wind had been harnessed for human purposes as early as the eighth century CE. Early windmills were capable of roughly the same amount of power as a team of horses. Steady improvements in design and construction increased their efficiency several times over, until they became the most powerful sources of energy available in areas without water power. Wind power was still used extensively in Northern Europe and the United States during the early stages of the Industrial Revolution.[22]

The history of water power, in turn, goes back at least to the first century BCE, when horizontal waterwheels were used to mill grain in Egypt.[23] According to the Domesday Book of 1086, thousands of stream-driven grist mills were then operating in parts of England.[24] Improvements in waterwheel technology led from the early horizontal version (wheel level with stream), through vertical arrangements (water running under or over an upright wheel), to radical water turbines of the early nineteenth century (water swirling under pressure toward directly facing blades). Waterwheels were instrumental in the engineering of deep underground mines, which required continuous power at a high level of reliability.

Another form of water power to be noted is that provided by fluctuating water levels in ocean tidal basins. The first tidal mill was built near Venice in 1044,[25] and more sophisticated versions were still being used in England some eight hundred years later. Because of siting limitations, the contribution of tidal mills to early industry was relatively minor.

Although these are the chief forms of energy associated with the developing technology of this era, several others deserve brief mention. One such is gunpowder, knowledge of which traces back at least to the ninth century CE.[26] Apart from its use in fireworks, the role of gunpowder in human history has to do primarily with violence and warfare. This, however, does not justify our overlooking the relatively peaceful role of blasting powder in the excavation of mines and quarries.[27]

Also worth mentioning is the fact that fossil fuels did not have to wait for the eighteenth century to play a part in human industry. As early as 200 BCE, the Chinese burned natural gas to evaporate brines.[28] Oil wells were being hand-drilled in Burma in the tenth century CE.[29] Most of the coal fields in England were being worked by the early 1600s, and by 1700 England and Wales had an output of about three million tons of coal a year.[30] These uses presaged the massive shift to fossil fuel that marked the Industrial Revolution itself.

As humankind moved into the industrial age, it had devised techniques to harness not only the muscle power of other animals but also the kinetic energy of wind and water, the gravitational energy of the tides, the solar energy stored in fossil fuels, and the chemical energy of gunpowder.

Over the course of this preindustrial period, as a result, improvements in technology had tripled the average energy consumption per person to approximately 38×10^9 joules. World population, in turn, had doubled several times to something in the neighborhood of six hundred to seven hundred million people.[31]

6.6 Industry and the Ascendancy of Fossil Fuel

Between 1750 and 1900, technologically advanced countries in Europe (led by England) moved gradually from agrarian economies to economies dominated by machine-driven industry. This transition, commonly known as the Industrial Revolution, was epitomized by the development of efficient steam engines in the 1760s and by the invention of the internal combustion engine roughly one hundred years later. By the beginning of the twentieth century, use of these new power sources had spread worldwide (making possible World War I), and the period of transition had given way to the industrial age, in which all countries to some extent participate today.

Although steam engines can also operate on biomass fuels (wood or charcoal), the efficient versions that got the Industrial Revolution under way depended primarily on coal. The first internal combustion engines ran on coal gas,[32] but commercially successful designs relied on fuels refined from oil (diesel or gasoline). Demand for fossil fuels increased with the harnessing of electricity as a commercial power source, inasmuch as most generators in operation by 1900 were run by steam turbines powered by coal.[33] The result of these combined influences was a massive increase in use of fossil fuels, accompanied by a proportionate decrease in reliance on renewable energy sources.

Independently of amounts consumed, the shift to fossil fuels resulted in dramatic changes in the methods by which consumer goods are produced. Prior to this period most goods were either crafted locally (the so-called cottage industry system) or produced in small mills usually driven by waterwheels. Given the superior power outputs of fossil fuel–driven steam engines, however, methods of production became organized in the form of centrally powered mills and factories. Many of the technological innovations made during this period had to do with machinery that could be driven by centralized power sources (e.g., Cartwright's power loom and Crompton's spinning mule).

Among the social consequences (many quite disruptive) of the resulting factory system in England was the amassing of large numbers of workers in workplaces where they could be assigned specific tasks that could be effectively supervised. Combined with standard economies of scale (fixed costs covering increasing amounts of goods produced), more efficient use of labor enabled goods to reach the market at prices considerably lower than had been commanded previously. Inasmuch as goods mass-produced in this fashion tended at first to be staple items, consumers of these products generally found less of their incomes being taken up by basic necessities. This tendency soon spread to Western Europe and North America.

As a result of this tendency, increasing numbers of people with ordinary incomes had money left over for what previously had been considered luxuries. More people were able to afford tea and spices imported from the Orient, more people were enabled to acquire fine clothing and dinnerware, and more had both leisure and means to spend vacations abroad. As the luxury market expanded, moreover, new ways were made available to dispose of extra income. The first passenger railway opened in England in 1825, an iron ship for passenger travel was launched in 1843, and privileged New Yorkers had the choice of electric lighting by 1882.[34] A defining moment of the Industrial Revolution was the Great Exhibition of 1851 in London's Hyde Park, in which many items displayed were intended for luxury consumption.[35]

In sum, whereas technological innovations during previous eras tended to provide better ways of doing things that people were doing already, technology during the Industrial Revolution was directed increasingly toward providing goods and services that were not previously available. Not only did the technology producing these benefits use up increasing amounts of fossil energy in its own right, but the machinery involved in delivering them (railways, ocean liners, electricity systems) became increasingly energy intensive. The benefits in question also became available to increasing numbers of people. The upshot of these tendencies in combination was a massive increase in energy consumption during the early stages of the Industrial Revolution.

By 1900, the end of this period, per capita energy consumption worldwide stood at about fifty-six billion joules a year.[36] World population at this time was in the neighborhood of 1.6 billion.[37]

6.7 Energy Use and Population

Estimates of annual per capita energy consumption for the periods treated thus far are shown in figure 6.1 in the form of a simple graph.

Figure 6.1. Growth in Per Capita Energy Consumption

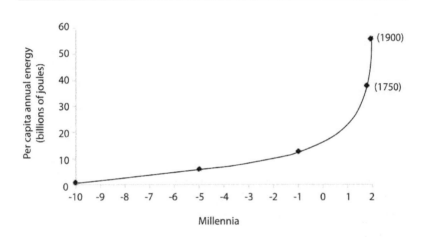

Points determining the shape of this curve are taken from estimates documented in sections 6.2 through 6.6. Despite the unavoidably tentative character of these estimates, the graph clearly exhibits the form of an exponential progression.

Growth in world population for the same stretch of time is shown on a separate graph (see figure 6.2).[38]

The curve in figure 6.2 is notably similar to the one in figure 6.1, although steeper from 1000 BCE to 1750 CE (in part because of recurrences of bubonic plague) and correspondingly straighter on either end.

Multiplying population by per capita energy use at any given point, of course, would give total energy consumption for the time in question. There are several reasons, however, for keeping the two sets of data separate, rather than presenting them in combination.

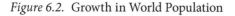

Figure 6.2. Growth in World Population

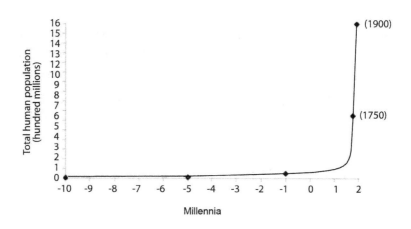

One reason is that total human energy consumption from the eigh-teenth century onward has been increasing at rates too steep to illustrate on a simple linear (nonexponential) scale. In point of fact, when the curve of figure 6.1 is imposed on that of figure 6.2, the result exhibits a rate of increase approximating that of a chemical explosion. Gun-powder, for instance, oxidizes (i.e., burns, producing heat) at a rate in-creasing exponentially with temperature. The positive feedback process involved progresses so rapidly that the resulting explosion appears in-stantaneous. What we see depicted in figures 6.1 and 6.2 together is an "explosion" in human use of energy. The severity of this exponential progression is more evident if its two major components are presented separately, thereby showing that each component is subject to exponen-tial increase itself.

Another reason for separate presentations is to counter the assump-tion that the increase in overall energy use during recent centuries is due primarily to increase in population. As the preceding discussion has shown, both sociological and technological factors have contributed sig-nificantly to the increasing amount of energy consumed by individual humans. While it undoubtedly is the case that human energy consump-tion would not have increased so dramatically over the past several centuries without accompanying increases in total population, an important condition for understanding our present dependence on fossil energy is to realize that it does not stem from population growth alone.

A third reason is that separate treatment of the growth processes depicted in figures 6.1 and 6.2 invites reflection on the manner in which these two processes interact. It seems clear, on one hand, that population growth in localized social groups often is accompanied by growth in social complexity. And growth in social complexity can provide resources, like administrative oversight and division of labor, that enable a group to make more energy available to its individual members. In brief, population growth can lead to increased per capita energy consumption. On the other hand, several authors have suggested ways in which increasing amounts of energy use per capita might lead to increasing numbers of people taking advantage of energy surpluses.[39] In this case, increased consumption leads to larger populations. We will look more carefully at these suggestions later, in connection with the relation between population and economic growth.

6.8 Energy Use in the Twentieth Century

The worldwide output of fossil fuels surpassed the total amount of biomass fuel consumption just before 1900.[40] While use of biomass energy continued to increase gradually during the twentieth century (by a factor roughly of 1.75), consumption of fossil fuel during that period increased many times over. The upshot of these trends is that over 90 percent of the fuel consumed by human activity today has fossil origins.

Since coal already was in extensive use before 1900, its continued use in the twentieth century expanded at a slower rate than that of either oil or natural gas. Annual worldwide use of coal grew from about eight hundred million metric tons in 1900 to about four thousand million in the 1990s.[41] This amounts to a fivefold increase over a period of a hundred years. World consumption of oil, on the other hand, increased from about twenty million metric tons in 1900 to nearly three thousand million in the 1990s, an increase of about 150 fold. And total annual consumption of natural gas during that period went from about seven billion cubic meters to about two thousand billion, amounting to more than a 280-fold increase.

When these three categories of fossil-fuel use are taken together, a remarkable pattern of growth emerges. By the end of the twentieth century, human use of fossil fuel increased to about 10 billion metric tons of oil equivalent.[42] Adding in roughly another half billion from increased use of renewable fuel, we arrive at an approximate 10.5 billion metric tons of total human energy consumption annually. At the end of the twentieth century, human population stood at about six billion. This comes to about 1.75 tons a person, equivalent to about seventy-nine billion joules.

Let us compare per capita growth during the twentieth century with figures for earlier periods cited previously. During the period from 1000 BCE to the beginning of the Industrial Revolution, per capita annual energy consumption grew from thirteen to thirty-eight billion joules. This amounts to an average increase of less than one billion joules per century. During the first 150 years of the industrial period, per capita annual consumption grew from thirty-eight to fifty-six billion joules, an increase of about twelve billion joules per century. During the twentieth century alone the figure increased another twenty-three billion joules. With reference to figure 6.1 (which ends with the year 1900), this shows that per capita consumption increased almost twice as rapidly from 1900 to 2000 as it did from 1750 to 1900.

Increase in total human energy consumption can be estimated in similar fashion. By 1750, some seven hundred million people were consuming 38 billion joules each. This gives a total of about (700×10^6 x $38 \times 10^9 =$) 27×10^{18} joules annually, up from (50×10^6 x $13 \times 10^9 =$) 0.650×10^{18} in 1000 BCE. The average increase for this (approximately) twenty-seven-century period, accordingly, was approximately 1 billion billion joules per century. By the year 1900 the total had increased to (1.6×10^9 x $56 \times 10^9 =$) 90×10^{18}, an increase on the average of 60×10^{18} joules per century since 1750. By the end of the twentieth century, the figure stood at (6×10^9 x $79 \times 10^9 =$) 474×10^{18} joules, up from 90×10^{18} a century earlier. This is an increase of 384×10^{18} for that century alone.

To say it again, total human energy consumption per year increased on the average about 1 billion billion joules per century between 1000 BCE and 1750 CE. From 1750 to 1900, the average increase was about 60 billion billion per century. And for the twentieth century alone, the increase was close to 384 billion billion.

In the previous section we observed that human energy use began increasing at an exponential rate as the industrial period got under way. With one significant difference, this pattern continues through the twentieth century up to the present. The difference is that during this most recent period the *rate* of increase itself has been growing at an exponential rate.

Ominous as it may appear, the explosion in human energy consumption brought on by the Industrial Revolution seems to be entering its final stage.

6.9 Entropy Increases with Use of Energy

Increasing energy use by humans is not necessarily a bad thing. It may be viewed merely as part of the general tendency for biological processes through the ages to take up ever larger portions of incoming solar energy. When life first appeared on Earth, the proportion of solar energy going into metabolic processes was negligible. During recent times, that proportion has increased to about 1 percent of incoming energy (see section 4.2). Given that humankind is only one among millions of species, the increases in human energy consumption documented in the previous section contribute only a small portion of that 1 percent.

The entropy resulting from this recent surge in human energy consumption, however, cannot be so easily dismissed. As stated in the Second Law of Thermodynamics, every unit of energy expended within an operating system gives rise to a corresponding unit of useless entropy (see chapter 1). In the case of the biosphere, this means that all energy consumed within it is eventually degraded into forms no longer capable of doing work. Whatever forms this entropy takes, it must be expelled for the biosphere to provide continued support for its top consumers (see section 4.8).

Several forms of entropy resulting from human activity have been examined earlier. One is low-grade heat, which humans produce in common with all living organisms. Low-grade heat is discharged from the biosphere by black-body radiation. Because of interference from other forms of entropy resulting from human activity, however, heat-removal mechanisms based on black-body radiation have ceased to function normally. As we have seen (in section 5.3), this has resulted in the current crisis of global warming.

Another form of entropy mentioned previously is material waste. Biological waste material is usually decomposed into compounds capable of providing nutrition to producer organisms (plants and algae), while the low-grade heat resulting from decomposition leaves the biosphere as black-body radiation. Increasingly prominent among waste products of human industry, however, are toxins that poison the decomposers and plastics that are not subject to decomposition. Once again we find circumstances in which humanly produced wastes impede the processes by which other wastes are removed from the biosphere.

The third form of entropy considered previously is degraded functional structure (see section 4.4). Functional degradation is occurring on a massive scale today as a result of species extinction brought about by human activity. Inasmuch as functional structure is nonmaterial (in the sense of section 4.4), its degraded form is not something that might be physically discharged from the biosphere (like low-grade heat). The humanly imposed impediment in this case thus is not an obstruction of the processes by which entropy is removed (as in the case of global warming), but rather a proliferation of *circumstances* in which higher levels of negentropy are essential for ecological stability. Removing entropy (disorder) in such circumstances boils down to regaining the structure (order) lost by the ecosystems in question.

This is a process requiring both time and energy. The time at issue is the time taken to regenerate the ecosystems affected, or more likely to replace them with other ecosystems supportive of human activity. And construction of ecosystems proceeds at an evolutionary pace, spanning decades or even centuries.[43] As far as energy is concerned, augmenting the functional interactions within a given ecosystem generally calls for increasing its number of participating species, which in turn requires increasing the amounts of nutritional energy provided by lower trophic levels within the system.[44] The energy required may not be readily available, inasmuch as the species to be replaced may have vanished from causes affecting species on lower trophic levels as well (consider the massive destruction of tropical rain forests).

Reflections along these lines show that the entropy stemming from human activity is not merely a negligible component of the large quantities of entropy generated within the biosphere overall. To the contrary, humanly generated entropy tends to take particular forms that actually disrupt the discharge of entropy produced by other biological processes. One lesson we learned in chapter 3 is that living organisms need to rid themselves of the entropy they inevitably produce in order to continue living. By impeding the processes by which other organisms discharge their entropy, human activity is undermining the structure of the biosphere itself.

6.10 Possible Remedies Anticipated

Before the Industrial Revolution, ecological damage caused by human activity remained mostly confined to its place of origin. Even when local populations abused their habitats in ways that made them uninhabitable (think of the Easter Islanders), the rest of the biosphere remained relatively unaffected. By the end of the nineteenth century, however, the disruptive by-products of industrialization had begun to spread far beyond their points of origin.[45] And by the late twentieth century, the entropy stemming from human activity—global warming, plastic junk, decreasing biodiversity—was adversely affecting biota in all parts of the globe.[46]

It is largely because of human intrusions like these that the biosphere has been degraded to the point of seriously threatening the well-being of its top consumers. Among people who have thought about the situation long and carefully (and without political agendas), there is general agreement that human life as we know it is increasingly in jeopardy. The urgent question at this juncture is what can be done about it.

One response that springs to mind immediately is that we should cut back on use of fossil fuels or even stop using them entirely. After all, fossil fuels contributed the lion's share of energy consumed the last century, and most of the noxious by-products now clogging up the biosphere are petrochemical in origin. To give up fossil fuels entirely, however, would result in misery and poverty for billions of people, which effectively removes it from the realm of acceptable solutions. Cutting back on the use of such fuels, on the other hand, seems both possible and socially desirable, although the severity of the cutbacks required remains unclear.

Continuing to explore the realm of the possible, we find various other approaches that deserve consideration. One falls under the general category of technical solutions. Given that most of our environmental problems are technological in origin, it has been argued that many of them can be solved by technological means. Another approach focuses on the advantages of "clean" energy. Observing that many ecological problems stem from consumption of fossil fuel, its advocates suggest that these problems could be overcome by increased use of solar and wind power. Yet another approach views our quandary from an economic perspective. Thinking of environmental integrity as a scarce commodity, its advocates contend that our predicament can be resolved by economic incentives.

Each approach is worth examining in detail. Doing so will enable us to weigh their strengths and weaknesses, as well as to gain further insight into the causes of our environmental plight. Knowing the cause is a prerequisite for prescribing remedies. Chapter 7 is taken up with a consideration of probable causes, and the remainder of part 2 is given over to an examination of possible remedies.

PART 2

Economics and Entropy

7

Economic Production and Its Ecological Consequences

7.1 Linking Ecological Damage to Economic Production

Chapter 6 recorded a roughly linear growth in human energy consumption from the hunter-gatherer era up to about one thousand years ago. The energy involved was almost exclusively renewable (in the form of biomass), and the by-products of human energy use were all biodegradable. A consequence was that most of the entropy resulting from this energy use was radiated back into space without widespread damage to other creatures on Earth. Up to this point in human history, humanity's involvement in the flow of energy through the biosphere did not differ significantly from that of other species.

This seemingly benign profile of human energy use began to change radically around the time of the Industrial Revolution. Humankind's rising energy consumption changed from a linear to an exponential progression (recall figures 6.1 and 6.2). An increasing proportion of energy consumed was fossil (nonrenewable) in origin (section 6.8). And increasing quantities of artifacts produced with this energy were not biodegradable. A result was that human energy use began to cause appreciable damage to the biosphere at large.

A general survey of this damage was given in chapter 5. One highly publicized form of damage is global warming, stemming from increasing amounts of low-grade heat building up in the atmosphere waiting to be discharged into space (sections 5.3, 5.4). Another form is the increasing presence in the biosphere of degraded material structure, typified both by plastic junk (section 5.5) and by breakdown of the ozone layer (section 5.6). A third is humanly induced loss of biodiversity (section 5.9), which amounts to a serious degradation of the biosphere itself. Damage of these and related sorts by now has become so severe that, in the sober judgment of many expert observers, the biosphere is on the verge of losing its capacity to support human society as we currently recognize it.

Several factors in the network of causes leading to this predicament have already been examined. One is the rapid increase in human population over the past few centuries (shown in figure 6.2). As long as per capita energy use holds constant, more people using energy obviously results in more energy being consumed overall. As might be expected, however, there are respects in which increased population follows from increased energy use no less than vice versa. Further consideration of the population factor is reserved for a subsequent chapter.

Another factor behind humanity's rising level of energy use is the rapid increase in per capita energy consumption shown in figure 6.1. Looking ahead to concerns of the present chapter, there are several things to note about this particular factor. One is that the increase in question probably could not have been achieved without extensive use of fossil fuel. As already indicated (section 6.8), fossil-fuel consumption grew many times more rapidly during the twentieth century than did use of renewable energy. Even if human population had been arrested at its 1900 level (the approximate point at which fossil fuel surpassed biomass fuel consumption), the increase in per capita energy use during that period would have required more energy than was available from renewable sources alone.[1]

This means that the exponential rise in per capita energy consumption following the Industrial Revolution played a substantial role in triggering many of the harmful effects of unexpelled entropy examined in chapter 5. Recall once again that global warming, destruction of the ozone layer, the accumulation of plastic junk, and the widespread poisoning of other species are particular examples of these effects. An obvious consequence is that our current level of per capita energy consumption is a primary source of ongoing damage to the biosphere on which human existence depends.

Yet another fact to bear in mind is that our present level of per capita consumption is due primarily to the economic production of the world's most highly developed countries. An average person in one of the less developed nations typically consumes only a small fraction of the energy used per capita in highly developed countries like the United States and Canada. Generally speaking, a country is classified as developed or undeveloped according to its per capita income.[2] This suggests that a country's per capita income is more than randomly associated with its per capita energy use.

A country's per capita energy use, of course, is its overall energy use divided by its population. These observations about per capita energy use are snippets from a larger picture relating economic production to energy consumption generally. The overarching fact of the matter is that a country's gross product tends to correlate positively with its fossil energy consumption. Although the correlation is not linear, countries with relatively high national incomes tend to generate relatively large amounts of ecologically damaging entropy.

The economic aspects of the big picture appear even more problematic in light of the common view among orthodox economists that economic health goes hand in hand with economic growth. According to the orthodox view, the sign of a properly functioning economy is its ability to maintain growth at a rate (say between 2 and 5 percent) that avoids both recession and inflation. But as just noted, growth in income generated is positively correlated with increased consumption of fossil energy. From an ecological perspective, accordingly, the long-term upshot is that the very process of maintaining the health of an economy (according to this view) makes the biosphere supporting that economy progressively unhealthy. This upshot, to say the least, appears counterproductive and as such calls for dispassionate examination.

Part 1 of this study (chapters 1 through 6) was taken up with an analysis of the ecological factors underlying our current environmental predicament. Part 2 (chapters 7 through 12) is given over to an investigation of relevant economic factors in turn. We begin in the present chapter with the connection between energy consumption and economic growth.

7.2 Energy Use Correlated with GNP

When economists speak of economic growth, they typically are talking about an increase in either Gross Domestic Product (GDP) or Gross National Product (GNP). Simply put, GDP is the total market value of goods and services produced within a country during a standard accounting period, and GNP is GDP plus income of domestic residents from abroad minus domestic income earned by nonresident foreigners. These measures are commonly regarded as indices of a nation's economic performance. Different reporting practices in different countries require that both measures be kept in use.

For some purposes, economists find it appropriate to compare economies in terms of GNP *per capita*. This more discriminating measure is often used in comparing economic conditions between developed and developing countries, inasmuch as the latter tend to be more heavily populated.

An often-reproduced set of statistics was released by the U.S. Office of Science and Technology (OST) in the early 1960s showing a close correlation between annual per capita commercial energy consumption and per capita GNP (in 1968 dollars) for more than forty countries worldwide. The chart in figure 7.1 shows the correlation for a representative sample of countries.[3]

The chart is divided into three zones, in order of increasing levels of energy consumption. Zone A includes countries with less than fifty million BTUs per capita annual energy consumption at that time and less than $1,000 per capita GNP. Zone B includes countries not in zone A with less than a hundred million BTUs energy consumption and less than $2,000 GNP. Zone C includes countries in neither zone A nor zone B, all of which have per capita energy consumption levels of over a hundred million BTUs. These zones will be discussed in reverse order.

Figure 7.1. Correlation of GNP and Energy Consumption

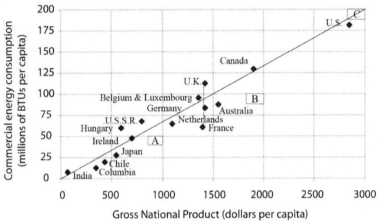

Zone C is occupied by the three countries then ranking highest in commercial energy consumption worldwide, the United States, Canada, and the United Kingdom. The United States is characterized by about 180 million BTUs of energy consumption per person and about $2,850 per capita GNP. This gives it an energy-in to GNP-out ratio of approximately 60,000 to 1 (rounded off to the nearest 5,000), which is to say that each dollar of goods and services produced requires an average consumption of about 60,000 BTUs of energy. Let us refer to this as the country's energy conversion ratio.[4] The corresponding ratios of Canada and the United Kingdom are roughly 70,000 and 80,000 respectively. The average per capita ratio for these three countries is thus about 70,000 BTUs for each dollar GNP.

Zone B contains fourteen countries figuring in the OST statistics, of which seven are shown in figure 7.1. The seven countries shown constitute the approximate center of an otherwise nebulous cluster. On the left side of the cluster are South Africa and Poland (neither shown) with conversion ratios of about 150,000 and 130,000 respectively. On the right are Finland and New Zealand (neither shown) with ratios respectively of about 35,000 and 30,000. The average ratio of the cluster of fourteen countries is close to 70,000, which is also the average ratio of the seven countries shown. In this respect, the countries shown are representative of the entire cluster.

Zone A includes twenty-seven countries, most of which were then designated "developing" by World Bank criteria. With a few exceptions, these are all close to the axis identified by the five countries shown in the chart. The average conversion ratio for the group shown is about 60,000 BTUs per dollar GNP, the same as the average of the twenty-seven countries in zone A taken together. The highest ratio in this zone, standing at approximately 80,000, is shared by Yugoslavia and Romania (neither shown). Ghana (not shown) has the lowest ratio, at about 20,000.

The solid line drawn from lower left to upper right corners of the chart indicates a ratio of 66.7, which in the context of the chart rounds off to 65,000 BTU for each dollar GNP. This line thus splits the difference, as it were, between the average conversion ratio of 70,000 for zones B and C and the 60,000 ratio of zone A. As a glance at the chart should reveal, the line pulls all the data represented on the chart together into a reasonably coherent pattern. On the average worldwide, production of one dollar's worth of goods and services in the early 1960s consumed about 65,000 BTUs of energy.

In one ordinary sense of the term, systems of production can be rated as more or less efficient according to the quantity of energy they require to produce a given amount of goods and services. The ratios of BTUs-in to GNP-out indicated above thus serve as measures of comparative efficiency of the productive systems concerned. A number of interesting conclusions in that regard are supported by the OST data upon which figure 7.1 is based.

One conclusion pertains to the dotted line drawn vertically between the entries on the chart for the United Kingdom and France. In the early 1960s, these two countries had about the same per capita GNP, an amount approximated by Germany and Belgium/Luxembourg as well. But France (with a ratio of 40,000) produced that amount with less energy and hence more efficiently than Germany (with 60,000), which in turn did so more efficiently than Belgium/Luxembourg (with 70,000), which did so more efficiently than the United Kingdom (with 80,000). Although the reasons for these disparities undoubtedly are many and various, they should be worth studying by anyone interested in energy conservation on a national level.

Another potentially interesting fact is that the twenty-seven countries occupying zone A had an average ratio of 60,000, in comparison with an average of 70,000 for the seventeen developed nations in zones B and C. This at least suggests that the process of development carries with it an overall loss in energy efficiency.[5] A contrasting observation, however, is that gain in overall production is not incompatible with increased energy efficiency, as witnessed by the comparative conversion ratios of the United States (60,000) and the United Kingdom (80,000).

It is also interesting to note the wide range in energy efficiencies among countries in zone B. The least efficient is South Africa, with a conversion ratio of about 150,000. The most efficient is New Zealand, with a ratio of about 30,000.

7.3 Factors Affecting a Country's Conversion Ratio (Energy-In/GNP-Out)

Figure 7.1 shows an average conversion ratio of 65,000 for the fifteen countries specified. In this respect, these countries were representative of the entire group of countries in the OST survey. This means that in the early 1960s the roughly forty countries involved consumed an average of about 65,000 BTUs of energy for each dollar of GNP their economies produced.

For present purposes, there is nothing special about the early 1960s other than its being a time for which such data are readily available. Neither is there anything especially significant about 65,000 as the average conversion ratio during that period.[6] Different ratios would be expected to emerge from analyses of other periods of recent economic history.

This is so for a variety of reasons. One obvious reason is fluctuation in energy costs. As the price of oil increases, for example, economies relying heavily on petroleum products will tend to use them in more efficient ways. Efficiency in this case amounts to more goods and services being produced with the expenditure of a given amount of energy.

Another reason for change in a country's conversion ratio is variation in the proportion of its GNP contributed by its service sector. As a general rule, services like education, health care, and entertainment are less energy intensive than industries dedicated to the manufacture of goods. It follows that an economy's conversion ratio will generally improve with an increasing proportion of services represented in its GNP.

A further reason has to do with changes in production techniques that take place as an economy evolves toward more fully developed status. Recall the fact indicated in figure 7.1 that developed countries tend to have higher conversion ratios (are less energy-efficient) than their counterparts in the developing world. This appears to be no accident. As labor-intensive productive practices give way to industrial modes of production, it should be expected that the economy in question becomes more energy-intensive. The price of increased energy intensity typically is an increase in energy (compared with labor) expended in producing a given amount of goods.

Despite the current drift toward increased industrialization on the part of developing countries, however, economic production worldwide has grown more rapidly over the past several decades than has consumption of energy for commercial purposes.[7] This is due in part to more efficient energy use on the part of developed economies. Another cause of this growing efficiency is a general shift from lower to higher quality energy sources, in particular a shift from coal to electricity and natural gas.[8]

Take the case of China, for example. In the approximately twenty years (1977–95) following the end of its revolutionary period, China's GNP grew annually at an average of over 9 percent. During the same period, however, its reported energy use per unit GNP fell a remarkable 55 percent (back to the 1957 level before its growth spurt began). This increase in energy efficiency is attributable in part to increasing imports of energy-intensive products. But it was a result also of advances in technology and of increased specialization. Another key factor was the rapid growth of its service sector in recent years.[9]

This recent pattern of economic production growing more rapidly than energy consumption worldwide suggests several attitudes toward energy use that might be adopted by economic planners in individual countries. A country might try to hold its GNP constant while decreasing its consumption of energy, or it might try to hold its energy use constant while increasing its GNP. As the case of China indicates, moreover, it might even aim to decrease its use of energy while increasing its GNP. Up to a point, any one of these strategies might conceivably be feasible.

But only up to a point. In this respect, attempts to get more and more economic production per unit energy by increased energy efficiency are like attempts by athletes to lower the record for the hundred-meter dash by improved diet and training. At some point not far from the present record, human beings just won't be able to run any faster. Similarly, at some point not far from the current performance of our more efficient economies, it will no longer be feasible to achieve increased economic output without expending more energy. From that point onward, continued economic growth will come only at a cost of increased energy expenditure.

Once again, there is no long-term significance in the fact that the average conversion ratio for some forty countries in 1961 was about 65,000 BTUs of energy per dollar GNP. What is significant for present purposes is the realization that producing a given unit of GNP always involves the expenditure of a nontrivial amount of energy. In all economies affected by the Industrial Revolution, regardless of their current state of development, the production of wealth invariably is tied to the consumption of energy. Given resources currently available, moreover, the energy consumed in economic production is bound to include a substantial component of fossil fuel.

7.4 The Entropic Residue of Economic Production

Let us take a broader look at the relation between energy consumption and economic production. Among other functions, a nation's economy generates benefits for the society it serves. These benefits include the goods and services that contribute to its GNP. In later chapters we will consider economic benefits that cannot be measured in quantitative terms. For now we are concerned only with benefits included under GNP (or GDP).

These benefits are produced out of, and with the help of, resources taken from the biosphere. Included among such resources are water, minerals like iron and copper, gases like oxygen and nitrogen, and various forms of energy. In the terminology introduced in chapter 2, these resources are all forms of negentropy. Let us refer to them generally as "ecological capital." Since expenditure of such resources is necessary in the production of economic benefits, moreover, we may speak of that expenditure as the "negentropic cost" of the benefits in question.

The negentropic cost of economic production has an entropic counterpart. As defined in chapter 2, negentropy is the opposite (the negative) of entropy. Expenditure of negentropy thus generates an equivalent in entropy. We may refer to this as its "entropic residue." Just as the negentropic costs of production include the minerals, fuels, and similar resources expended in the process, so the entropic residue includes all the various forms of entropy generated by that expenditure. In accord with the Second Law of Thermodynamics, negentropy entering the economic process exits in the form of entropy.

By way of parallel, recall the thermodynamic description of living organisms laid out in chapter 3. An organism maintains its life process by "sucking up" negentropy from its environment and by passing off the resulting entropy into the environment for disposal. In the case of individual organisms, the negentropy in question consists mainly of energy and nutrition, whereas the expelled entropy consists largely of waste materials and low-grade heat.

When human beings band together and engage in economic activity, however, other forms of entropy and negentropy become involved. Negentropy "sucked up" from the environment comes to include energy in forms other than sunlight and foodstuffs (notably fossil), along with minerals and other natural resources. At this present stage in economic history, unfortunately, the resulting entropy has come to include global warming, a decimated ozone layer, and the disruption of natural cycles essential to life (chapter 5).

There are parallels with the flow of energy through ecosystems as well. A healthy (normally functioning) ecosystem replaces negentropy consumed by its constituent organisms by new supplies from outside sources (mainly solar radiation), and passes off the resulting waste products (mainly waste heat) for more comprehensive ecosystems to dispose of. An ecosystem tends to remain healthy as long as the negentropy it takes in and the entropy it discharges remain in approximate balance.

By definition, the biosphere is the most comprehensive ecosystem of all. The biosphere, accordingly, is the system into which all other ecosystems discharge their entropy and from which the entropy accumulated from all life processes on Earth is passed off into space. The health of the biosphere over the long term depends upon its ability to maintain an approximate balance between the negentropy consumed by its resident organisms and the entropy resulting from this consumption that is radiated into space as low-grade heat. Entropy that must be disposed of in this fashion includes both that generated by biological processes and, in recent centuries, that generated by economic activity.

As long as the biosphere remains healthy, it is able to get rid of most of the entropy stemming from these two sources as a matter of course. In preindustrial times, the entropic residues of economic production were not exorbitant, and the biosphere was able to accommodate them without undue stress. Within the span of the twentieth century, however, constantly expanding economic production resulted in an accumulation of vast amounts of entropy that the biosphere has been unable to discharge into space. The inevitable result is a disruption of its thermodynamic balance, which throws into doubt the biosphere's continued ability to support its top consumers (described in chapter 4).

Conventional economic wisdom takes it for granted that economic health requires constant growth in the production of goods and services. If so, then economic health and the health of the biosphere cannot be maintained simultaneously. We will continue to be concerned with this conflict and ways of moderating it throughout part 2 of this study. By way of preparation, let us consolidate the results of the present chapter.

7.5 A Disregarded Economic Principle

During the middle decades of the twentieth century the correlation between energy consumption and GNP remained controversial, with mainstream economists generally reluctant to admit more than a co-incidental relation.[10] Recent economic studies, however, have demonstrated a remarkably close interaction between the two. The results of these studies support my conclusion in section 7.3 that production of any given unit of GNP requires the expenditure of significant amounts of energy.

One study published in 1986, for example, showed a tight year-by-year correlation between GNP and energy use in the United States from 1870 to 1981, a correlation that extended even to minor yearly fluctuations.[11] Another study, reported in David Stern's 2003 paper "Energy and Economic Growth," concluded that "there is a very strong link between energy use and both . . . economic activity and economic growth."[12] With reference to the world economy generally, it has been remarked in a report published by the Institution of Engineering and Technology that there "has always been a tight coupling between energy consumption and economic growth."[13]

A consequence of this tight correlation between economic production and energy consumption is that the production of economic goods unavoidably generates entropy. This consequence is formally demonstrated by the following inference:

(1) The production of economic goods consumes energy.
(2) The consumption of energy generates entropy.
Therefore, (3) the production of economic goods generates entropy.

In 1 and 3, the expression "economic goods" is intended to cover services as well as material products. The general sense of the expression here is that of economic benefit or economic utility.

This inference is obviously valid, which means that the truth of conclusion 3 follows necessarily from the truth of premises 1 and 2. The truth of 1 is shown by considerations discussed in section 7.3, as well as by studies like those mentioned earlier in the present section. The truth of 2 follows from the Second Law of Thermodynamics. Since the inference is valid and its premises are true, the inference establishes the truth of its conclusion.

By itself, the conclusion that economic production gives rise to entropy seems unremarkable. Entropy results whenever energy is consumed. The conclusion is cast in a more ominous light, however, by the fact that most of the energy used in economic activity today comes from fossil fuels.[14] In ways discussed at length in chapter 5, entropy deriving from fossil fuels often shows up as some form of ecological damage (global warming, habitat destruction, etc.). Under these circumstances, the consequence to bear in mind is not just that economic production results in entropy, but that it results in various forms of ecological damage.

The previous inference should be augmented accordingly. Given the present stage of global industrialization, involving massive consumption of fossil fuel,

> (4) Entropy generated by economic production by and large is ecologically damaging.

Therefore, (5) the production of economic goods by and large is ecologically damaging.

The truth of 5 follows from the truth of 3 and 4. Thesis 3 is established by the previous inference, and thesis 4 rests on considerations such as those examined at length in chapter 5. Taken together, these two inferences establish the conclusion that, under present circumstances, economic goods come at a cost of ecological degradation.

This result can be expressed in various ways, highlighting different aspects of the ecological expense in question. Interpreted in terms of costs and benefits, the result says that, under current conditions, the ecological cost of economic benefits carries with it a severe entropic liability. In terms of utilities and disutilities, it says that economic utility implicates extensive ecological disutility on the part of the supporting environment. In thermodynamic terms, it means that economic activity transforms *ecological* negentropy (energy and structure) into *economic* negentropy (goods and services), leaving a residue of highly damaging entropy (low-grade heat and structural disorder) for the biosphere to cope with. However put, the upshot is that economic production at this point in time results in appreciable damage to the biosphere.

Let us repeat the conditions under which this conclusion applies. We are not talking about economic growth in all times and circumstances. In eras when only renewable energy was used in economic production, and when goods produced were all biodegradable, economic activity generated relatively little long-term ecological damage. Entropy originating under such conditions is passed off into space at roughly the same rate as it is produced within the biosphere. Lasting ecological damage results only when entropy is produced more rapidly than it can be dispelled.

The conclusion also needs to be more specifically stated. Economic production, of course, is quantitatively measurable (e.g., so many dollars worth of goods produced per person). But nothing has been said so far about quantities of ecological damage. In this regard, it should be recalled that the damage in question notably includes low-grade heat trapped within the atmosphere, accumulation of plastic junk, and decreased species diversity in the biosphere. All these are forms of entropy subject to quantitative measurement.

Ecologically damaging entropy is also illustrated by breakdown of functional structures, such as the biosphere's nitrogen and carbon cycles (section 5.7). Just as functional structure is a form of order, so breakdown of functional structure is a form of disorder. Although not gauged in calories, pounds, or inches, structural order and disorder nonetheless are measurable quantities. (One relevant quantitative measure is provided in the appendix to chapter 2.)

With this in mind, we may state conclusion 5 in somewhat more concrete terms. With applicable qualifications taken into account, the conclusion may be reformulated as follows:

(P) The production of a given quantity of economic goods introduces a corresponding quantity of ecological degradation.

The corresponding amount of degradation will vary with case and circumstances. And in individual cases, it will often be hard to measure exactly. But given our massive reliance on fossil fuel, that amount in any given case most certainly will not be negligible.

Result P has the status of an economic principle. For convenience, we may refer to it as the Entropy Principle of Economics. It is economic in the sense that it applies to industrialized economies. It has the status of a general principle because it applies to such economies generally, regardless of their geography or stage of development. The fact that nothing like it can be found in standard economic textbooks does not deprive it of this status.

7.6 Shifting the Focus to Economics

Principle P states that a certain measure of ecological capital must be expended for any given quantity of economic production. A direct consequence is that growth in economic production results in growing amounts of ecological degradation. Part 1 of the present study has argued that continued human welfare requires that ecological damage be curtailed. This seems to indicate that economic production must be curtailed as well.

Curtailing production, however, runs directly counter to the view of mainstream economics that a healthy economy requires continuing growth. This puts mainstream economics on the spot. If human welfare is to continue, it appears that the orthodox view favoring continuing growth must be relinquished. The only alternative is to find some way of fostering growth that will not cause increasing ecological damage.

In point of fact, two strategies are currently being pursued for curtailing environmental damage brought about by economic activity. One addresses the various damaging effects of economic production on a case-by-case basis. Thus we might try to alleviate global warming, for example, by cutting back on CO_2 emissions, try to reduce damage to the ozone layer by eliminating use of CFCs (recall the Montreal Protocol), and try to deal with other problems likewise on a piecemeal basis. In the analogous case of trying to cut back on lung disease caused by smoking, this is like isolating various types of disease (emphysema, pulmonary fibrosis, cancer) and devising treatments for each one separately. Since this strategy seems compatible with continued economic growth, it has gained widespread support among mainstream economists.

The other strategy is to break the link between economic production and ecological degradation. This would be analogous to breaking the connection between lung disease and smoking by eliminating harmful ingredients from cigarettes. In the present case, one way of breaking the link would be extensive substitution of "clean energy" for fossil fuels. This also is an approach many economists find attractive.

It should be evident that these approaches are not mutually exclusive. An entirely feasible joint strategy is to replace fossil energy sources where possible by wind and solar and to rely on abatement technology (like catalytic converters) to counter damage from fossil fuels still in use. To assess the strengths and weaknesses of these two main strategies, however, it is better to examine them separately. This sets the agenda for chapters 8 and 9.

A third general strategy for curtailing ecological damage of economic origin, needless to say, is to cut back on economic activity itself. In the analogy of lung disease caused by smoking, this is like cutting back on use of tobacco products ("An ounce of prevention is worth a pound of cure"). Since cutting back on economic activity amounts to curtailing growth, however, this approach is anathema to mainstream economists. We turn to the topic of economic growth and its curtailment in chapters 10 and 11, after evaluating the shortcomings of the other two strategies.

With the remarks of the present chapter on the deleterious impact of economic activity, it may appear that undue blame is being cast on the discipline of economics itself. For economists to find offense in this appearance, however, would be like criminologists finding offense in a discussion of crime. The cause of our environmental crisis is not economics itself, but rather the real-life activity that economists study.

At the same time, it must be admitted that certain branches of economics have been more influential than others in charting the course along which real-life economies are currently heading. Unfortunately, these branches of economics are in the ascendancy; hence the description "mainstream economics." The final chapter of part 2 distinguishes the mainstream from the other branches of economics not beholden to the doctrine of perpetual growth.

8

Technological Solutions to Ecological Problems

8.1 Specific Fixes for Specific Problems

Three general strategies for alleviating the environmental impact of economic production were identified at the end of chapter 7. First mentioned was the strategy of identifying specific environmental problems and then devising suitable technology to correct them. The purpose of the present chapter is to examine the strengths and weaknesses of this general strategy.

A brief review of chapter 5 will verify that most of the ecological problems it discusses are technological in origin. Global warming stems largely from the emission of greenhouse gases from factories, power plants, and petroleum-powered vehicles. Holes in the ozone layer are caused mostly by synthetic gases known as CFCs. Agricultural destruction of habitats is due in large part to use of petroleum-based pesticides, and so forth. Given their technological origin, it is not unreasonable to think that such problems admit technological solutions in turn. Inasmuch as human ingenuity is responsible for these problems in the first place, that is to say, it seems natural to rely on human ingenuity in seeking effective remedies.

This approach is attractive from an economic viewpoint because remedial technology almost always involves the development of products and services that can be sold on the market. Another attractive feature is that, in certain cases at least, technology appears to work quite successfully. Let us begin a brief survey of apparent successes.

8.2 Encouraging Precedents

One frequently cited success story is the replacement of CFCs (chlorofluorocarbons) with environmentally less damaging substitutes. For decades after their invention in the 1930s, CFCs were used widely as refrigerants, solvents, aerosol propellants, fire retardants, and foam-extrusion agents. In the mid-1980s it was discovered that escaped CFCs were causing ruptures in the earth's stratospheric ozone layer, which shields the biosphere from potentially damaging ultraviolet radiation. By this time, ozone depletion was progressing detectably year by year, with corresponding increases in damage to plant life (notably marine phytoplankton) and to human beings (in the forms of skin cancer and ophthalmic disease).

An international conference was convened in 1987 to address this problem, resulting in a treaty known as the Montreal Protocol. Among its main provisions, the protocol called for a phasing-out of the most damaging forms of CFC by 1996 and for an elimination of all ozone-depleting chemicals by 2030. As a result, worldwide production of CFCs dropped by a factor of five between 1989 and 1996, most of the remainder being due to continued use by developing countries. The plight of developing countries in this regard was acknowledged with the establishment of the Multilateral Fund to help them phase out ozone-depleting substances without compromising their economic growth.[1]

By 2005 the vast majority of the world's industrial nations had become parties to the treaty. This joint effort was enabled in large part by the availability of other chemical products to replace the most damaging CFCs. Transitional replacements included HCFCs (hydrochloro-fluorocarbons), which, although still containing chlorine, have less effect on the ozone layer. More promising substitutes in the long run are HFCs (hydrofluorocarbons), which are relatively inactive and have little ozone-depleting effect. While efforts to find other replacements continue, the replacement of CFCs with HFCs provides a striking illustration of technologically induced problems alleviated by technological remediation.

Another instructive illustration of this approach is the use of desalination technology when adequate supplies of fresh water are unavailable. The first large-scale desalination plant was built by Kuwait in the 1960s, providing water for agriculture as well as for human consumption.[2] By the turn of the century, more than 7,500 desalination plants were in operation worldwide, a majority of which are located in the Middle East.[3] As of 2005, the largest plant in operation (in Ashkelon, Israel) was capable of producing a hundred million cubic meters of salt-free water per year. Desalination of seawater is a proven technology for augmenting the world's diminishing supplies of potable water.

Consider also the use of smokestack scrubbers and similar devices to reduce emission of sulfur compounds from the burning of fossil fuels. Sulfur dioxide, for instance, combines with water in the atmosphere to form sulfuric acid, which returns to earth as acid rain. Although its presence had been recognized for well over a century, acid rain became notorious in the late 1960s for its destructive effects on lakes and forests. Particularly affected were large areas in the northeastern United States and Canada, along with parts of Germany, Russia, and China. An abatement technique now common in various countries, including the United States, is to mix hot gases in smokestacks with a limestone slurry that reacts with sulfur dioxide to form calcium sulfate. The latter then can be recovered and used in the manufacture of gypsum products. Here, as in previous cases, technology has proved effective in the mitigation of environmental problems originally stemming from the use of technology.

Yet another illustration that might be suggested is the use of genetic engineering to reduce the need for chemical pesticides. As noted in chapter 5, the application of commercial insecticides and herbicides to control agricultural pests can have effects beyond those on the targeted species. Not only can pesticides cause harm to workers applying them, but they pose dangers to ecologically beneficial life-forms as well. Agricultural poisons kill off honeybees needed to pollinate crops, reduce populations of birds that eat noxious insects like mosquitoes, and destroy earthworms and other decomposers essential to healthy soil.

Viewed from a perspective not dominated by profitability, use of pesticides is a serious environmental problem.

One anticipated benefit of genetically engineered crops was that they would require less pesticide than their naturally occurring counterparts. Prominent instances are Bt crops, whose seeds contain genes of *Bacillus thuringiensis,* a soil bacterium that attacks insects that prey on naturally occurring strains. Bt corn that is resistant to common corn borers has become commercially available in recent years. Bt cotton also has found widespread use because of its ability to control several varieties of budworm and bollworm. At least in theory, genetic engineering appears to provide access to farming methods less dependent on ecologically harmful pesticides than methods used now.

Apparent successes like these lead to optimism about technological solutions to other ecological problems of our industrial age. Many such problems have their origin in commercially successful technology. Against this background, it seems natural to assume that technological ingenuity will be able to provide workable solutions for these other problems as well. We have next to consider whether this assumption is justified.

8.3 Problems Unresolved by Piecemeal Fixes

Initial confidence in the piecemeal approach is likely to be dampened by the realization that the remedial technologies described in the previous section have proven less successful than those brief descriptions indicate. In some instances, use of technological remedies has actually aggravated the problems they were intended to remedy. A case in point is the use of genetically modified seed to reduce the need for agricultural pesticides.

According to a report by agronomists from Cornell University,[4] Chinese cotton growers were among the first to plant Bt cotton, which reduced pesticide use during the first three or four years by more than 70 percent. Within less than a decade, however, Bt farmers had to apply just as much pesticides as conventional farmers because of a proliferation of other insect pests that replaced the bollworms killed by the toxins. Inasmuch as Bt seed is much more expensive than unmodified varieties, farmers who continued to use it generated appreciably less income than their neighbors using conventional farming techniques.

The same tendency has been observed in crops grown from seeds genetically modified for herbicide tolerance (HT crops). Most HT crops are engineered to tolerate glyphosate, a herbicide marketed under the trade name "Roundup." Whereas HT crops generally required less pesticide during the first few years after the technology became widespread, however, in recent years (since about 2000) emergence of HT-tolerant weeds has resulted in use of over one million pounds more herbicides on genetically engineered than on conventionally grown crops.[5] According to a recent report released by a prominent agricultural economist,[6] farmers in the United States now use more pesticides per acre on the top three genetically engineered crops (corn, soybeans, and cotton) than on conventional varieties.

As far as smokestack scrubbers are concerned, one should not dispute the view that preventing large amounts of sulfur dioxide from being emitted into the atmosphere is all in all a good thing. But we have a tendency to think that once scrubbers have been installed in a coal-burning plant's smokestacks, further emissions from the plant are environmentally benign. Thinking this way overlooks the fact that a typical five-hundred-megawatt power plant (enough to power a city of about 140,000 people) also produces yearly over 10,000 tons of nitrogen oxide, almost 4 million tons of carbon dioxide, about 700 tons of carbon monoxide, over 300 pounds of arsenic and toxic heavy metals, and more than 190,000 tons of commercially worthless sludge from its scrubbers that end up in local landfills.[7] The typical plant also uses some 2 billion gallons of water, which is raised about 16° Fahrenheit in temperature before being discharged into a nearby lake or river (for the effects of this, see section 5.2).

Environmental drawbacks of desalination processes, in turn, include the generation of waste brine with a salinity about 1.8 times that of seawater, in amounts greater than those of the potable water produced.[8] Whether discharged on land or in water, this brine has devastating effects on local ecosystems. The desalination process is also very energy intensive, requiring several times more electricity per gallon produced than water pumped directly from lakes and rivers.[9]

Even in that much celebrated case of new technology coming to the rescue, the substitution of HCFCs for ozone-depleting CFCs, the full story is not one of unqualified success. On the positive side are the facts that replacement technology was readily available, that most countries were able to embrace it without economic disadvantage, and that amounts of ozone-depleting gases escaping into the atmosphere at first declined dramatically as a result. Experts now predict that the ozone layer might actually heal itself by 2065.[10]

Nonetheless, although HCFCs are only 5 percent as active in depleting ozone as CFCs, their use in rapidly increasing quantities has resulted in continued damage to the ozone layer.[11] Another problem is that HCFCs are greenhouse gases. While their contribution to global warming is substantially less than that of CO_2, it is far from negligible. No solution to one problem is fully successful if it exacerbates another that is equally serious.

A further negative aspect of this case is that replacement of CFCs by other coolants, however expeditiously accomplished, is a remedy whose curative effects will be delayed for several decades. Ruptures in the ozone layer continue to expand as a result of chlorine already in the stratosphere, and time is required for accumulation of additional ozone to replenish areas where ruptures have already developed. It may indeed turn out that the ozone layer regains its integrity sometime later in this century. But in the meantime, increasingly large numbers of people will develop skin cancers and cataracts, and the ocean's supply of phytoplankton will be progressively depleted.

The Montreal Protocol stands as a model of international cooperation in the face of an environmental problem by which all nations are in some way affected. But the shift from CFCs to HFCs by way of response should not be held up as an exemplary instance of environmental problems being solved by technological means. It is instead a case of engineers being able to replace an ecologically damaging technology with another that is less damaging but no less profitable economically. The main lesson to be learned here is that abatement technology often has unintended effects that can cause additional damage in their own right.

8.4 Ecological Problems Have Multiple Causes

A common feature of the ecological problems to which these abatement strategies are addressed is that each is influenced by a variety of causes. Chlorine in the upper atmosphere is not the only factor of human origin contributing to ozone depletion. Another chemical agent with this effect is bromine, stemming from halons commonly used in fire extinguishers and from agricultural fumigants containing methyl bromide. Although less bromine than chlorine reaches the stratosphere overall, bromine is 60 percent more destructive of ozone on an atom-by-atom basis.[12]

Desalination technology provides local solutions to widespread shortages of water for drinking and agriculture. These shortages are attributable to a multitude of causes. Some have to do with human consumption. In addition to increasing numbers of people requiring increasing amounts of water, there is a growing demand for water in industry (mining, electricity generation), recreation (golf courses, swimming pools), and lifestyle preference (green lawns, frequent bathing). Others have to do with pollution from human sources. Inland bodies of water are subject to eutrophication by agricultural runoff, and water from springs and rivers is rendered undrinkable by bacteria from animal farming and urban sewage. Natural causes include shifting weather patterns and ocean currents, often associated in some fashion with global warming (see section 5.4).

Consider next the matter of acid rain. Although its main cause is airborne sulfur dioxide, acid rain can result from nitrogen and chlorine compounds and to some extent from CO_2 as well. To some extent, all these compounds were present in the environment before the advent of heavy industry. Sulfur dioxide and hydrogen chloride are typically discharged in volcanic eruptions, and dimethyl sulfide is emitted by phytoplankton in the ocean. Forests and lakes were subjected to acidic precipitation from time to time as a matter of course, but usually at levels to which their resident populations could at least partially adapt.

Acid rain became an ecological problem with the additional accumulations of sulfur and nitrogen oxides in the air resulting from the increasing use of fossil fuel (mostly coal) in factories and power plants. By the late 1960s, scientists had begun to associate the collapse of freshwater ecosystems and the decline of northern forests with noxious emissions from upwind smokestacks. Here was a clear case of other creatures being destroyed by human use of fossil fuel. Moreover, human beings themselves were increasingly at risk from the corrosive effect of these chemicals on their respiratory systems.

The basic problem is not simply that acid rain is falling on our lakes and forests. This has been happening for countless millennia. The problem is that human technology has increased the intensity of this natural phenomenon to a point where many organisms affected can no longer cope with it. Installing scrubbers in factory chimneys curtails one source of the problem. But the problem is not resolved merely by cutting back on one of its multiple sources.

Genetic engineering of crops was cited in section 8.2 with reference to its potentially beneficial effect of reducing the use of chemical pesticides in agriculture. As already noted, Bt corn and cotton contain genes lethal to natural predators, and the built-in resistance of HT crops to commercial herbicides makes it easier to control nearby weeds with less concentrated doses of poison. Although reduction of pesticide use probably was not a motive in their development (companies responsible for genetically modified seeds sell pesticides as well), it seemed theoretically possible that these products might help alleviate the baneful effects of commercial poisons throughout the biosphere.

The problem of a biosphere choking with noxious chemicals, however, is much too general to be relieved by cutbacks in one or another kind of pesticide. From an ecological perspective, reduction in use of any pesticide is unquestionably a good thing. But the sources of noxious substances in the biosphere are so many and so varied that the problem is not likely to be solved on a piecemeal basis.

By and large, the ecological problems we confront today are too pervasive to be attributable to simple causes. A consequence is that they cannot be resolved by technology designed to lessen their impact in specific circumstances. Attempting to solve ecological problems of this dimension with specialized technological fixes is like trying to cure lung cancer by giving up a particular brand of cigarettes: when there are multiple sources of a pernicious condition, focusing on a single source in isolation is a feckless expedient.

8.5 Systemic Malfunctions Are Not Fixed by Technological Remedies

The idea of fixing ecological disorders of technological origin by further technology may strike one initially as an appealing concept. Other things being equal, it is better to take advantage of technological innovations when they prove helpful than to eschew new technology entirely. But new technology often turns out to have adverse effects that were not anticipated. And the ecological disorders in question are often too complex to be alleviated by technology addressed to specific manifestations. These points have been argued in the preceding sections.

Let us take our analysis of this approach another step forward. The basic shortcoming of this piecemeal approach is not merely that technology often has unexpected side effects nor that the ecological problems in questions have multiple sources. The underlying shortcoming is that these problems show up as malfunctions on the part of complex adaptive systems and that the proposed remedies do not correct these functional failures.

Return to the case of desertification. Precipitation over a given land mass occurs as a result of complex interactions involving evaporation from neighboring bodies of water, prevailing winds in the upper atmosphere, temperature gradients set up by convection currents near the surface, and the atmospheric drift of dust particles around which moisture can condense, along with various other factors (see section 5.4). Dry spells typically result when one or more of these factors behave abnormally, such as prolonged extension of equatorial convection currents as a result of global warming. And the affected land mass turns arid when dry spells persist.

It is obvious that climatic abnormalities of this sort are not neutralized by desalination plants installed in Saudi Arabia (or anywhere else) to supply drinking water not available through local rainfall. While this technology can help mitigate the effects of desertification in specific locales, it does nothing to restore regularity to the water cycles on which countless ecosystems depend in other affected areas. These cycles are parts of complex systems of interacting climatic variables that extend over vast areas of the earth's surface. Increasing local supplies of drinking water by technological means does not help bring these complex systems back into equilibrium.

Consider also the case of ozone depletion. When intact, the layer of ozone in the stratosphere blocks ultraviolet radiation that is harmful to many forms of life, including plants that convert lower wavelengths of sunlight into biomass.[13] This ozone layer is held in equilibrium by a set of complex chemical interactions involving high-level winds, changing seasons, oxygen molecules migrating from the troposphere, ultraviolet radiation, and so forth (see section 5.6). As long as these interactions remain in balance, the quantity of ozone they produce remains relatively stable and almost all of the incoming ultraviolet energy is absorbed in the process.

This equilibrium was seriously interrupted during the last century, when chlorine compounds used as air-conditioner coolants and spray-can propellants began entering the stratosphere in substantial quantities (as discussed in sections 5.6, 8.3). As mentioned previously, the industrial world's response to this threat was the Montreal Protocol of 1987. This initial agreement called for a phased reduction in use of CFCs by developed nations (with special provisions for developing countries), which it was hoped would lead to a total cessation by 1996. Subsequent amendments allowed continued production of CFCs at 15 percent of the 1986 baseline to help developing countries meet their basic needs, and mandated a complete phaseout of HCFCs by 2030. An important proviso is that all parties can continue to use the compounds in question when necessary for health or safety and when there are no technologically or economically feasible alternatives.[14]

Even if these extended deadlines prove impracticable (which seems likely), the Montreal agreement is admirable both as an example of international cooperation and as a response to a serious environmental problem. But let us look carefully at the nature of the response. Ozone-depleting substances (e.g., from volcanoes) have been in the stratosphere for eons and were part of the balance that remained basically stable until the introduction of CFCs produced by human industry. The basic stability of this balance enabled the development of life as we know it. Introduction of CFCs and related substances early in the twentieth century upset this balance, allowing appreciable amounts of harmful radiation to reach the earth's surface in subsequent decades. The proposed solution is just a partial reduction in use of the chemical that caused the problem in the first place.

What we need to realize is that even a *complete* phaseout of ozone-depleting chemicals will not automatically restore the balance. Once the stability of a complex system of interacting processes is upset, it cannot be reinstated merely by removing the aggravating cause. Ozone-depleting agents now in the stratosphere will remain active for decades, and it will take the ozone layer an unknown period of time (probably longer than currently predicted) to regain its equilibrium even after these disruptive chemicals are dissipated.

To a limited extent, technology can compensate for the unfortunate effects of natural systems that have gone out of kilter. We can build desalination plants on desert seashores, and we can wear sunscreens while waiting for the ozone layer to regain stability. But the fact remains that complex natural systems that have become dysfunctional cannot be set aright by technological means.

For yet another example of massive system malfunction, recall the underlying causes of global warming laid out in chapter 5. Global warming results from a disturbance of the complex feedback mechanisms by which tropospheric air temperature is maintained within a relatively constant range. These mechanisms are based on the tendency of low-lying cumulus clouds to reflect sunlight, keeping it from reaching the earth's surface, and the tendency of higher cirrus clouds to absorb heat, keeping it from leaving the atmosphere. When some sector of the air mass becomes abnormally hot, moisture evaporates from the earth's surface, and the proportion of cumulus to cirrus clouds increases. The net effect is a decrease in temperature of the air mass affected. When a given sector of the troposphere becomes unusually cold, conversely, the proportion of cumulus to cirrus clouds decreases. This amounts to fewer low-lying clouds to block incoming sunlight and more high-level clouds to absorb outgoing heat, causing the affected air mass to undergo a temperature increase.

As noted previously (see section 5.3), greenhouse gases cooperate with upper-level water vapor in absorbing heat energy before it can leave the atmosphere, thus leading to global warming. Greenhouse gases (notably CO_2) have been present in the atmosphere since plant life began, and were instrumental in establishing a stable range of ambient temperatures in which current life-forms could develop.

The problem stemming from the introduction of additional greenhouse gases by fossil-fueled industry is that more heat is now building up in the atmosphere than its natural heat-control mechanisms can handle. A dire consequence is that greenhouse gases from human industry have incapacitated these feedback mechanisms to a point where the very ecosystems supporting that industry are severely threatened. Both the problem and its sources have been amply documented.

It goes without saying that every effort should be made to reduce the amount of greenhouse gas produced by human industry to the lowest level possible. What we need to realize, however, is that even radical cutbacks in greenhouse-gas emissions will not restore the temperature-control mechanisms of the atmosphere to their preindustrial effectiveness. As is the case with the water cycle and the ozone layer, once natural systems of this magnitude are rendered dysfunctional by outside influences, they cannot be restored merely by reducing the presence of the disturbing factors. In particular, they cannot be restored by technological intervention.

8.6 Emissions Trading, Carbon Offsets, and Similar Matters

Technological remedies are popular within the business community because they provide a new kind of commodity to buy and sell. An example is equipment manufactured to reduce pollution. At the turn of the century, compliance with existing air pollution standards in the United States would have required a ten-million-ton reduction in sulfur dioxide emissions. Using smokestack scrubbers to achieve that goal, at over $300 a ton, would have cost an estimated $4.5 billion.[15] Purchase of equipment accounts for a major portion of this estimate.

Anticipating a bonanza in pollution-abatement markets, Japan moved into the new century aiming to be the world's leading vendor of abatement technology.[16] This led to the development of equipment for eliminating both sulfur and nitrogen emissions, which it is now selling to other countries. Another emphasis was the development of low-polluting automobile engines, which led to Japan's being recognized a decade later as the world's leading manufacturer of hybrid cars.

Another lucrative commodity was inspired by the Kyoto Protocol of 1997, which authorized the buying and selling of carbon emission credits. This agreement required the countries involved to reduce their emission by a certain amount (averaging 5.2 percent below their 1990 baseline) over the next decade, which those countries may convert into emission caps assigned to individual factories or industries. Companies with emissions exceeding their caps then are enabled to buy carbon credits from others with emissions below their limits. This scheme allows polluting industries to decide between buying expensive abatement equipment and paying other business entities for permission to continue polluting.

Once emission credits became established as tradable instruments, financial investors started buying them to sell at a profit. Open trading in SO_2 emission rights, for example, got under way in 1995, and by 2000, rights to emit SO_2 sold for about $260 a ton.[17] By 2007, markets for trading emission credits had been established in several countries, including the United Kingdom, Canada, and Japan, and non–Kyoto-based markets had been set up in Illinois, California, and Oregon.

Closely related to these emission-credit schemes is the business of carbon offsetting. Like carbon credits, carbon offsets involve paying someone else to compensate for your carbon emissions. A typical transaction of this sort involves a conscience-ridden individual paying an enterprising organization to plant trees that will absorb CO_2 in an amount supposedly matching that produced by the individual's air travel.

By some calculations, planting nine hundred trees is enough to remove as much CO_2 as an average U.S. citizen generates each year. At a typical going rate of $1 for each ten trees planted, this average citizen can continue to drive his or her SUV with an environmentally clear conscience for less than $100 annually.[18]

A major difference between emission credits and carbon offsets is that trading the former is regulated by a strict legal framework, whereas carbon offset transactions are generally unregulated. One source of abuse is to base offsets on emission-abatement operations of dubious effectiveness. Offsetting schemes currently in business include reducing methane seepage from farm manure and throwing iron particles into the ocean to bring phytoplankton to the surface.[19]

In contrast with the initial impact of carbon-credit trading,[20] the environmental effectiveness of these offsetting schemes remains unclear. While there is no doubt about trees taking up CO_2, for example, the cooling effect of trees planted as carbon sinks in temperate climates may be largely canceled out by the heat produced as they absorb winter sunlight that would otherwise be reflected back into space by an open snow cover.

Another source of uncertainty is that many of the tree-planting projects on which carbon offsets are based would probably have been initiated anyway for other reasons, such as producing pulp for paper mills or lumber for construction. When this is the case, money paid for offsets is merely a bonus for the producer and accomplishes little by way of CO_2 reduction.

About the only thing certain about such schemes is that carbon offsetting has developed into a very profitable business. Recent data indicate that over $280 million of offsets were sold in 2006 alone.[21]

The point to bear in mind, by way of summary, is that carbon offsetting, emission credit trading, and abatement technology are all piecemeal responses to a complex crisis. In their emphasis on technology and profit, they are basically extensions of the economic practices that led us into this predicament initially. Attractive as they may be from a business point of view, they offer no additional credibility to the dubious prospect that our environmental crisis can be resolved by technological means.

9

Replacing Fossil Fuel with Clean Energy

9.1 The General Strategy

Chapter 7 documented a close connection between recent economic activity and widespread ecological degradation. It also laid out reasons for thinking that ecological damage resulting from current economic activity is due largely to use of fossil fuel. There is increasing awareness among scientists and other cognizant people that these damaging side effects of fossil fuel must be curtailed.

Three strategies were identified in chapter 8 by which these harmful side effects might conceivably be reduced. One is to confront the many undesirable consequences of fossil-fuel use on a piecemeal basis. This strategy was examined in the course of that chapter and found unreliable as a general cure of our environmental problems.

Another strategy calls for replacing fossil fuel with so-called clean energy, hoping thereby to break the link between economic production and ecological destruction. Like the piecemeal strategy of chapter 8, this approach is compatible with continued economic growth. Also like the previous strategy, accordingly, it is viewed with general favor both by mainstream economists and by many members of the business community. Because of this, significant steps have already been taken in the development of clean-energy sources.

This is the main reason for dealing with the replacement strategy in the present chapter, following the piecemeal strategy already treated in chapter 8. Both are strategies already under way. When the shortcomings of these two current approaches have been assessed, we will be in a better position to consider the strengths and weaknesses of the remaining strategy calling for curtailment of economic growth. Treatment of this distinctly less popular strategy is set aside for the next two chapters.

9.2 Clean Energy

To understand the significance of the replacement strategy, we need a working definition of the term "clean energy." Commonly recognized examples of clean energy include energy obtained directly from solar radiation (by means other than photosynthesis), from wind, from moving water, and perhaps a few other sources (such as geothermal). What these types of energy have in common to merit the description "clean" is that they can be harnessed and used without direct damage to the biosphere.

One reason clean energy is environmentally benign is that it is available without major disruption of ongoing natural processes. Solar radiation is naturally present throughout the biosphere, along with wind and moving water. Although technology often comes into play in harnessing such energy for human use, the technology involved need not be invasive. Windmills and waterwheels can often be put to use without interrupting the balance of local ecosystems.

For the most part, sources of clean energy are also renewable. An energy source is renewable if it can be replenished without technological assistance. Although technology is sometimes (not always) required to make solar energy available for human use, the radiation providing this energy is continuously renewed without human help. Energy from wind and water is also renewable in this respect. While not always available in quantities or forms desired, energy from such sources is never completely used up.

Other forms of energy frequently touted as clean and renewable are nuclear and biofuel.[1] Thinking of biofuel simply as fuel derived from biomass, we can acknowledge a sense in which its source is renewable. Plant life is constantly being replaced on a natural basis, regardless of its use for human purposes. But commercial biofuels themselves are not renewable. Once consumed, they would never be replenished without extensive involvement of human technology.

Biofuels, moreover, are far from clean. This is the case particularly with ethanol produced from corn. One reason is that ethanol itself is the product of a technological process that releases various noxious chemicals.[2] Another is that most corn used to produce ethanol today is grown with commercial fertilizers and pesticides, which are environmentally harmful in many ways.[3] Yet another reason is that burning ethanol as a fuel generates carbon emissions in amounts only marginally less than those produced by fossil fuels.[4]

As far as nuclear power is concerned, the best to be said in its behalf from an environmental perspective is that nuclear reactors do not emit greenhouse gases. Nonetheless, significant greenhouse gas emissions are involved in the mining and enrichment of nuclear fuel, and ozone-depleting gases are discharged during enrichment as well.[5] For the record, it should be noted that nuclear sources are not renewable either. The only way such sources can be replenished is through complicated, expensive, and dangerous technology.

The overarching environmental objection to nuclear power, of course, is that radioactive waste from the reactors must remain within the biosphere for indefinite periods of time before it loses its capacity for severe ecological damage. Today, more than sixty years after Hiroshima, the nuclear industry has yet to devise more than temporary storage facilities for its radioactive wastes. The time may come when the waste generated in developing the world's nuclear arsenals becomes no less dangerous to the biosphere than the weaponry itself. Be that as it may, nuclear energy is neither clean nor renewable. Most certainly it is not a solution to our clean-energy needs.

9.3 Putting the Strategy in Perspective

The question posed for this chapter is whether substituting clean energy for fossil fuel is an adequate strategy for resolving our environmental crisis. A scenario that comes to mind in this regard is dismantling all power plants that run on fossil fuel and replacing them with banks of solar collectors, fields of wind turbines, and other clean-energy installations. Instead of smokestacks belching soot and greenhouse gases, we envisage power stations that operate with zero emissions.

At first blush, it may appear that this strategy deserves support without further debate. As previously defined, clean energy is energy that can be harnessed and used without harm to the biosphere. Given that our current crisis consists largely of ecological damage caused by fossil fuels, replacing those fuels with harmless forms of energy might seem a natural expedient for eliminating the damage.

On reflection, however, it becomes evident that things are not nearly that simple. One complication is that fossil resources currently serve in many capacities other than sources of energy. This is true of oil in particular, which is converted into products ranging from lubricants, solvents, and automobile waxes to fertilizers, paints, and road materials. And oil and natural gas are used to produce synthetic fabrics, plastics, and pesticides. Although estimates vary, the percentages of oil and natural gas put to nonfuel uses are far from negligible.[6]

The complication in this regard is that clean energy cannot replace fossil sources in their nonfuel uses. Energy from solar collectors cannot be used as feedstock in the production of pesticides, energy from wind turbines cannot provide raw material for the manufacture of plastics, and so forth. As we know all too well, several of the more severe forms of ecological damage under consideration are tied in with these nonfuel applications. This means that several major forms of ecological damage would not be alleviated by the substitution of clean energy for fossil fuel.

Among uses of fossil sources for fuel, moreover, there are many that do not involve electricity. Electricity is obviously involved when coal or natural gas is used to generate electric power conveyed to customers over high-voltage power lines. Electricity is also involved in diesel-powered railway engines, for example, which convert diesel fuel into electricity for driving their wheels. A large portion of fossil-fuel consumption, however, comes from uses in which electricity is not directly involved.[7] Salient examples of nonelectricity uses are oil and natural gas for heating, gasoline and diesel for ground transportation, and a close cousin of diesel for aviation.

These nonelectricity uses pose a further difficulty for the clean-energy strategy because power from clean sources is delivered mostly in the form of electricity. There are exceptions, of course. Windmills can be used to pump water, waterwheels to drive mill stones, and passive solar collectors to heat buildings. In such uses, electricity typically is not involved. For the most part, however, clean-energy installations are designed to produce electricity. Power from these sources can be directly substituted for electricity generated from coal and oil.

The difficulty here is that clean energy cannot be directly substituted for fossil fuel in applications where electricity is not involved. In some cases it is possible to put electricity to work in tasks previously fueled by other forms of energy. A straightforward example would be shifting from natural gas to electricity in heating one's home. If this electricity were produced from a clean-energy source, this would count as a replacement of fossil fuel by clean energy.

Another (and less realistic) example would be to replace most of our petroleum-fueled transportation fleet with electric vehicles. If this were to happen, then the electricity used to power these vehicles in theory could be produced from renewable sources. One reason this scenario is fanciful is the sheer improbability that enough clean energy could be produced to operate such a prodigious number of vehicles. Another is that a wholesale conversion of our petroleum-fueled transportation fleet to electricity would be exceedingly expensive. Furthermore, a conversion of this sort would involve changes in infrastructure more extensive than society could tolerate.[8] From a practical perspective, it is hard to imagine more than a small portion of the world's transportation needs being served by clean energy.

Even if conversions to electricity like these were made whenever possible, moreover, there would still be many applications in which fossil fuel would remain irreplaceable. Science fiction aside, neither cargo ships nor airplanes can run exclusively on electricity. Heavy-duty construction equipment generally requires fossil energy, as does most of the farm machinery involved in food production. Also included among types of equipment dependent on fossil energy are the mobile armaments employed in modern warfare.

Summing up, we find that roughly 10 percent of our natural gas and oil production goes into nonfuel uses[9] and that about 30 percent is converted into electricity.[10] As an optimistic estimate, let us assume that another 10 percent is used for purposes electricity could serve as well.[11] These last two, totaling 40 percent, are the only categories of use in which clean energy could be substituted for fossil fuel. The remaining portion of fuel uses (50 percent) are such that the fossil resources involved could not be replaced by clean energy.

The significance of these percentages may be more apparent if they are presented in graphic form. Think of the line in figure 9.1 as divided into ten equal segments, each representing one-tenth of the world's total consumption of fossil resources.

Figure 9.1. Fossil Resources Replaceable by Clean Energy

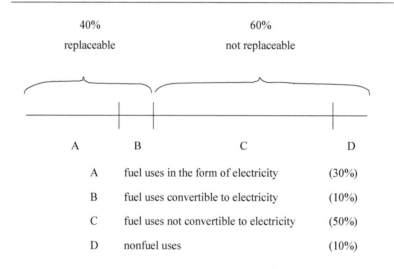

	A	fuel uses in the form of electricity	(30%)
	B	fuel uses convertible to electricity	(10%)
	C	fuel uses not convertible to electricity	(50%)
	D	nonfuel uses	(10%)

Needless to say, these percentages are rough estimates at best. But the proportions they represent still give us cause to believe that most of the fossil resources on which industrialized economies now rely could not be replaced by supplies of clean energy.

The reason for this conclusion is not just that clean-energy supplies are limited, which indeed they are (see section 9.6). The main reason is that most fossil resources consumed by contemporary society are employed in uses to which clean energy cannot be applied.

9.4 Positive Results of the Replacement Strategy

The fact that fossil sources are replaceable by clean energy in less than half of their uses (40 percent in figure 9.1) does not mean that replacement is not a good strategy in those particular uses. What it means is that the strategy is feasible in some cases but not in many others. One way of putting the strategy to the test is to return to the several forms of ecological damage distinguished in chapter 5 and to assess its potential impact on these individual problems.

For this purpose, let us assume that fossil fuel has been replaced when feasible by clean energy. More specifically, we will assume that clean energy has been substituted for coal and natural gas in the production of all electricity used in category A of figure 9.1. We will also assume that all uses in category B have been converted to electricity, which likewise is produced from clean-energy sources. In effect, we are assuming the best possible case, which is that clean energy has replaced fossil fuel to the fullest extent practicable.

Although it is probably unrealistic to assume that fossil fuel can be totally eliminated as a source of electricity, the assumption nonetheless will help us test the strategy in question. It will enable us to determine the most environmentally beneficial results to be achieved by the replacement strategy. If the replacement of fossil sources when feasible is only partial (contrary to the best-case assumption), then the results that follow perforce will be less beneficial.

The first ecologically harmful effect of fossil fuel discussed in chapter 5 is the thermal pollution of lakes and streams supplying water for cooling in the process of generating electricity. Installations for producing electricity from renewable sources (solar radiation, wind, moving water) typically do not require cooling.[12] Replacement of coal and natural gas by such sources, accordingly, would remove this particular cause of thermal pollution.

A more serious and pervasive form of thermal pollution is global warming. As currently understood, this condition is caused by greenhouse gases, of which CO_2 is a prime example. Carbon dioxide is one by-product of burning fossil fuels and thus is produced by power plants, factories, and internal combustion engines. If fossil fuels were replaced in these uses by clean-energy sources, which produce no carbon emissions, this source of greenhouse gas would be eliminated.

Elimination of carbon emissions from power plants and internal combustion engines obviously would have beneficial results. As we shall see in the following section, however, even a total shift away from fuel in these applications would not do away with the problem of global warming. The unfortunate fact of the matter is that there are many sources of greenhouse gases other than exhaust pipes and smokestacks.

Beyond thermal pollution of our waterways and atmosphere, there are a few more problems that would be at least somewhat alleviated if the replacement strategy were systematically pursued. One worth mentioning is the increasing acidification of ocean waters that is destroying coral reefs and their associated ecosystems (discussed in section 5.4). Marginal decreases in atmospheric CO_2 presumably would be followed by marginal decreases in rate of acidification. Another matter to bear in mind is the considerable quantities of sludge produced by scrubbers in the smokestacks of fossil-fueled power plants (see section 8.3). Solid waste such as this would be eliminated if these power plants were replaced by clean-energy installations.

9.5 Global Warming Not Eliminated by Use of Clean Energy

Despite these undoubted benefits, the replacement strategy leaves several of the ecological problems distinguished in chapter 5 relatively untouched. Let us begin with global warming. Although a complete replacement of fossil energy in its uses where electricity could also be used (categories A and B of figure 9.1) would be a significant step in the right direction, global warming almost certainly would continue even if this step were successfully taken.

One reason is that many other fuel-related uses of fossil resources would remain (the 50 percent of category C) that continue to produce greenhouse gases. A second reason is that there are many sources of CO_2 other than combustion of fossil fuels. As far as human responsibility is concerned, another major source is the intentional burning of tropical rain forests. Add to this the fact that CO_2 is only one of several greenhouse gases. Another is CH_4 (methane), large quantities of which are produced in the guts of animals raised for food. Replacing fossil fuel by clean energy in the generation of electricity would do nothing to remove these (and various other) sources in which people are directly implicated.

Shifting attention to nonhuman sources, we note that the respiratory and digestive systems of most large animals (not just beef cattle) are effective sources of greenhouse gases as well. Perhaps the most significant nonhuman source under present circumstances, however, is the thawing of permafrost as a result of global warming already under way. Thawing permafrost under lakes releases CH_4, whereas carbon entrapped by dry thermafrost emerges as CO_2. As more of these newly released gases reach the atmosphere, global warming intensifies, thus hastening the melting of larger expanses of thermafrost, and so on in a continuing spiral.[13] This positive feedback process has the potential to accelerate indefinitely, wiping out any advantage gained by reductions in human consumption of fossil fuel.

Another factor to bear in mind is that once CO_2 enters the atmosphere it can stay there for many decades.[14] Even if humanity gave up fossil fuels overnight, greenhouse gases previously aloft would continue to work mischief for years to come. The process of global warming already has reached the point at which even a complete shift to clean energy would not bring it to a halt.

Apart from these ongoing sources of greenhouse gases, it seems likely that extensive use of clean energy could contribute to global warming in its own right. This is true of solar energy in particular. One reason has to do with a natural tendency to situate solar collectors in relatively barren areas like deserts, where a substantial portion of incoming solar radiation is reradiated directly back into space without being converted into low-grade heat. Solar radiation converted to electricity, on the other hand, generally undergoes conversion into heat before being returned to space. Let us look at this effect more closely.

Of the large amounts of solar radiation reaching the earth, roughly 35 percent is reflected back into space directly (the albedo effect[15]), leaving about 65 percent to be absorbed and reradiated back subsequently. Making up the former figure, 24 percent is reflected by clouds, 7 percent by the atmosphere, and 4 percent by the earth's surface (mostly by snow and light soil like desert sand). Making up the latter 65 percent is 5 percent reradiated at its original wavelengths (mostly higher than infrared) and 60 percent reradiated in the infrared range after being absorbed by the atmosphere or by the earth's surface. This 60 percent is the solar radiation that contributes heat directly (i.e., without human intervention) to global warming.

With regard to heat contributed by human activity, it seems evident that use of electricity produced from solar energy presently contributing to this 60 percent will not increase the amount of low-grade heat to be discharged through the atmosphere. Heat from this 60 percent has to exit through the atmosphere anyway, and converting some of this energy to electricity will not add to the total amount of heat in question. In other words, heat from the use of electricity derived from this 60 percent will not increase global warming.

But this is not the end of the matter. As just mentioned, there is a tendency to situate solar collectors in relatively barren lands like deserts, which reflect most of the solar radiation falling on them directly back into space. And since radiation reflected back directly does not convert to low-grade heat, it does not contribute appreciably to global warming. Generating electricity from collectors situated in areas like these, accordingly, decreases the amount of solar energy that does not contribute to global warming and increases the amount that has to pass through the atmosphere as low-grade heat.

In this somewhat indirect manner, extensive use of solar energy could have a potentially substantial effect on global warming. For a hypothetical example, if one-fourth of the 4 percent currently reflected back into space from the earth's surface were converted into electricity instead, then the amount of total solar radiation ending up as low-grade heat would increase from 60 percent to 61 percent. The upshot would be close to a 2 percent increase in the amount of low-grade heat waiting to exit through the atmosphere. It seems not unreasonable to expect that any increase in the neighborhood of 1 to 2 percent would have a significant impact on global warming.

9.6 Other Problems the Strategy Leaves Unresolved

Global warming is not the only ecological problem left unresolved by the replacement strategy. Another problem left untouched is the accumulation of plastic junk building up in our oceans and terrestrial landfills. Until recently, plastic artifacts were all made of petrochemical polymers, which are materials manufactured from fossil resources. Traditional plastics (celluloid, rayon, cellophane, Styrofoam) are nonbiodegradable,[16] which means that refuse from such products now cluttering up the environment may be expected to stay there indefinitely. Inasmuch as manufacture of plastics falls under the nonfuel use of fossil resources (D of figure 9.1), however, replacement of fossil fuel with clean energy would do nothing to alleviate this problem. No matter how zealously the replacement strategy is pursued, plastic junk in the environment will continue to accumulate.

Another nonfuel (category D) use of petroleum is the manufacture of such ozone-depleting chemicals as CFCs (see section 5.6). Despite the initial success of the Montreal Protocol (discussed in section 8.2), the ozone layer has continued to deteriorate under the influence of replacement petrochemicals such as HCFCs (see section 8.3). Moreover, since these chemicals disperse only slowly once they reach the upper atmosphere, the ozone layer is not expected to regain its integrity before the end of the century. In the long run, this problem may prove even more severe than global warming. And the replacement strategy will not help resolve it.

Equally regrettable is that this strategy shows little promise of re-
lieving the very extensive ecological damage inflicted by industrial agriculture (described in section 5.8). For purposes of brief review this
damage can be classified under four or five headings. Of these, the most
widespread may be the damage caused by commercial fertilizers.

Ecological damage caused by use of nitrogen-based fertilizers includes eutrophication of lakes and other waterways, dead zones at the
mouths of rivers caused by fertilizer runoff, release of nitrous oxide
(a greenhouse gas) into the atmosphere by denitrification, and their contribution to acid rain by release of ammonia. Production of commercial
fertilizer is a nonfuel (category D) use of fossil resources involved. A
consequence is that such damage cannot be relieved by substituting
clean energy for fossil fuel.

Another ecologically pernicious practice of industrialized farming
is the extensive use of pesticides. Although designed to kill competing
plants ("weeds") and unwelcome insects, these poisons endanger ecologically beneficial life-forms as well. Pesticides kill honeybees instrumental in pollinating crops, decimate populations of birds that control
noxious insects like mosquitoes, and destroy decomposers (earthworms,
beetles, etc.) essential for healthy soil. Manufacture of pesticides, once
again, is a nonfuel (category D) use of fossil resources. And once again,
the damage they cause would not be affected by a shift to clean energy.

A third class of damage caused by factory farming pertains to loss
of species diversity. Although agriculture by nature involves giving
special care to individual plant species, industrial farming tends to focus
on crops that are particularly profitable (corn, wheat, soybeans, etc). As
noted in chapter 5, whereas humans once used roughly 3,000 species of
plants for food, only about 150 of these are grown commercially today
and most of the rest have gone extinct. By declining to cultivate unprof-
itable species, the monoculture practiced by factory farming has radically reduced the variety of plants available for human consumption.

Species loss among sustenance crops has been exacerbated by a
growing reliance on genetically modified (GM) seeds. Within the first
four years of the present century, for example, the proportion of GM
corn grown in the United States doubled to 45 percent (see section 5.8).
It seems not inconceivable that GM varieties will completely dominate

the world's food markets in a matter of decades, while the natural varieties they replace will simply disappear. This would constitute a dramatic loss of species diversity.

Whether or not loss from genetic modification is to be treated as a separate class of ecological damage, species loss itself is beyond reach of the replacement strategy. To whatever extent petrochemicals are a contributing factor, the effect stems from a nonfuel (category D) use of the fossil resources involved. The substitution of clean-energy sources for fossil fuel would do nothing to stem the loss of species diversity.

The last kind of damage to be considered is caused by use of heavy farm equipment that cannot be powered by electricity. Included are tractors, combines, and bulldozers, along with trucks used for harvesting and for delivery of products. Use of fossil fuel to power such machines belongs to category C, the category of fuel use not convertible to renewable energy. The associated problems will not be helped by the replacement strategy.

On balance, the observations of these last two sections are not encouraging. While there are types of ecological damage that would be somewhat alleviated by a shift to clean energy, even in these cases the damage would not be wholly eliminated. And since most forms of damage we have surveyed are due to petrochemicals, meaning fossil resources put to nonfuel uses, most of the damage they cause is beyond reach of the strategy in question. Sobering as it may be, the only reasonable conclusion is that replacing fossil fuel with clean energy would contribute far less than is needed to resolve our environmental crisis.

9.7 Ecological Problems Initiated by Use of Clean Energy

Not only will the replacement strategy leave most of our current ecological problems unresolved, but other problems arise with use of clean energy itself. Although details are hard to predict, the more serious of these will probably center on the quantities of space required for its production and distribution. Currently available technologies for producing clean energy, as will be recalled, include solar collectors, wind turbines, and generators run by moving water. It is instructive to compare siting requirements of these technologies with those of the chlorophyll-bearing organisms responsible for almost all our renewable energy before the industrial era.

Before the industrial era, the role of converting solar energy into forms usable by other creatures fell exclusively to plants. Ecosystems supporting consumer organisms developed only where there was adequate plant life to support them. As a consequence, most ecosystems had adequate supplies of energy in their immediate vicinities.

This held true of ecosystems supporting human existence in particular. Early human society had neither need nor opportunity to import energy from distant places. A hallmark of modern industrial society, on the other hand, is that it requires technology for moving energy supplies over long distances. Railways carry coal, ships carry oil, pipelines carry natural gas, and so forth. While local industry can thrive with local sources of energy, the fact that energy sources are found only in localized areas means that society at large must move energy from one place to another.

This is the case with clean energy no less than with fossil fuels. Although solar energy is available almost anywhere on earth, converting it to electricity is feasible only under fairly restrictive conditions. Collection by photovoltaic cells, for instance, is inefficient in regions with low insolation as well as in regions with frequent cloud cover. This means that use of solar energy for commercial purposes would be impractical, among other places, in much of Canada and northern Eurasia.

Siting of wind turbines, in turn, is commercially feasible only in areas where average wind velocity exceeds a certain minimum level.[17] For this reason, wind turbines most often are located offshore (where wind flows freely), on mountain ridges and other high elevations, or on open plains. These restrictions have limited commercial development of wind power to a relatively small portion of the globe. To date, wind farms account for less than 1 percent of worldwide energy production, with concentrations in Germany, Spain, and the United States.[18]

As far as hydropower is concerned, siting is limited mostly to offshore tidal fields and to areas featuring dams and natural waterfalls. Use of inland streams to produce electricity is generally impractical. Although hydroelectric installations currently supply almost one-fifth of the world's electricity,[19] further expansion is unlikely save in a few especially suitable locations.

A consequence of these several limitations is that if clean energy were to take over from fossil fuel in production of the world's electricity, a lion's share would have to be conveyed considerable distances from origin to point of consumption. This would necessitate vast networks of expensive power lines traversing every inhabited region of the earth.[20] The ecological consequences of these massive delivery systems would be far more intrusive than our current criss-cross of high-tension power lines (think of the region around Niagara Falls), which are responsible for widespread ecological damage as matters stand.[21]

Another ecological drawback of clean-energy technologies is the sheer space required for their installation. Take wind-turbine technology as an example. For optimal aerodynamic efficiency, individual turbines must be placed considerable distances from each other. A terrestrial wind farm of twenty turbines, for example, might occupy an area of one square kilometer.[22] Although the spaces between towers could be used for grazing, and sometimes farming, the rotating blades would rule out ecosystems supporting avian wildlife.

A point in favor of solar technology is that photovoltaic cells (currently the most practical way of converting sunlight to electricity) can be installed adjacent to each other. Whereas the spacing among turbines allows wind farms to be used for other purposes, the close juxtaposition of collectors on solar farms typically prevents much sunlight from reaching the ground underneath. A consequence is that areas set aside for solar power generation tend to be ecologically impoverished and generally unavailable for other human uses. Al Gore has estimated that, given the conversion efficiency of current photovoltaic collectors, about 10,000 square miles of land would be required to supply the power needs of the United States with solar energy.[23] This means that if the country were to go completely solar (ignoring the distinctions in figure 9.1), a total land area roughly double the size of Connecticut would be totally given over to the production of electricity.

Yet another shortcoming of the clean-energy sources we have been considering is that they rely on humanly designed and fabricated machinery. Unlike solar conversion processes in the plant world, which if left undisturbed can continue to supply energy in useful forms indefinitely, machinery wears out and needs replacing. Manufacturers' estimates suggest that a wind turbine, if properly lubricated and maintained, should last about twenty years. Comparable estimates for photovoltaic collectors range from twenty to fifty years.[24] Apart from conventional generators in existing hydroelectric plants, with life expectancies of a hundred years or so,[25] technologies for converting motion of tidal waters into electricity are too new for informed estimates of longevity. But a lifetime of more than a few decades seems unlikely.

Assuming a generous average of thirty to forty years of operating life for a typical clean-energy installation, the upshot is that the machinery providing a country's supply of clean energy will require continuous updating and replacement. Old equipment will have to be dismantled and added to the debris in our crowded landfills. And replacement equipment constantly will be under construction, with the ecological disadvantages attendant upon the various manufacturing processes involved. In one sense of the term, technologies of this sort provide a source of renewable energy. Although the energy itself is renewable, however, the technology for harnessing it is not.

9.8 Summary Assessment of the Replacement Strategy

Like the piecemeal approach of chapter 8, the replacement strategy is already being implemented in many industrial economies. One reason for its ready acceptance, again like the previous strategy, is that it involves new products (e.g., solar installations, wind turbines) on which businesses can make a profit. A more compelling reason for acceptance among mainstream economists is that both strategies appear to be compatible with continuing growth. The general idea seems to be that if specific problems are handled piecemeal when they appear, and if clean energy is substituted for fossil fuel, then economic growth can continue indefinitely.

In the case of the replacement strategy, at least, the appearance of compatibility may prove illusory. Although fossil resources are replaceable in less than half of their uses (categories A and B of figure 9.1), the strategy calls for replacement in such uses rather than cutting back on the amount of energy consumed. A consequence is that demand for clean energy will tend to increase with an expanding economy.

Despite clean energy's being generally renewable, however, there are practical limits on its availability. There are limits on the siting of clean-energy installations, limits on the size of delivery systems that society will tolerate, and limits on the amount of clean-energy technology in which society will find it profitable to invest. As the economy continues to expand, these limits might make use of clean energy increasingly unattractive. In effect, continued expansion might actually derail the replacement strategy.

We have seen that clean energy can be substituted for fossil resources in only about 40 percent of their current uses. Although substitution when possible may be expected to have some moderately beneficial results, we have seen that it would have little impact on many of our environmental problems. We have seen, furthermore, that extensive use of clean-energy technology would create environmental problems in its own right. These results are enough to show that the replacement of fossil fuel by clean energy will not do much to alleviate our environmental crisis.

It should be emphasized that these results do *not* show that current efforts to phase out fossil fuel in favor of clean energy are themselves counterproductive. All in all, such efforts are a good thing and should be continued to the fullest extent feasible. As far as our environmental crisis is concerned, nonetheless, we should realize by now that it will not be resolved by strategies pursued in conjunction with continuing growth. Economic growth itself is the source of the crisis. The next two chapters are given over to an examination of the remaining strategy. Unpalatable as it may be in certain sectors of society, this third strategy calls for a curtailment of economic growth.

10

History and Theory of Economic Growth

10.1 The Remaining Strategy

In chapter 7, the ecological damage caused by present-day economic activity was likened to health problems caused by smoking cigarettes. Three possible strategies were identified for counteracting this damage. One is to devise technologies for reducing the damage in individual cases (e.g., smokestack scrubbers), analogous to devising distinct medical treatments for emphysema, for lung cancer, and so on. A second strategy is to replace fossil fuel with clean energy as a way of breaking the link between economic production and ecological damage. This is analogous to replacing nicotine in cigarettes with less harmful ingredients.

These strategies were examined and found wanting in chapters 8 and 9 respectively. While both are capable of beneficial results, and should be implemented whenever possible, by themselves these strategies show little promise of restoring the biosphere to a healthy condition. Dysfunctional conditions such as global warming and ozone depletion result from disruption of natural feedback systems that keep crucial environmental variables in balance. Natural control systems of this sort cannot be repaired by specific technologies, to the discredit of the first strategy in question. Dysfunctionality in such circumstances, moreover, often results from uses in which fossil resources cannot be replaced by clean energy. In such cases, the second strategy is not even applicable.

We come now to the remaining strategy identified in chapter 7. This final strategy boils down to cutting back on the sheer volume of human energy consumption that is primarily responsible for the massive damage humanity has inflicted on the biosphere. Taking the analogy with nicotine consumption one step further, this is akin to curtailing the use of addictive tobacco products. In one way or another, we shall be occupied with this third strategy and its ramifications throughout the remainder of the study.

Considered apart from context, this strategy seems obvious and relatively straightforward. If consuming overly much of a given item is causing problems, the obvious remedy is to curtail consumption. If the item in question happens to be energy, the countermeasure that comes immediately to mind is simply to cut back on use of energy. A straightforward solution to any problem of excess is to adopt alternatives that are not excessive.

As the comparison with tobacco use suggests, however, solutions that seem straightforward in the abstract might turn out to be considerably more complex in practice. Like nicotine intake for the individual smoker, profligate use of energy can be addictive for a consumer society. Advising a society of consumers to cut back on its use of energy is comparable to advising an addicted smoker to cut back on nicotine intake. In both cases, the advice would probably be ignored. Even when the behavior in question is recognized as harmful, cutting back on consumption of an addictive substance is all too often a difficult matter.

The point of the analogy here is not that excessive consumption of energy is addictive in its own right. We are not addicted simply to using lots of energy. What makes our current pattern of energy use so difficult to reverse is the close correlation between energy use and economic production. As shown in chapter 7, and as some economists have known all along, high levels of production are directly dependent upon correspondingly high levels of energy consumption. The addiction in question is a result of this dependency.[1]

It is a monumental misfortune, both for the biosphere at large and for the societies it supports, that the high levels of energy use involved in economic production are also a major source of ecological destruction. It is additionally unfortunate that habits of excessive energy use have become addictive for the societies concerned. The starkly unfavorable upshot is that human society must find a way of breaking its energy addiction in order to have any chance of staving off the fatal consequences of a collapsing biosphere.

As stated previously, the remaining strategy for counteracting the damage to our ecological support system caused by expanding economic production is to cut back significantly on the amounts of energy involved. The remaining strategy thus stands in direct opposition to conventional economic wisdom that social well-being requires continual economic growth. In an effort to understand the attitude of mainstream economists toward growth and its importance, let us briefly consider the history of the concept of economic growth.

10.2 Early History of the Concept of Economic Growth

Our term "economics" comes from the Greek *oikonomia,* which originally meant the art of household management. One of the first treatises on economics was the *Oeconomicus* by Xenophon (fourth century BCE), a discourse concerned with specific skills such as furnishing houses, training servants, and procuring food. Economics in this original sense was not far removed from the subject designated (redundantly) "home economics" in high school curricula of previous decades.

Economics in this original sense seems at first to be quite different from the subject taught by economics professors in present-day colleges and universities. An obvious difference is that the "household" whose management is studied by economists today has been extended from the individual family to a more diffuse and extensive group of people. The sense of "home" (Greek *oikos*) pertaining to contemporary economics is close to that figuring in our terms "homeland" and "home state."

Another difference is that when someone completes the required course of studies and becomes an economist today, that person is not thereby qualified to manage an actual economy. Xenophon's treatise in effect was a textbook of household management. By contrast, the skill of a trained economist today is not primarily one of practice but rather one of understanding the principles and dynamics by which economies operate.

There is one key respect, however, in which mainstream economics today retains continuity with its ancient Greek prototype. In both contexts, the mark of a successful economic regime is its ability to maintain a consistent pattern of growth. The goal of Xenophon's economics went beyond merely running the household efficiently on a daily basis. Another aspect of running a household successfully was being able to increase its wealth over an extended period.[2] A well-run household was one that kept on growing, while stagnation was a portent of impending failure.

The notion of what constitutes economic growth underwent considerable change between Xenophon's time and recent centuries. For Xenophon, growth was primarily a matter of augmenting resources (living accommodations, provender, servants) by which the needs of a household could be met. There was no expectation that the household itself should produce these resources, only that resources made available externally (by farming, commerce, conquest, etc.) be brought into its service.

With the shift in focus from the domestic family to larger political organizations, however, functions of production and consumption began to merge within the context of a comprehensive system of interacting economic interests. Needed commodities might still be acquired from sources outside the system, either by exchange or by military action. But the items of value (precious metals, currency, crops) for which outside goods were exchanged, along with the resources by which standing armies were maintained, typically were produced within the system itself. Increasingly complex interactive networks of this more comprehensive sort presumably evolved into prototypes of our modern economic systems.

Once an economic system of this rudimentary sort has taken shape, it can be characterized with respect to the parties (rulers, merchants, workers, etc.) whose interests play significant roles within it. It can also be characterized with respect to the kinds of benefits (monetary wealth, land, tithes in kind) it brings to bear in serving those interests. Generally speaking, for an economic system to undergo growth is for it to gain added capacity to make such benefits available.

Although economics did not emerge as a discipline with a distinct subject matter until the latter half of the eighteenth century, historians have ventured to characterize earlier periods of economic activity in terms of more recent economic theories. According to one source,[3] the ancient Roman economy was an agrarian system based on slave labor, geared to feeding its vast numbers of soldiers and citizens and, ultimately, to keeping its ruling class in power. Economic growth was a matter of acquiring additional territories for growing grain and additional forces of slaves to work them. Insofar as new territory was acquired by conquest, economic expansion led to increasing numbers of soldiers and administrators who needed to be fed. Once the empire ran out of additional territory to conquer (around the second century CE), it began a period of decline and of inevitable collapse.

A similar story can be told about the feudal period in Europe.[4] For centuries leading up to 1500, feudal economies were based on land, which typically was devoted to producing food through the labor of serfs. Economic production was aimed at serving the needs of the lord and his manor, leaving the serfs a bare minimum for daily subsistence. To the extent that economic growth was possible, it usually amounted to increasing productivity of the fields through minimal improvements in farming methods (e.g., replacement of oxen by horses) or to demanding larger tithes for use by the lords. Feudalism broke down with the rise of monetary economies, in part because paid workers proved more productive than bondaged labor.

Increasing emphasis on money as a means of exchange led to what has come to be known as the "mercantile system," prevalent in Europe from the sixteenth to the eighteenth century. During this period, economic growth constituted increasing amounts of negotiable wealth at the disposal of a nation-state, usually in the form of currencies and the precious metals on which they were based.[5] Growth was achieved by several methods, one of which was the exploitation or conquest of gold-rich territories in previously "unexplored" parts of the world (think of Cortez and his incursions into Mexico).

As its name suggests, the mercantile system was based on trade. Another means of gaining wealth was for a nation to maintain a positive balance of trade with other nations. Governments would attempt to foster growth by discouraging imports (money leaving the country) and encouraging exports (goods leaving the country). Imports were discouraged by imposing tariffs. Exports were encouraged by granting monopolies intended to make favored merchants more competitive in foreign markets. This enabled monopoly holders like the Dutch East Indian Company and the Hudson's Bay Company to manipulate prices in hopes of eliminating competition from other countries. Once a company's monopoly became international, it could raise prices and reap profit from its original investment.

In this competition among nations for foreign markets, the stakes were high enough to lead frequently to armed conflict. Several wars between England and the Netherlands in the seventeenth century were precipitated by mercantilist ambitions. So too, in large part, was the American Revolution. As dedicated markets for British goods, and as sources of raw materials, the American colonies were compelled to trade only with England and on English terms. This arrangement the increasingly prosperous colonists found increasingly intolerable. A result of the revolution was that the United States soon became a formidable trading power in its own right, as witnessed by the infamous "triangular trade" of the late eighteenth century (slaves from West Africa to the Caribbean, sugar from there to New England, and rum from New England to West Africa). The wealth generated by this commerce helped lay the basis for nineteenth-century American capitalism.[6]

Given its emphasis on production of goods for export and on price manipulation, the mercantilist system did not favor the working population. Local consumption was not a factor in economic prosperity, with the consequence that laborers and farmers generally lived on a level of bare subsistence. As in the previous period of feudalism, the people primarily responsible for economic growth received few of its benefits. The main beneficiaries of mercantilism were the trading companies and the governments that supported their operations.

10.3 Adam Smith's Theory of Economic Growth

Adam Smith's *The Wealth of Nations* is usually considered to be the founding work of modern economics.[7] Considerable portions of this work are spent rebutting the policies of mercantilism, with its emphasis on the importance of accumulating bullion. For nations operating under mercantilist principles, gold and silver had inherent value and served as media of foreign trade. In Smith's view,[8] by contrast, bullion had the status of other commodities with value varying according to supply and demand.

Another line of criticism he directed against the mercantilist assumption that commerce is a zero-sum game (i.e., that every commercial transaction has a winner and a loser). As suggested previously, this assumption was a factor in the trade wars that troubled Europe during the sixteenth and seventeenth centuries. Smith's position was that all parties benefit from well-informed transactions when freely undertaken.[9] If Portugal is better at making wine and England at producing cloth, then each country should be able to benefit by purchasing the other's goods.

A more radical divergence from mercantilism was Smith's view that economic prosperity should benefit society as a whole. According to mercantilist ideology, the working classes had no rightful expectation of leisure time, education, or extra money to buy more than bare necessities. Such amenities were the prerogative of financiers and merchants, who brought wealth into the nation's coffers. The role of working people was to produce goods for the consumption of others, rather than themselves to enjoy the goods produced.

Smith's rebuttal amounted to a fundamental rethinking of the mechanisms by which economies operate. By nature, he believed, human beings tend to act for personal gain. Although this may not be a good thing generally, in the context of a free and well-ordered economy self-interested action can work for the general well-being. If producers and consumers are able to choose freely what they sell and buy, the marketplace will distribute goods at prices that are beneficial to the entire community.

Smith describes this tendency of the unfettered market in terms of the now famous metaphor of the "invisible hand."[10] As he puts it in *The Wealth of Nations,* when a man conducts his business with the motive of personal gain he is "led by an invisible hand to promote an end which was no part of his intention."[11] That end, we are given to understand, is an optimal distribution of economic goods across all levels of society. With respect to the relationship between landlord and laborer in particular, he remarks that, although the owner seeks only to gratify his own desires, he is "led by an invisible hand to make nearly the same distribution"[12] of necessities among the poor as would have resulted if all parties had been allotted equal portions of land. Rightly or wrongly, this rationale has been used by economists subsequently to justify modern free-market capitalism.

For the invisible hand to operate in this manner, it is necessary that the producers of economic goods function in the role of consumers as well. In the mercantile system, Smith observes, "the interest of the consumer is almost constantly sacrificed to that of the producer." Smith, on the other hand, considers it "self-evident" that "consumption is the sole end and purpose of all production."[13] His most consequential departure from mercantilism may have been his account of growth, in which expansion of consumption is necessary for an expanding economy.

Smith's account of economic growth begins in the first chapter of *The Wealth of Nations* with a discussion of the division of labor. When a complex productive task can be broken down into simple components, and each member of a workforce assigned a specific subtask, production can be achieved more efficiently and cheaply than when each member is directly responsible for the finished product. Smith's own example is the production of pins, which at that point involved a considerable number of distinct operations. By his estimate, a single worker would have had difficulty making twenty pins a day, which would have amounted to fewer than two hundred pins for a ten-person workforce. When each person specializes in a particular operation, however, the same work force could produce close to forty-eight thousand pins a day.[14]

Division of labor thus enables the production of larger numbers of goods, which then can be sold at cheaper prices. Yet unless there is ample demand for his goods on the market, no manufacturer will be motivated to expand his productive capacity. To make this point, Smith changes his example from pins to nails. Consider a nail maker who produces one thousand nails a day and works three hundred days a year. If the nail maker lives in one of "the remote and inland parts of the Highlands of Scotland," the market for nails might be so small that only one day of a year's labor would turn out to be profitable. The resulting constraint is summed up in the title of chapter 3 (book 1): "The Division of Labor Is Limited by the Extent of the Market."[15]

Large markets are required to absorb the large volumes of goods made possible by division of labor. This means that the class of consumers has to be expanded beyond the elite few who benefited from the mercantile system. Expansion of the consumer class is enabled in part by lower prices resulting from the cheaper production of goods by specialized labor. Another factor is the ability of factory owners to share profits more generously with a productive workforce. A combination of these two factors brought consumer goods within reach of the common wage earner. The upshot is that workers responsible for producing goods joined the ranks of consumers by whom those goods are purchased.[16]

Division of labor increases labor productivity. Increased productivity leads to higher wages. Higher labor income swells the consumer market. With an expanded market comes a demand for more consumer products. This demand then is met with further increases in productivity, resulting in part from more efficient division of labor, and so forth. Working together, this set of dynamics constitutes a positive feedback loop (see section 3.5). Set off by a tendency toward worker specialization, increased productivity leads to yet higher productivity, improved worker income leads to yet higher income, and larger markets lead to yet larger markets.

This, in a nutshell, is Adam Smith's theory of economic expansion. Although the upward spiral can be joined at various points, his view seems to be that the cycle is driven primarily by increases in consumption—thus his dictum that consumption is the sole end and purpose of production. In line with our characterization of earlier theories in section 10.2, we may say that, for Adam Smith, economic growth amounts to an increase in consumption that benefits all sectors of the economic community.

It should be noted in passing that there is nothing in Smith's account suggesting that economic growth can continue indefinitely. Growth could be curtailed by government policies attempting to enhance profits or to hold down wages, such as the creation of monopolies or an excessive taxation of income.[17] Given his view on the importance of land as a source of food and revenue, moreover, along with the obvious fact that land is limited, Smith may well have realized that economic growth is constrained ultimately by environmental factors.[18]

10.4 Classical Growth Theory Following Adam Smith

Beginning with Adam Smith, the development of classical economic theory continues with the work of Thomas Malthus, David Ricardo, and John Stuart Mill. All three shared Smith's belief in private property, competition, and markets free from government supervision. They also shared his confidence in the public benefit to be derived from the pursuit of private gain, which Smith epitomized in his metaphor of the invisible hand.

Malthus is best known for his book *An Essay on the Principle of Population,* published in 1798, in which he points out the dangers of unchecked population growth. In briefest form, the problem he anticipated is that, while food supply would increase arithmetically at best, population would grow geometrically. The result would be mass starvation, which, along with war, crime, and epidemic, would reduce human population to sustainable levels. Malthus's concern with the effects of overpopulation contributed to the reputation of economics as "the dismal science."

Adam Smith had argued that expanding consumer markets led to increased production and higher wages, which in turn would result in yet further consumption. For this positive feedback effect to continue, Malthus pointed out, the workforce would have to remain more or less fully employed. But population increase tends to result in underemployment. Another problem with Smith's doctrine, according to Malthus, had to do with what has come to be known as the "law of diminishing returns."

In point of fact, Malthus and Ricardo arrived at this principle independently.[19] As it applies to production systems with variable inputs and outputs, the general idea is that increasing inputs leads to progressively smaller additional (so-called marginal) outputs. A simple example has to do with increasing the amount of seed applied to a fixed piece of land. If one pound of seed per acre yields one hundred pounds of produce, it would be unreasonable to expect that each time the amount of seed is doubled the amount of produce would double as well. At some stage the acreage would become overplanted, and the produce returned from additional plantings would begin to diminish.

With regard to Smith's agrarian-based production system, Malthus observed that increasing population would increase the supply of labor. But inasmuch as productivity depends on land (a fixed quantity) as well as labor, continual increase in supply of labor will not lead to continuously increasing levels of productivity. In this case, diminishing returns show up as diminishing supplies of food in proportion to an increasingly hungry population.

In various publications from 1815 to 1821, Ricardo extended the principle of diminishing returns to other factors involved in land productivity. One factor is the amount of capital applied per laborer. Up to a point, return on capital will increase with amount of capital expended. But bringing more capital to bear, other things being equal, requires bringing more land under cultivation. Given limitations in fertile land available, this leads to cultivation of progressively less productive land, and accordingly to decreasing returns on capital.

Another factor is technological progress in farm equipment. On one hand, improvements in technology can make a given piece of land more productive and thus allow for more growth. On the other hand, Ricardo realized, introduction of labor-saving machinery tends to reduce employment. While this does not automatically decrease returns on capital, it leads to a decrease in consumer spending, which Adam Smith thought must increase for economic growth to occur.

As a result of Ricardo's work, the model economy studied by theoretical economists provided roles for three classes of participants. One is that of workers, who spend most of their wages on necessities and whose income is under constant threat of erosion by pressure from the other classes. Another is landowners, who, inasmuch as they follow the principle of self-interest, spend most of their revenue on luxuries. Third are the capitalists, who tend to reinvest their profits in hopes of increasingly higher rates of return. This opened the door to "class-warfare" theoreticians like Karl Marx, who, in his *Communist Manifesto* of 1848, focused on the conflict between the capitalists and the working class.

The way this plays out historically, according to Marx, is for capitalists to pay their workers subsistence wages while retaining for themselves as much profit as the market allows. In sociological terms, this amounts to powerful (capital) interests exercising force and fraud in taking advantage of the weak (workers). In economic terms, it amounts to undercutting the purchasing power of the consumer on whom Adam Smith's dynamics of growth was based. In Marx's view, the fatal flaw of capitalist economies is their internal contradiction between improving technological efficiency, which drives up profits, and declining purchasing power of the so-called proletariat, who consume the products of an expanding economy.

The summary work of classical economic theory was John Stuart Mill's *Principles of Political Economy*. Although Mill is best know outside economics for his more broadly philosophical works, *Utilitarianism* and *On Liberty*, his *Principles* remained the most widely used textbook in economics for some forty years after its publication in 1848. In this treatise, Mill synthesized and expanded on the contributions of Smith and Ricardo to the theory of free markets, and he added original work of his own on taxation, foreign trade, and the distribution of income.

In the preface to the first edition of his *Principles*, Mill expressed hope that its contents would impress themselves "strongly on the minds of men of the world and of legislatures." This is comparable in intent to a desire that the ethical prescriptions laid out in his philosophical works actually be adopted by human society generally. In retrospect, it is not clear that a full-scale implementation of the dictates of either *Utilitarianism* or the *Principles* would have had a more salutary effect overall than that of Marx's economic philosophy. Further consideration of Mill's influence is set aside for chapter 14.

10.5 Neoclassical Growth Models

The transition from classical to neoclassical economics began late in the nineteenth century. A brief story of the transition might focus on a variety of themes. One theme is the emergence of a highly abstract conception of the individual consumer. According to this conception, individuals maximize utilities in much the manner that firms maximize profits. As with firms, individuals have preferences among possible outcomes of their economic activity that can be optimized on a rational basis. The sense is that economic behavior is governed by innate mechanisms operating according to the principles of Bentham's hedonic calculus.[20] As Thorstein Veblen put it, the "hedonistic conception of man" in question "is that of a lightning calculator of pleasures and pains."[21]

Another transitional theme was an attempt to make economics "scientific" in the manner of physical science. In this context, a discipline was deemed scientific to the extent that its subject matter could be formalized in mathematical terms, enabling it to proceed on an axiomatic basis. An integral part of this approach was the previously mentioned conception of an individual consumer's preference being guided by algorithmic calculations of pleasures and pains. Other axioms brought into play as this approach developed have to do with the interplay between supply and demand and with various factors influencing return on capital.

The cause of mathematical economics was advanced by the work of John Maynard Keynes, who began his career as a mathematician. *A Treatise on Probability* (1921) has been touted as no less original than his towering economic work, *The General Theory of Employment, Interest, and Money* (1936).[22] A key theme of *The General Theory* is that mass unemployment cannot be explained by high wages or high prices alone, although both tend to weaken demand and hence decrease sales and jobs. In Keynes's view, unemployment is a function primarily of aggregate demand, which is demand on the part of consumers, investors, and governmental bodies alike. The general idea is that sales and jobs fluctuate with changes in aggregate demand: high demand leads to general prosperity, low demand to depression and unemployment.

Keynes thus followed classical economics in associating economic vitality with consumer activity. However, he departed from the laissez-faire doctrines of Smith and his followers in advocating governmental intervention to sustain market equilibrium. Because of its contribution to aggregate demand, government can and should act as a counterpoise to vacillating business cycles. When the economy is depressed, government should encourage private investment and expand its own expenditures, even if this leads to budget deficits. When the economy is flourishing, on the other hand, government should reign in its expenditures and in the process make up its budget losses. The influence of Keynesian economics is evident in the New Deal politics of the Roosevelt era, in part through the involvement of Milton Friedman.[23]

From a Keynesian perspective, growth rates of an economy vary (as do sales and jobs) with aggregate demand, to which firms react by producing more or less goods for the consumer market. Pursuant to Keynes's emphasis on growth as a matter of increased production, economics began to focus on the phenomena of economic growth specifically, with attention both to its causes and to how it is measured. Since early in the twentieth century, the conventional measure of an economy's growth has been its GDP (Gross Domestic Product). Due to the prevailing assumption that GDP per capita is directly correlated with standard of living, it is generally taken for granted that growth in economic production is a desirable thing. Research into the causes of growth during recent decades has been motivated by the aim of maintaining long-term growth in production and of moderating the effects of short-term recessions.

The basic model of economic growth during this period was articulated in a paper by Robert Solow,[24] for which he won the 1987 Nobel Prize in economics. In keeping with its classical antecedents, Solow's model deals with the interaction between capital and productive output. It assumes a fixed growth rate of labor and a fixed proportion of depreciation in capital stock over time. This depreciation is compensated in varying degree by savings that change in constant proportion with overall income. The model assumes that rate of return on capital decreases as capital investment increases (the principle of diminishing return on capital). It also assumes a constant progression of technological innovation, which means that technology is external ("exogenous") to the model.

When saving equals depreciation, the capital/labor ratio is constant. The dynamics of the model hinges around this point of stability. When saving is greater than depreciation, and when capital per worker is relatively low, capital investment generates a relatively large increase in future income and yields a relatively high rate of return. A consequence is that capital continues to rise. Because of diminishing returns on capital, however, additional increments of capital generate decreasing amounts of additional income and thus a falling rate of return on investment. By the time depreciation catches up with saving (i.e., the capital/labor ratio returns to its steady state), the rate of return on investment will have declined to a point where there is no incentive to accumulate more capital.[25]

In Solow's model, stability is a state of zero growth in income per capita. Because of diminishing returns on capital, there is no additional investment, and economic growth comes to a halt. This means, in effect, that no economy can grow indefinitely merely by accumulating capital.

One factor that can induce additional growth once an economy has reached equilibrium is technological progress. As the level of technological sophistication increases, a given quantity of input can yield greater quantities, or improved qualities, of output. In effect, technological innovation raises the rate of return on capital, thus counteracting the diminished rate of return that otherwise would lead to economic stagnation.

As already noted, Solow's model treats technological innovation as exogenous. This is one of the model's obvious limitations in today's economic climate. Economists working in the 1980s undertook to rectify this shortcoming by developing growth models that included mathematical explanations of technological advancement. In addition to making technology thus "endogenous," this more recent approach to growth theory also downplays the principle of diminishing returns on capital. The sense is that increasing productive efficiency stemming from technological innovation can work instead to enhance the marginal product of capital investment. This supposedly enables continuing growth as capital increases.

10.6 Growth Not Vindicated by Economic Growth Theory

Let us review the circumstances that called for this brief survey of the history of economic growth. After examining current strategies for curbing the severe ecological damage resulting from two centuries of excessive energy use by industrial society (chapters 8 and 9), we concluded that the only effective remedy would be to cut back radically on the energy consumption involved. This latter strategy runs directly counter to the widely held belief that continued growth is essential to economic health. If economic well-being indeed is inseparable from perpetual growth, then humanity faces a dilemma it might be unable to resolve. Either it retreats into a state of economic calamity, or it faces collapse of the biosphere that supports its existence. The foregoing account of economic growth over recent millennia was undertaken in an effort to determine why mainstream economists assume that unending growth is essential to economic well-being.

One thing our survey indicates is that large-scale economies in fact have tended to expand up to the point of ultimate collapse. This is true especially of the economies of the Roman Empire, of feudal Europe, and of the mercantilist period of expansion into colonial markets. In each of these cases, economic growth was bound up with warfare and some form of slave labor. This pattern continued through the twentieth century with its global wars, albeit with a gradual shift from slave labor to poorly paid workers. Generally speaking, it seems that the world's largest economies tend to get bigger and bigger until they either run out of resources or are defeated militarily.

By and large, the economies of developing regions tend to follow the same path of expansion. In recent decades, to be sure, some emerging economies have been expanding more rapidly than their previously developed counterparts. Notable instances are the ballooning economies of China and India.

It should be noted, however, that our brief survey says nothing about innumerable small economies that may have functioned adequately for extended periods without significant expansion. It says nothing about any number of relatively stable economies in Europe, Asia, and Africa that must have coexisted with their neighbors (not always peaceably) for centuries before being swallowed up by the Roman Empire. It says nothing about the presumably numerous agrarian societies that existed on other continents while wars raged in feudal Europe. And it says nothing about the many traditional economies that may have flourished in previously isolated territories before being subjugated in the expansion of European colonialism.

In the unlikely event that a full history were written of the world's less ambitious (and hence less noteworthy) societies, we might expect it to reveal a large portion of cases in which economic stability had been maintained for considerable periods without constant expansion. History might teach us that economic growth is necessary to support political and military expansion or to maintain borders against growing threats from outside enemies. But there is scant historical evidence to back up the common assumption among mainstream economists that continued economic health invariably depends upon continuing growth.

As far as I am aware, no economic historian would claim otherwise. When mainstream economists discuss the topic of growth, it usually is in the context of axiomatic growth theory of the sort reviewed previously (in section 10.5). What mainstream growth theory amounts to, by and large, is an abstract consideration of the likely effects of various factors that contribute to growth.

Thus Solow's model, for example, implies that increases in capital investment will yield growth only up to a point of diminishing returns, after which technological innovation is necessary for growth to continue. By making innovation endogenous, more recent growth theorists have explored ways in which technology can increase the effect of additional investment on increasing production and so forth. Mainstream growth theory is concerned not with vindicating growth but rather with laying out possible ways in which growth can be maintained.

Contrary to our initial hopes, accordingly, we find that technical growth theory offers little help toward understanding why continual growth should be thought essential to a thriving economy. If growth theory shows that stratagems A, B, and C can lead to increased economic output, this result provides incentive to implement those approaches for policymakers already committed to economic growth. But as far as the theory itself is concerned, it might also provide incentive to avoid those approaches in the view of other people convinced that continued growth at this point in human history may not be a good thing.

As matters stand, it appears that neither economic history nor standard growth theory offers persuasive reasons why human society should continue along the path of continued growth that has caused such extensive damage to its supporting ecosystems. Given the deep involvement of economics in other facets of human affairs, however, there are arguments to be heard from other quarters to the effect that growth is indispensable for human well-being. We consider several such arguments in the following chapter.

11

Why Economic Growth Is Considered a Good Thing

11.1 Mainstream Policy Calls for Continuing Growth

Among mainstream economists, there is little doubt about the desirability of continued economic growth. Indeed, their literature tends to convey the impression that commitment to growth is a defining mark of economic orthodoxy.[1] To call the desirability of growth into question is commonly viewed as defection from the profession.

This puts mainstream economics at loggerheads with our earlier conclusion (in chapter 7) that ecological damage tends to increase with increasing economic production. The conflict is intensified by the findings that our environmental crisis can be resolved neither by technological remedies for isolated problems (discussed in chapter 8) nor by substitution of clean for fossil energy (discussed in chapter 9). Unpalatable as the remaining alternative may be, it appears that the only way out of the crisis involves cutting back on our consumption of energy. And this in turn involves cutting back on our economic production.

On one hand, our argument to this point indicates that return to ecological health requires a substantial curtailment of economic growth. On the other, economic orthodoxy maintains that continuing growth is essential to economic health. Which counsel should prevail?

Despite the unhelpfulness of economic growth theory in this regard (discussed in chapter 10), it should be noted that this commitment to growth is not without supporting arguments. Although these arguments are usually formulated in ways suggesting that their outcome is predetermined, numerous reasons can be found in relevant literature purporting to show why continuing economic growth is desirable. Economic growth increases tax revenues, stimulates employment, generates additional goods and services, advances standards of living, and so forth. The general thrust of these arguments is that economic growth leads to results that are socially beneficial. Not only is it good for the economy at large, but growth also contributes to social well-being.

The cumulative argument of chapters 7 through 9 indicates that worldwide economic production must be curtailed for the survival of present-day society. Defenders of the mainstream counter with various arguments suggesting that continued growth is necessary for continued social well-being. The purpose of the present chapter is to consider a few of the more prominent among these arguments and to show that they fail their intended purpose.

While not denying that economic growth often leads to beneficial results, our general conclusion will be that its benefits fall short of those claimed by its mainstream advocates. On balance, as we shall see, the social benefits of continued growth fail to outweigh its mounting costs in environmental degradation.

11.2 An Argument Tying Growth to a Growing Population

According to much-published statistics, world population quadrupled in the twentieth century. During the year 2000, it reached six billion. The U.S. Census Bureau predicts two billion more people will be added before 2030. It seems obvious that increasing amounts of goods and services will be needed worldwide to sustain all these additional people. This provides an initially plausible argument for the desirability of continued economic growth. Economic growth, the argument claims, will always be needed to take care of the world's growing population.

Before we respond to this argument, it should be acknowledged that growth in world population has far-reaching economic ramifications. Some of these will be considered later in this chapter. But the prospect of continuing population growth does not automatically convert into a rationale for continued economic expansion. One factor that needs to be taken into account is that population is increasing more rapidly in developing countries than in countries with established industrial sectors. For example, in 2007 the populations of Angola and Ethiopia increased at rates of about 2.2 and 2.3 percent, respectively, compared with increase in the United States and the United Kingdom of less than 1 percent.[2] Such figures can be read as indicating that continued economic growth is called for in developing countries. With regard to continued growth in other countries with relatively stable populations, however, the fact that population is increasing in developing countries seems to have no direct implication.

Another relevant consideration is the large disparity in economic production between richer and poorer nations. In the year 2008, for example, the U.S. GDP per capita was fifty-two times greater than that of Angola, and that of the United Kingdom sixty times greater than that of Somalia.[3] Someone concerned with inequities of this sort could plausibly argue that population growth calls for a more equitable distribution of income rather than continued expansion of the overall world economy.

While keeping population growth and its ramifications in view, let us shift attention from national income to the energy required to produce that income. In chapter 6, we estimated that during the hunter-gatherer era, around 10,000 BCE, a typical human consumed approximately one billion joules of energy per year. By 5000 BCE, annual consumption per person had increased to about six billion joules, and by 1000 BCE to about sixteen billion. Average per capita consumption had increased to about thirty-eight billion joules by 1750 CE (marking the beginning of the Industrial Revolution) and had spurted ahead to fifty-six billion joules by 1900. By 2000, average annual energy consumption per person stood at about seventy-nine billion joules. As already noted, world population in 2000 had reached six billion, with another two billion predicted by 2030. Let us consider these figures from various angles with the contention in view that continuing economic growth is required to accommodate a growing population.

Suppose first that per capita energy consumption holds constant at somewhere near the 2000 level, while population grows to eight billion, as projected for 2030. Eight billion people each consuming 79 billion joules of energy would result in a worldwide consumption of 632×10^{18} joules, a 33 percent increase over the 2000 level of 474×10^{18}. Given that the biosphere already is in serious trouble with the entropy stemming from human energy consumption at the present level, increasing that consumption by a third would only hasten the impending ecological catastrophe. If per capita consumption were actually to increase during this period, as some observers envisage, the likelihood of disastrous results would be even higher.

Another abstract possibility is that *total* (not per capita) world energy consumption holds constant at the 2000 level (474×10^{18} joules), while world population increases to eight billion as predicted. This would provide an average of about 59 billion joules per person, more than our estimate of 56 billion for the year 1900. The early 1900s comprised an era in which a sizeable portion of the human race (certainly not all) appeared to live fairly comfortably. Given subsequent improvements in productive techniques, along with more equitable access to natural resources, per capita energy consumption at the 1900 level might leave most of the eight billion people alive in 2030 reasonably well off. This option seems clearly more desirable than the first possibility.

The brute reality of the matter, however, is that the total amount of energy consumed by the human race in the year 2000 was already far too high. Suppose that by 2030 total (not per capita) world consumption were reduced to the level estimated for 1900, a time before massive side effects like global warming and ozone depletion became evident. In 1900, world population stood at about 1.6 billion people, consuming an average of 56 billion joules per person. This amounts to a worldwide consumption of roughly 90×10^{18} joules per year. If that amount were distributed among the 8 billion people predicted for 2030, it would provide each about 11 billion joules of energy, which is only slightly less than the amount per capita at the beginning of the Roman civilization (see section 6.4).

Suppose even that world consumption by 2030 were cut back to the level at the beginning of the Industrial Revolution, before humanity began producing more entropy than the biosphere could absorb. In 1750, approximately seven hundred million people consumed about 38 billion joules per person, resulting in a total consumption of around 27 billion billion joules worldwide. Distributed among eight billion people (as of 2030), this amount would allow each person about 3.4 billion joules of energy. Although this would not support many lives of luxury, it is still several times more than the 1 billion joules needed to sustain the poorest of the world's population today (see section 6.2). If humankind were in a position to choose between this relatively impoverished state and the imminent extinction threatened by continued energy growth, bare subsistence would seem the better choice.

Needless to say, merely citing figures like these will not have much influence on the amount of energy actually consumed by the world economy in the year 2030. The point of these figures is to show that humanity could probably get along reasonably well on only a fraction of the total amount of energy we consume today. In terms of consumer dollars, this means that the world's gross economic product could (in theory) be cut back considerably without dire effect on most of the world's population for decades to come.

What these figures show more specifically is that the argument with which this section began is basically wrong-headed. The fact that world population continues to expand does not constitute a justification for continued growth in economic production. Given that economic growth requires increasing energy consumption, an expanding population argues for reduced per capita production instead. Our present ecological predicament is sufficiently dire that some degree of economic retrenchment would be called for even if world population held constant at the present level.

11.3 An Argument Proposing Growth as a Means of Reducing Population

Another argument involving population, but with a different thrust, is suggested by the previously noted fact that population is increasing more rapidly in developing than in developed countries. In the form we shall consider, the argument is based on two assumptions. One is the assumption that these differences in population growth rates are due primarily to differences in affluence.[4] A supposed consequence is that population will grow more slowly in developing countries as their per capita GDP increases. Second is the assumption that an overall reduction in world population would be an effective means of mitigating humanity's adverse effect upon the environment. This, of course, is the approach advocated in Paul Ehrlich's high-profile book *The Population Bomb*. On the basis of these premises, the argument concludes that economic growth should be encouraged as a means of reducing the number of people the biosphere overall has to support.

One problem with this argument is that its initial premise is oversimplified at best. While relative poverty may be one factor influencing population growth in developing countries, there surely are other factors as well. Among factors often suggested are a need for larger labor forces, lack of educational opportunities for women, lack of occupations other than "homemaker" open to women, lack of access by women to information about family planning, and social practices that subjugate lower-class women by keeping them preoccupied with child rearing.

Reservations are also in order regarding the second premise. It is of course clear that population level is one factor influencing the amount of degraded energy the human race inflicts upon the biosphere. Inasmuch as total human energy consumption equals per capita energy consumption multiplied by total world population, a cutback in population obviously would result in reduced energy consumption *if* per capita consumption itself did not increase. The problem here is that per capita consumption has been increasing dramatically during recent centuries (described in chapter 6) and probably would continue to increase even if world population were reduced to a lower level. We must bear in mind that the ecological degradation we are dealing with is a consequence of the amounts of entropy produced by human activity at large, rather than the number of people producing that entropy.[5]

A further shortcoming of the argument is its underlying supposition that population level can be controlled by deliberate intervention, particularly of an economic nature. Prior to the last century or two, world population was held in check largely by pestilence, famine, and high rates of infant mortality. The main factor behind dramatic increases in population during recent times is a disengagement of these natural controls by technological developments in medicine and agriculture. Human ingenuity was a major source of our current high population levels in the first place, and now we find ourselves proposing further bursts of ingenuity in search of an economic remedy.

Attempts to regulate population inevitably are problematic, regardless of the techniques employed—in some cases problematic on ethical grounds. For example, many people find the most directly effective techniques (infanticide, abortion, even contraception) morally unacceptable. Given the conclusions of chapter 7, similar misgivings might be raised about more indirect techniques featuring economic growth. Another issue is the extent to which such techniques are really effective.[6] Other ramifications of the so-called population bomb aside, any lasting solution to our environmental predicament is more likely to rely on curtailed energy use than on humanly contrived techniques of population control.

Given the problematic character of its premises, the argument that economic growth is needed as a means of population control seems dubious at best.

11.4 An Argument Based on Disparities in Wealth among Nations

At the beginning of the present century, there were about twenty countries producing less than $1,000 a year per person, compared with over $27,000 for the twenty most prosperous nations. This means that richer countries had about twenty-seven times more income to support their individual citizens than did poorer countries in underdeveloped areas. Apart from tiny Luxembourg, with $36,400 per capita, the world's wealthiest country in 2000 was the United States with $36,200. The average U.S. citizen enjoys amenities stemming from per capita income approximately twenty-four times greater than those of nearly fifty less prosperous countries.[7]

In the view of many mainstream economists, the primary reason some countries are more affluent than others is that the economies of wealthy countries have been able to grow more efficiently over longer periods of time. The way to bring underdeveloped countries up to speed, accordingly, is to provide them with the same advantages that have done the job for their more affluent neighbors. These advantages include up-to-date technology (especially in agriculture and manufacturing), abundant energy (mainly fossil fuels), and the financial wherewithal to acquire these and other resources when needed. Technology and energy are available in the open market, and the money to acquire them can be borrowed from organizations like the World Bank and the International Monetary Fund (underwritten by more affluent countries).

When a laggard economy has been jump-started in this fashion, the argument continues, it will enter a period of economic growth. In the process, it will generate enough income to pay off its debts and raise its country's inhabitants to a higher standard of living. If all goes well, the eventual result is that the country will work itself into the ranks of developed nations. Countries like Zambia, for instance, with per capita GNP in 2000 of less than $900, would no longer count as underdeveloped if their productive output began to approach the $25,000 characteristic of Austria and Japan.

Here, in outline, is another familiar argument for economic growth. As poorer countries undergo development, their growth in income will remove the inequities they presently suffer in comparison with more prosperous countries. Economic growth thus serves as an instrument for equity, earning a country effective membership in the society of developed nations.

This argument encounters difficulties on several fronts. One obvious problem is that it supports economic growth only in poorer countries. Since the inequities in question are due to the wealth of some countries no less than to the poverty of others, an argument much like this could also be marshaled in behalf of curtailed growth in wealthier countries. Even if it were successful in other respects, the argument would do nothing to justify continued economic expansion on the part of already developed nations.

Another problem is that the argument overlooks the close correlation between economic production and energy consumption stressed elsewhere in this book (e.g., sections 7.2 and 11.2). Any process of development leading to parity in production of wealth would lead toward parity in energy consumption as well. And parity in energy consumption at levels current in wealthier countries would be catastrophic for the biosphere at large.

The extent of this problem can be illustrated by some statistics provided by energy scholar Vaclav Smil, which, although not entirely up to date, are indicative of a presumably continuing pattern.[8] According to Smil, whereas rich countries contain only about one-quarter of the world's population, they account for about four-fifths of its total energy consumption. If consumption in underdeveloped countries were raised to the same level (i.e., the lower three-quarter sections each raised from one-fifteenth to four-fifths of the present level), energy use worldwide would more than triple for a given year: $4/5 + (3 \times 4/5) = 16/5$. Since energy currently being consumed by industrially developed countries already is producing more entropy than the biosphere can handle, the prospect of this entropy load being tripled over the next several decades cannot be viewed as a desirable goal.

Yet another major shortcoming of the argument is that the strategy behind it usually requires poor countries to borrow money for the capital improvements that are supposed to raise their standards of living. To the misfortune of millions of Third World people, these loans often have the opposite effect. Not only have many projects undertaken with borrowed money had socially disruptive consequences (e.g., the Aswan dam in Egypt), but time after time the poor recipients have been unable to meet the interest payments on their debts (let alone repay the principle). It has been estimated that by the mid-1980s Third World countries had run up a total debt of about $1 trillion, interest payments on which have been taking up increasing shares of the earnings those loans enabled. All too often, the result has been lower standards of living in these countries than prevailed before the debt was incurred. After successive loans from private Western banks and the International Monetary Fund in the 1980s, for instance, real wages in Zaire were only

10 percent of their level in 1960, and 80 percent of its population was in abject poverty.[9]

The equity problem is compounded by the fact that the main beneficiaries of these loan programs often turn out to be the First World lending institutions that provide the financing, along with the private corporations supplying the goods and services the money is used to purchase.[10] The practice of rich countries providing loans that their poor recipients cannot repay is uncomfortably similar to that of financial institutions urging credit cards on people with marginal incomes: the primary beneficiaries are the institutions providing the credit.

11.5 Arguments Relating Economic Growth to Quality of Life

A favorite argument among advocates of economic growth recommends growth as an avenue to social well-being. Commonly cited benefits of economic growth include longer and healthier lives, wider choices of lifestyles, more leisure time, more jobs at better wages, and more money available for health care and education. A recent book by a prominent academic economist even advocates economic growth as a basis for moral improvement.[11]

In one way or another, benefits like these are thought to be linked with a society's economic prosperity. Thus it seems natural to rely on parameters like per capita GDP as measures of a social group's standard of living.[12] Regardless of how measured, the mainstream view is that quality of life increases with per capita GDP. This correlation has served as a powerful argument among the faithful for economic growth.

A dissonant note over recent years has been a spate of news accounts about how economic growth has brought increasingly disproportionate amounts of wealth to sectors of society that are already prosperous.[13] This raises questions of equity in personal income, which we shall have occasion to consider in a moment. Our present concern is with how the growth extolled by mainstream economists affects society at large, including its less affluent members.

One often noted by-product of economic growth is the creation of new jobs. In public discussions of economic policy, the availability of ample job opportunities is generally assumed to be a good thing. Promise of new jobs is a common ploy of political candidates, and loss of jobs is usually a hazard for political incumbents. The perceived association of economic growth with the creation of jobs thus appears to support the conclusion that growth itself should be encouraged.

Effective as an argument like this might be in a political setting, however, the premise linking job opportunities to economic growth is more problematic than at first might appear. The problem is that creating the conditions of growth on the part of large corporations often leads to a cutback of existing jobs instead. In parlance of the 1990s, this effect is known as "downsizing."

The overall dynamics of this effect are roughly as follows. Economic growth is tantamount to increased production. In modern industrialized economies, increased production typically entails increased use of automation. Inasmuch as automation is a way of doing things without people, increased automation leads to decreased use of human labor. The upshot is that economic growth is often accompanied by loss of employment opportunities.

There of course is no ironclad rule that this downsizing effect will invariably occur. Under certain circumstances, it might turn out that the number of jobs added by economic expansion exceeds the number lost to automation. This seems to be typical of industrialized economies during wartime, when an urgent need for war materials overrides considerations of corporate efficiency. It may also apply in cases of a massive shift from one type of technology to another, such as the recent transition from machine skills to data processing in most advanced economies. From a broad perspective, however, such circumstances probably count more often as exceptions than as illustrations of the rule. The high incidence of job loss in expanding economies effectively undercuts the argument that economic growth is desirable because of the jobs it creates.[14]

One way in which economic growth might bring about genuine improvement in social well-being, nonetheless, is through the provision of social services such as health care, education, public transportation, and recreational facilities. Such services bring benefits to rich and poor alike. Although improvement in social services typically requires increased government expenditures, the additional funds can sometimes be made available by a larger tax base without a significant increase in prevailing tax rates. Given that economic expansion generally goes hand in hand with an increase in taxable holdings and activities, this converts into another frequently heard argument for economic growth. In short, economic growth enables an expansion of social services, bringing benefits to all sectors of the society involved.

The problem with this argument is that it runs contrary to empirical data. As might be expected, social services in developed countries tend to be more extensive than in countries with relatively low per capita income.[15] Within the category of developed countries, however, there appears to be no distinct correlation between per capita GDP and expenditure on social services. In a recent tabulation (2001), the five countries ranking highest in per capita GDP (the United States, the Netherlands, Ireland, Iceland, and Norway, averaging $32,000) spent an average of 17.9 percent of their total GDP on social services. On the other hand, the five countries with highest percentage of GDP spent on social services (Denmark, Sweden, France, Germany, and Belgium, averaging 28.2 percent) had an average per capita GDP of $23,600. In brief, the five countries with the highest per capita GDP spent about 57 percent less of their total GDP on social services than did those ranking highest in expenditure on social services, while the latter had per capita GDPs about 27 percent lower than those of the wealthier group.

While figures like these change from year to year, what these particular statistics indicate is that greater national wealth does not translate into larger expenditures on social services. Although increased national income makes enhanced social services possible in theory, in practice those additional resources flow in other directions. Although enhanced services may provide a theoretical justification for economic growth, accordingly, the justification fails when confronted with actual data.

Another factor that throws doubt on the supposed link between economic performance and social well-being is the growing evidence that economic growth may tend to erode various social amenities that contribute significantly to overall quality of life. Recent analysis of statistics from the United Nations System of National Accounts has identified significant components of economic well-being that are not measured by GNP, including crime, family breakdown, suicide rates, poverty, and loss of leisure time.[16] And the Dutch Council for Health Research issued a study in 2006 of why people in leading economies appear increasingly less happy in spite of their increased affluence, with the summary conclusion that "what drives economic growth is not necessarily good for . . . mental health."[17]

Yet another relevant factor concerns the relative satisfaction people feel with their economic circumstances. To the extent that personal satisfaction depends on economic considerations in the first place, there is evidence that people judge their well-being in comparison with others rather than on an absolute basis.[18] In practical terms, this means that individuals with a certain income (say, $10,000 per year) in a society with a lower average per capita income (say, $5,000) will tend to be more content than other people with the same income in a society with higher average income (say, $20,000). Once again, we find a disconnect between quality of life and per capita GDP in absolute terms. Here is one more reason why economic growth cannot be justified with reference to the social benefits that are supposed to follow in its train.

11.6 The So-Called Trickle-Down Effect

The observation that people judge their economic well-being in relative terms ties in with the so-called trickle-down principle that has figured in recent debates about U.S. economic policy. By way of justifying the very considerable gains in corporate profits and executive salaries during recent years of economic upturn, conservative politicians have argued that these earnings tend to percolate downwards through the economy in the form of higher wages benefiting people in lower income groups as well. This reasoning is deceptive. It suggests that workers benefit merely by bringing home higher wages, whereas what would really be beneficial is an increase in actual purchasing power instead.

To distinguish the two, let us first consider the effects of inflation. If inflation results in both take-home pay and the price of goods increasing at the same rate, then take-home pay increases while purchasing power remains constant. Apart from other drawbacks of inflation, this causes no direct economic damage to the average wage earner. As long as wages keep pace with inflation, the wage earner is able to purchase the same amount of goods and services as before.

But the trickle-down process does not work like inflation. As a result of trickle-down, someone's purchasing power can actually decrease simultaneously with an increase in wages. When this happens, the increase in wages occurs under conditions that are detrimental to the workers involved.

To illustrate how this can happen, we need a standard unit with reference to which purchasing power can be measured. Let us take as a unit the wherewithal to purchase an item that, lacking a better name, we will call a "widget" (something most people need and will purchase if they can). A person who can buy just one of these items has a purchasing power of just one widget. To keep things simple, let the going price for widgets be set initially at $1,000.

Now imagine two people with different incomes, each a multiple of this standard unit. Person A is an executive whose current annual salary is $100,000, giving A a purchasing power of a hundred widgets. Person B is a worker with an annual income of $10,000, and thus a purchasing power of just ten widgets. At first, A's purchasing power exceeds B's by a gap of ninety widgets. The difference in lifestyles available to A and to B is the difference made by being able to afford this additional number of widgets.

Let us also imagine that the economy has been subject for several years to an annual inflation rate of 10 percent. The incomes of A and B have changed accordingly, along with the going price of widgets. After one year of inflation, A's income is $110,000, B's is $11,000, and the price of widgets has increased to $1,100. A can still buy a hundred widgets compared to B's ten, leaving the same gap in purchasing power as before. After four years, A earns $146,410, B earns $14,641, and the price of the unit commodity is $1,464. At this point, A can still buy ninety more widgets than B. And so forth and so on. As long as incomes and commodity prices increase at the same rate, the relative purchasing powers of A and B hold constant.

To get the trickle-down process under way, let us now suppose that A's corporation continues to make good profits year after year (as a result of steady growth in the economy), and that A is awarded a 10 percent salary increase annually. As a best-case scenario, assume that enough money has percolated down to provide B an annual 10 percent increase as well. These figures are adjusted for inflation, as is the price of widgets itself. This means that widgets still cost an amount equivalent in value to $1,000 during the initial year.

Under these circumstances, it can be seen that the purchasing power of A relative to B increases significantly each year. Whereas A could buy 90 more widgets than B initially, during the next year A can afford 110 compared with B's 11, a difference of 99 standard units. After four years, A can buy 146 widgets to B's 15, a difference of 131. If the process continues for another five years, the difference in purchasing power between A and B will stand at over 200 standard units.

Not only does the gap in purchasing power between A and B widen year by year, but it widens more rapidly as the years continue. Even under the generous assumption that B received a 10 percent raise nine years in a row, by the end of that period the gap between them has more than doubled. In order to keep the gap in purchasing power between them constant, B would have to receive a 100 percent increase the first year, an additional 55 percent the second, and well over 10 percent increases for each year following (in comparison with A's steady rate of 10 percent). While A's income slightly more than doubles over the nine-year period, B's income would have to increase more than twentyfold to avoid falling further behind. There is not the remotest possibility that the trickle-down process would be that generous.

Conceived in terms of purchasing power, this is the way trickle-down works. This is the way the rich get richer in an expanding economy—at the inevitable expense of the poor getting poorer. Even when substantial amounts of wealth generated by a growing economy are passed on to people in lower income groups, these people usually suffer in relative purchasing power as a result of the process.

11.7 The Management of Growth in Free-Market Economies

We have examined a number of arguments commonly used to *justify* economic growth and found them wanting. Given the way free-market economies operate, however, failure to justify growth will not keep it from continuing. As we shall see in chapter 13, free-market economies are driven by desire for profit. And insofar as increase in economic production is usually accompanied by increase in profit, such economies are managed accordingly. Justified or not, market managers consider it part of their job to sustain conditions of continuing growth.

It is important to realize that, despite the sense of self-determination conveyed by its label, the free-market economy is subject to a considerable degree of deliberate control. In economic parlance, the free market is usually contrasted to managed economies (like those of recent communist states) where governments play a direct role in setting prices and allotting resources. Although this distinction is genuine, there is a sense in which free-market economies are managed as well. They are managed by various governmental and private entities that set the conditions under which the market operates.

In the paradigmatically free-market U.S. economy, for example, Congress enacts laws controlling taxation and tariffs, and the executive branch enforces (or neglects) those laws with an eye toward their economic effect. The Federal Reserve Board regulates commercial interest rates, hoping to steer a middle course between inflation and recession. The Interstate Commerce Commission sets parameters for the conduct of interstate business. And so forth. Indirect as their effects might be, each of these governmental entities plays a substantial role in directing the nation's economic activity.[19]

On the private front, the nation's stock exchanges provide channels through which investors monitor the financial performances of profit-making corporations, which they influence by buying and selling shares. Corporations themselves exercise influence by lobbying Congress and by funding political campaigns of candidates sympathetic to business interests. Not to be overlooked are the large numbers of mainstream economists hired to advise these interested parties and to provide assistance in their shaping of market activity.

Each in its own way, these groups all have an interest in maintaining a consistent pattern of economic growth. For corporations, growth means expanding markets and increasing profits. For investors and corporate executives, it means increased dividends and personal incomes. Another group with a stake in growth includes countless participants in retirement plans, mutual funds, and savings accounts, many of whom experience declining assets when the economy goes stagnant. As far as mainstream economists are concerned, continued growth represents a vindication of their particular theoretical commitments. To the extent that the interests of these highly influential groups prevail, the free-market economy will be abetted in its inherent tendency toward continued growth.

From an ecological perspective, of course, the problem with this whole dynamic is that the free-market economy generates wealth out of resources taken from its supporting environment. The more wealth produced by the economy, the more severe the resulting ecological damage. (To show this was the burden of chapters 5 through 7.) The unfortunate fact of the matter is that what is good for the free-market economy is bad for the biosphere at large.

11.8 The Inherent Anomaly of Continual Growth

From an ecological perspective, to be sure, the very notion of economies that must continue growing to remain healthy is an unprecedented anomaly. Consider the analogy between an economy and a biological organism. Like any living organism, an economy takes organic materials and minerals (negentropy) from its immediate environment, converts these into useful products, and discharges the inevitable wastes (entropy) back into the environment. It undergoes periods of early growth and maturation ("development") during which there is an increase both in the resources it takes from the environment and in the wastes that result from its productive operations.

But at this point the comparison begins to falter. According to mainstream economics, an economy must continue to grow even after it reaches maturity (becomes "developed") and faces the threat of failure if its growth is impeded. For a living organism to keep growing past maturity, on the other hand, would be an unmistakable omen of failing health. When a healthy organism reaches maturity, it typically shifts its energies from growing larger to simply maintaining a stable state. An organism that keeps growing beyond maturity becomes dysfunctional and enters a terminal period of deterioration and decay.

Unpalatable as it may seem, the best biological analogy for a continually growing economy may be that of a cancer tumor that grows at an ever-increasing rate.[20] In its more malignant forms, cancer continues to grow either until it is removed or until it destroys the afflicted organism. An economy geared for growth, in like fashion, faces two possible futures. Either it finds a way to root out its malignancy, or it ends up destroying the very environment on which it feeds.

Given the complicity of mainstream economics in this dire predicament, it seems imperative that we look elsewhere for economic guidance. The purpose of the next chapter is to canvass other approaches to economics that are not beholden to the dogma of continued growth.

12

Economics without Continuing Growth

12.1 Dissenting Voices

Neoclassical economics is commonly dubbed "mainstream" because of its dominance in the economics programs of most major English-speaking universities.[1] Graduates of these programs become faculty for many other universities and colleges as well as professional staff for large numbers of businesses and governmental bodies. Given that mainstream economics is premised on the desirability of continual growth, this accounts for the widespread assumption within the economic establishment that growth is necessary for a healthy economy (discussed in section 11.1).

But while neoclassical economics was building up its hegemony, its path was never entirely free from dissonant voices. Some objected to what was perceived as an excessive use of mathematics in economic theory, and others to the general methodology of establishing a social science on an axiomatic basis. Yet others expressed concern that axiomatic economics had little to say about disparities in wealth between rich and poor nations or about the growing income gap between management and labor.

The most insistent challenge to mainstream economics, however, is probably that issued by an increasingly influential group of economists and ecologists working in the field known as "ecological economics." Their charge is that the mainstream is systematically oblivious to the ecological impact of economic activity.[2] The main purpose of this chapter is to gain an overview of this relatively new branch of economics, with particular attention to its message regarding economic growth. This done, we shall attempt to evaluate its contribution to the solution of our environmental problems.

Before turning to ecological economics, however, it will be useful to look briefly at a recent offshoot of the mainstream, with which it is sometimes confused. This is generally referred to as "environmental economics." As a tributary of the mainstream, environmental economics is not in a position to call continued growth into question. It nonetheless puts forward some useful proposals for mitigating the ecological damage caused by free-market activity.

12.2 Environmental Economics

Through most of its history, neoclassical economics viewed environmental resources such as clean air and fresh water as essentially limitless in supply. While it was recognized that these resources contribute significantly to human well-being, the general perception that they are free and unlimited led to their having no market value. A consequence is that they did not figure in the theoretical models by which economists studied the dynamics of the free-market system.

A further consequence is that environmental factors did not play a role in the operation of actual markets. Thus there were no market mechanisms for factoring pollution and other environmental by-products of production into the pricing of commodities. In economic parlance, this means that harmful environmental effects of production were treated as "externalities." The main thrust of environmental economics is to make such externalities part of the pricing process and then to allow the market to assign them costs in monetary terms.

From the perspective of environmental economics, the fact that externalities like these previously had no role in the pricing process was viewed as a market failure. By one definition at least, a "market failure" occurs when the market concerned does not reflect the full social costs of goods or services it provides.[3] When this happens, the goods and services in question are said to be allocated inefficiently, meaning that consumers would buy fewer units of a commodity at a price reflecting its true social cost. Another way of characterizing environmental economics, accordingly, is that it advocates the correction of market failures pertaining to the costs of environmental resources.

Although pollution is a major concern among environmental economists, there are other kinds of externalities that figure in their discussions as well. One is the environment's ability to absorb potentially disrupting by-products of industrial activity, for example the ability of vegetation to absorb carbon dioxide. Another broad category of externalities is the earth's stock of mineral resources that are subject to depletion. Because of this relatively broad range of interests, the branch of economics in question is sometimes labeled "environmental and natural resource economics."[4]

Various measures have been proposed for factoring environmental costs into the pricing process of the free-market system. One obvious approach would be to regulate the emission of polluting by-products, with the imposition of financial penalties when maximum thresholds are exceeded. Even if such regulations were enacted, however, they would be difficult to enforce, given a general antipathy to financial disincentives in a free-market context.

Other measures rely on more effective motivations. One such is the implementation of pollution quotas, with permits that can be traded on the open market. When tradable quotas of this variety are available, a firm can compare the cost of reducing its own pollution with that of buying permits from another company capable of accomplishing the same reduction more efficiently and then make the choice it finds less costly. As already observed in section 8.6, a sulfur dioxide trading program of this sort has been in place since 1995 and to date has proved at least moderately successful.

Another market-based approach is to provide tax incentives for installing pollution-control equipment (such as smokestack scrubbers; see section 8.2). An added economic advantage of this approach is the considerable money to be made in manufacturing the equipment involved. According to one source, profits in the pollution-control industry topped $3 billion in 2004, with growth to $15 billion projected for 2015.[5]

A further approach often associated with environmental economics is the use of cost-benefit analysis to compare the value of specific sites in providing essential ecological services (e.g., storm and flood protection) with that of their being exploited for other purposes. It is not unusual to find that currently undeveloped areas in various parts of the world are economically more valuable if left in their natural state. According to a recent report out of Cambridge University,[6] for example, logging of a certain forest in Malaysia would cut its economic value by 50 percent.

Although particular approaches like these have obvious merits on their own, the fact that environmental economics remains dependent on its neoclassical roots compromises its usefulness as an approach to ecological survival. One problem already noted is that it leaves unchallenged the doctrine that economic growth is tantamount to social improvement. Bringing environmental costs and benefits into the pricing process by itself is not enough to keep an economy from expanding to the point where it overwhelms its ecological support system.

Also problematic is the methodological requirement that qualitative states of the environment be measured in economic terms. Apart from the grossest kind of approximation, there is no way in which an intact ozone layer—or a normally functioning Gulf Stream or an optimal level of CO_2 in the atmosphere—can be assigned a credible dollar value. One cannot help suspecting that if realistic prices were put on these basic ecological resources, then the market would assign prices to products made from them that few customers would be willing to pay. In effect, the pricing mechanisms of the market would not work for such products.

Ecological economics avoids these problems by dissociating itself from the mainstream paradigm. This parting of the ways was gradual and seems not to have emerged as an accomplished fact until sometime in the 1980s. To appreciate the significance of this departure, let us trace out the origins of ecological economics in the 1960s and 1970s.

12.3 Antecedents of Ecological Economics

Public concern about environmental problems was galvanized by con-servationist Aldo Leopold's *A Sand County Almanac* (1949) and biolo-gist Rachel Carson's *Silent Spring* (1962). Among economists, the most prominent pioneer in this movement was probably Kenneth Boulding, author of "The Economics of the Coming Spaceship Earth" (1966). Boulding's status as a respected member of the profession is indicated by his having been elected president of the American Economics Asso-ciation in 1968.

Boulding's article has become famous for its contrast between the "cowboy" or open economy and the "spaceship" economy, which is es-sentially closed. As he explained the metaphors, a cowboy is symbolic of the "illimitable plains" and is also associated with reckless and exploi-tative behavior, while a spaceship symbolizes the planet earth "without unlimited reservoirs of anything, either for extraction or pollution." In the cowboy economy, he goes on to say, both production and consump-tion are regarded as good things, and the success of any economy is measured in terms of raw materials cycled through the system and dis-charged as waste products (which he labels "throughput"). In the space-ship economy, by contrast, the primary concern is with maintaining stocks (e.g., of raw materials and pollution sinks), and success is a matter of achieving maximum human good with a minimum of throughput.

Boulding is aware that most economists will find "very strange" the idea that production and consumption are not good things in them-selves. Another departure from traditional economics is his explicit use of concepts from thermodynamics.[7] As he puts it, there is "no escape from the grim second law of thermodynamics," especially with respect to our rapidly depleting supply of fossil fuel. Boulding also makes effec-tive use of the contrast between entropic and negentropic processes (he uses the term "anti-entropic"). He illustrates the former by the diffusion of once concentrated materials over the earth's surface, the latter by the concentration of previously diffuse materials. Similar concepts are em-ployed by later writers on ecological economics.

The much maligned but highly influential volume *The Limits of Growth,* by Donella H. Meadows, Dennis L. Meadows, Jorgen Randers, and William W. Behrens III, was first made public in 1972. As far as methodology is concerned, the genius behind the work was Jay Forrester, founder of the computer simulation technique known as "system dynamics" and then professor in MIT's Sloan School of Management. In bare essentials, the computer model ("World 3") used in *The Limits of Growth* study was a system of several dozen interdependent variables that could be either altered or held constant in order to explore their effects on each other. The primary contribution of the volume's authors, all of whom had extensive experience in the study of complex systems, was to select the variables to be manipulated and to make educated guesses about how they might be empirically related.

Variables were selected to represent five areas of major global concern: industrialization, population growth, food shortage, depletion of nonrenewable resources, and environmental degradation. Runs of the model based on estimates of then current trends predicted that natural resources would soon be exhausted, that pollution was about to reach life-threatening levels, that massive famine was only a few decades away, and that catastrophic collapse of the whole system would occur by 2050. Ultimate collapse in the model could be delayed by holding one or a few of the key variables constant (e.g., population or the use of natural resources). Given the systematic interconnections among variables, however, the only way to avoid eventual collapse involved arresting or reversing existing trends in all five areas. Piecemeal approaches focusing on population control or food production in isolation would not assure a sustainable future and might even backfire.

Needless to say, *The Limits to Growth* received scathing reviews from mainstream economists. Not only were its results radically at odds with the doctrine that continual growth is necessary for a healthy economy, but the methods of computer simulation employed in the study were fields apart from the mathematical models favored by mainstream economists. One might surmise that the appearance of this study was a defining moment in the fixation of mainstream orthodoxy; henceforth to take this kind of study seriously was to invite excommunication from the fold.

Despite derision from the mainstream, however, by the year 2000 *The Limits to Growth* had sold over three million copies in at least thirty-one languages.[8] Its importance lies not in the accuracy of its predications, which were not intended to be definitive in the first place, but rather in its having alerted the world to reasons why economic growth cannot continue indefinitely. In this manner it helped clear the way for the distinctly unorthodox approach represented by ecological economics.

Other trail-breaking works by economists during this period were E. F. Schumacher's *Small Is Beautiful: Economics As If People Mattered* (1973) and Herman Daly's *Steady-State Economics* (1977). The title of Schumacher's book reflects his ideas that large organizations should operate like networks of smaller groups and that the technology used by such groups should be "intermediate" in the sense of designed for the job satisfaction of self-employed workers. For Schumacher, quality of work experience is a basic economic good, to be stressed even at the expense of economic growth.

Daly's book also argues for a deemphasis of economic growth. In place of the standard economic model focusing on the flow of products to consumers and the return flow of income from consumer to producer, which ignores the economy's dependence on environmental resources, Daly advocates a model focused on the throughput (Boulding's term) of low-entropy resources resulting in high-entropy by-products. His basic idea is that an economy is sustainable only if it maintains a stable balance (steady state) between the natural resources it consumes and those the environment can replenish at a sustainable rate. Daly was a student of Nicholas Georgescu-Roegen, whose contributions we consider in section 12.4.

Among important contributions from outside the field of economics were ecologist Howard T. Odum's *Environment, Power, and Society* (1971) and geophysicist Earl Cook's *Man, Energy, Society* (1976). Odum's book was an application of general systems theory to the interaction between ecological and economic systems in particular. One of its basic themes is that all wealth stems from the environment and that the value of commodities ought to be based on the energetic resources needed to produce them (rather than on, e.g., the consumer's willingness to pay).

Cook's book is a thoughtful and sober examination, using the Second Law of Thermodynamics as backdrop, of the various forms in which human society appropriates energy from the natural environment and of the inefficiencies with which that energy is often put to use.

12.4 The Conceptual Reorientation Brought About by Georgescu-Roegen

Kenneth Boulding may have been the first prominent economist to have applied the Second Law of Thermodynamics explicitly to the economic process, but his use of the principle involved little technical sophistication. Other attempts to make this principle relevant to economics were made by physical scientists, including Frederick Soddy and Earl Cook, but their works were seldom read by mainstream economists. Within the field of economics proper, the first technically rigorous treatment of the Second Law and its economic implications came with Nicholas Georgescu-Roegen's *The Entropy Law and the Economic Process* (1971).

Georgescu-Roegen (G-R) was born in Romania, received a doctoral degree in mathematical statistics from the Sorbonne, and in 1934 came to the United States, where he studied economics under the influence of Joseph Schumpeter at Harvard. He subsequently moved to Vanderbilt University, where he gained a reputation among mainstream economists for his work on mathematical models of consumer preference and market equilibria. When he turned his mathematical skills from mathematical modeling to thermodynamics and its economic applications, the shift was perceived as a revolt against the mainstream, to which he himself had made substantial contributions.

G-R's departure from neoclassical economics can be described in terms of what, following Schumpeter,[9] might be called a "conflict of preanalytic visions." The preliminary conception of a functioning economy that stands behind the standard neoclassical approach posits an isolated system in which exchange value of one or another sort is passed back and forth between firms and households. More concretely, an economy is conceived as a closed loop in which products are exchanged for money, which then enables the generation of more products, and so on indefinitely. The system is isolated in the sense of being cut off from its environment. No account is taken either of raw materials and energy coming into the system or of the resulting waste products discharged back into the environment.

In place of the familiar interchange between producers and consumers, G-R substituted what amounts to a one-way flow beginning with resources taken from the environment and ending with a discharge of wastes.[10] The one-way flow is roughly equivalent to the quantity termed "throughput" by Boulding and others. In G-R's own terms, it consists of "an input flow of low entropy" joined to "an output flow of high entropy."[11] So-called "low entropy" is what we have been calling "negentropy" (G-R also uses this term occasionally), and "high entropy" is energy or structure too far degraded for further use.

Another way of getting at the disparity between G-R's approach and that of the mainstream has to do with their respective postures toward classical (Newtonian) mechanics. The equations of classical mechanics have no temporal parameters, which is to say that the processes to which they apply are theoretically reversible. The equations would apply even in a world where, so to speak, time ran in the opposite direction. This ties in with the atemporal character of the neoclassical flow model, with exchange value circulating endlessly between producer and consumer. In G-R's way of thinking, by contrast, the economic process is essentially temporal and not mechanical.[12] Because of the Second Law, it is physically impossible for the economy to reverse itself and to begin operating so as to convert a high-entropy input into a low-entropy output.

A concrete result of G-R's entropy-flow model (low entropy in, high entropy out) is his rejection of the neoclassical notion that increases in capital can make up for decreases in natural resources. According to the standard conception, economic output is a joint function of given quantities of capital, labor, and natural resources, such that a greater quantity of one factor can sometimes compensate for a lesser quantity of another.[13] In G-R's analysis, this notion trades on a confusion between funds and flows. Whereas natural resources constitute flows passing through the economy, both capital and labor are funds (agencies) that transform resources into products. In terms of a favorite analogy of

Herman Daly's, to think that capital can effectively substitute for resources is like thinking that the same house could be built with half the lumber by using twice the number of saws.[14]

G-R is generally recognized as the theoretical progenitor of ecological economics (although he seems not to have used the label himself). Within the last few decades, ecological economics has emerged as a major source of dissent against mainstream orthodoxy.

12.5 Ecological Economics (EE)

EE took on a distinct identity with the founding of the International Society for Ecological Economics in 1987 and the establishment of a journal bearing its name in 1989. Founding editors of the latter were Robert Costanza and Herman Daly (author of *Steady-State Economics* mentioned earlier), who had collaborated previously in editing a special issue of the biology journal *Ecological Modeling* (1987) on topics germane to the emerging field of EE. This volume still serves as a useful introduction to the field, along with several more recent books explicitly directed to that purpose.[15]

This new field is interdisciplinary in the following sense. Whereas mainline economics is concerned with the operation of economic systems, and mainline ecology with that of ecosystems, EE is concerned with the *interaction* between systems of these two different sorts. Moreover, the manner in which it addresses this interaction is distinct from the way in which the interdisciplinary science of biophysics, for example, addresses biological phenomena from the perspective of physical theory. Strictly speaking, EE is neither the application of mainline economic theory to ecological phenomena (environmental economics is closer to this description) nor the application of ecological theory to economic phenomena (this is closer to bioeconomics[16]).

This is not to say that individuals working in the field of EE lack training in these two previously established disciplines. From its founding period to the present, EE has included both economists and ecologists who bring their specific skills to bear on phenomena of shared interest. Thus there are ecological economists without specialized training in economics and others without specialized training in ecology. This underscores the fact that EE as such is not an extension of some previously established discipline (the way biophysics is an extension of physics). It is interdisciplinary in the sense of including researchers from different disciplines engaged in a collaborative effort, rather than a merger of the two disciplines involved.

For these reasons, EE not only lacks an overarching theoretical framework but also has no generally shared methodology. In the words of one prominent representative, EE is "a large umbrella under which economists, ecologists, and other scholars" engage in searching for "a better understanding of the interrelations between people and their environment." Taken on its own, it is "methodologically eclectic" without integrating models or distinct theoretical paradigms.[17]

Given its methodological and theoretical eclecticism, EE is held together by a set of shared attitudes and perceptions regarding economic systems and their natural environments. Probably without exception, anyone identifying with the field would accept G-R's preanalytic vision that an economy is an integral part of a more comprehensive ecosystem. Along with this view goes the understanding that an economy takes low-entropic resources from its environment, uses up these resources in its internal processes of production and consumption, and discharges the resulting high-entropic wastes back into the environment. Another part of this preanalytic vision is an awareness that the environment is limited in its ability to sustain economic activity, because of limitations both in the low-entropic resources it makes available and in its capacity for assimilating high-entropic wastes.

Tied in with this conception of a limited physical environment is an understanding that economic growth cannot continue indefinitely. Continuing growth involves taking increasingly greater quantities of resources from the environment and discharging increasingly greater amounts of wastes for the environment to assimilate. Inasmuch as the environment is limited in both respects, continuing growth hastens the time when the economy in question will simply stop functioning.

From this perspective, it seems obvious that there will come a time (if it is not here already) when growth will have to be curtailed for the world economy to continue functioning. But curtailing growth does not require that the economic quality of human life will necessarily diminish. Herman Daly has made this point repeatedly, relying on a distinction between growth and development.[18] Economic growth is a matter of increasing quantities of materials flowing through the economy (which Daly, following Boulding, calls "throughput"). Economic development, on the other hand, is qualitative in character, having to do with increasing economic services achieved with a given amount of throughput.

Daly's proposal in this regard is that economic planners shift their aim from quantitative growth to what he calls "sustainable development." For him this means qualitative improvement in economic services, while avoiding levels of quantitative throughput that exceed the carrying capacity of the ecosystem.[19] While not all interested parties believe such a shift in emphasis would be realistic in the long run,[20] Daly at least has pointed out a kind of economic expansion that would not necessarily be environmentally destructive.

Another departure of EE from standard economics is its focus on environmental quality as such, in contrast with environmental economics' focus on human benefits. As a branch of the mainstream, environmental economics relies on cost-benefit analysis in policy decisions and evaluates costs and benefits in terms of individual (and/or group) utilities and disutilities. This means that factors of environmental quality and impairment are evaluated with respect to human preferences. EE, however, evaluates these factors in terms directly relevant to their ecological functions, such as productivity (of biomass), stability and resilience, and biodiversity. In brief, whereas standard economics focuses on the environment as a source of human services, EE treats ecological systems as having value in their own right.

This illustrates yet another departure of EE from standard economics, in that EE, unlike the latter, makes no pretence of being value-neutral.[21] To be sure, proponents of EE often are quite upfront in their concern for justice, equity, and human dignity.[22] Some even discuss values of an overtly religious nature.[23] EE is also described by its advocates as a field especially suited for taking the needs of future generations into account,[24] a concern which conventional economics, with its emphasis on actual preferences, is ill-suited to address.

This brief discussion is enough to show that, from an ecological perspective, EE's approach to the interaction between economy and environment is clearly preferable to anything mainstream economics has to offer. A further question is how, and to what extent, EE can help us find a way out of our current environmental predicament.

12.6 EE's Contribution to the Resolution of Our Environmental Predicament

Our discussion of economics in its various manifestations began in chapter 10 with the observation that mainstream economics is committed to economic growth. Since economic growth generally requires increased energy consumption, and since increasing energy consumption over recent centuries has caused increasing damage to the biosphere, the influence of mainstream economics impedes the otherwise promising strategy of trying to alleviate our environmental predicament by substantial reductions in energy consumption. What help does EE offer in pursuing this strategy?

In reading the literature of EE, one gets the sense that its advocates are putting it forward as a workable alternative to mainstream economics. One encounters talk specifically about paradigm shifts, like that from classical (Newtonian) to quantum mechanics, suggesting that the paradigms of neoclassical economics are due for replacement by those of its ecological counterpart.[25] Although a shift of this sort clearly would be beneficial to the environment, there are reasons for thinking that such a transition will never actually take place.

These reasons have to do with differences in status between the two types of theory involved. For one thing, in comparison with EE, neoclassical economics is typified by a more or less integrated theory held in place by a central paradigm. Although formulated differently by different authors, its dominant paradigm is that of a perfectly competitive market in which all participants make rational and fully informed choices intended to maximize their own interests. As already noted, however, EE is theoretically eclectic and possesses no single paradigm with respect to which its concerns can be integrated. A consequence is that a transition from neoclassical to ecological economics would be less a paradigm shift within the same discipline than a change to another discipline of a rather different sort.

Perhaps a more substantial point of difference is that neoclassical theory has descriptive power that EE lacks. The fact that mainstream economics makes extensive use of mathematical modeling does not prevent it from being an empirical science. Its subject matter covers factual topics pertaining to the distribution of scarce resources within organized societies. While the gap between fact and theory is often considerable, people actually learn things about these topics when they study neoclassical economics. Knowledge gained through this study enables trained economists to find influential jobs in government and business. Mainstream economics is descriptive in that its practitioners are trained to make factual observations about the operation of actual economies.

Given that EE focuses on the interaction between economy and environment, rather than on economies by themselves, it does not provide its practitioners with the same kind of factual expertise. A study of the interaction between economy and environment does not prepare people for economic positions in government and industry. (It is hard to imagine an ecological economist as chair of the Federal Reserve Board, for instance.) This is not to deny that ecological economists are knowledgeable in the operation of actual economies; undoubtedly many are. When this is the case, however, it is because of their experience with standard economics rather than their study of EE as such.

As far as its consideration of operating economies is concerned, EE tends to be more prescriptive than descriptive.[26] Rather than featuring information about how economies actually operate, it is a primary source of advice about how they *should* operate (including advice about policy). To mention two of many proposals by way of illustration, Herman Daly has urged that we (1) stop counting the consumption of natural capital as income and (2) maximize the productivity of natural capital in the short run and invest in increasing its supply in the long run.[27] If these and similar proposals were implemented, our environmental circumstances undoubtedly would begin improving as a result. The point to be emphasized, however, is that they are only proposals and that under current political circumstances they stand little chance of being implemented.

As matters stand, EE has neither the technical expertise nor the political clout to change the economic process in ways it thinks would be environmentally desirable. Take the case of economic growth in particular. EE has been loud and clear in its insistence that upward-spiraling growth is inflicting increasingly severe damage upon the environment. But most governments and financial institutions see growth as essential to full employment and price stability. This means that calls from EE (and elsewhere) for no-growth policies go largely unheeded. EE can advocate such policies until the end of the day, but as matters stand it lacks the clout to put these policies into effect.

The present section began with the question of how EE might help reduce the quantities of energy consumed by industrialized countries. This boils down to the question of how it can help curtail economic growth. By way of answer, we observed initially that EE is not the kind of theory that could replace mainstream economics in the deliberations of industry and government. The basic reason is that it lacks a unified vision that might supplant the paradigm of the competitive market. We observed further that EE addresses the economic process in a prescriptive rather than a descriptive mode. Its descriptive efforts are aimed at the interaction between economy and environment, which puts it in a position to recommend opportune changes in economic policy and practice. But it has neither the technical expertise nor the political leverage to bring such changes about.

Nonetheless, the fact that EE is primarily prescriptive does not prevent it from being of appreciable help in resolving our environmental predicament. Any change from the status quo that is not merely happenstance relies upon insights of a prescriptive sort. Prescriptive advice is advice about how things might be done in ways more in keeping with humanly significant values and aspirations. Preeminent among these are values pertaining to an environment capable of supporting human endeavors, now and into the indefinite future. There is no reasonable doubt that EE has made valuable contributions toward articulating such values, regardless of how ultimately they might be put into effect.

12.7 Curtailment of Growth a Matter of Values

Part 2 of this study has been concerned with the role of economics in our gathering ecological crisis. Chapter 7 examined the relation between energy consumption and economic productivity, culminating with the general principle that the production of a given amount of economic goods introduces a corresponding measure of ecological degradation. The upshot, in effect, is that economic goods come with ecological price tags and that our current crisis amounts to an ecological deficit resulting from an overproduction of economic goods.

Three strategies were identified for defusing the crisis. The first was to isolate particular sources of ecological damage (e.g., sulfur dioxide emissions) and then to devise technologies (e.g., smokestack scrubbers) to eliminate them on a piecemeal basis. Chapter 8 evaluated this approach and found it wanting. One major shortcoming, among several, is that ecological degradation often takes the form of massive system malfunctions (like rupture of the ozone layer), which cannot be remedied by technological means.

The second strategy addresses the fact that the problem stems largely from humankind's excessive use of fossil fuels during the past two centuries. An initially obvious countermeasure is to replace fossil fuel with energy available from renewable sources (e.g., solar, wind, hydroelectric). This strategy is criticized and judged inadequate in chapter 9. The result, in summary, is that while human industry should eliminate its use of fossil fuel to whatever extent feasible, we must be aware that fossil resources are used in many ways for which there are no clean-energy replacements. Even if clean energy were substituted whenever possible, ecological damage on many fronts would continue unabated.

This result sets up the final three chapters of part 2 dealing with economic growth. The third and most direct strategy for combating ecological degradation is to avoid using energy in amounts producing more entropy than the biosphere can cope with. Inasmuch as this strategy calls for levels of energy consumption far lower than those occurring presently, it runs directly counter to the mainstream doc-trine that economic health requires continuing growth. In order to un-derstand the mesmerizing force of this doctrine, we first looked into the history of the concept of economic growth in chapter 10. We then ex-amined various rationales offered in support of the doctrine in chap-ter 11 and found them generally inadequate.

Not only does the mainstream commitment to growth stand at cross-purposes to the general strategy of decreasing energy use, but we should note further that this commitment is a primary cause of the environmental predicament we are concerned to alleviate in the first place. In effect, the third strategy is to break free from the influence of this commitment. Given the very low likelihood that mainstream eco-nomics will relinquish this doctrine under its own initiative, we turned in the present chapter to explore the possibility that mainstream eco-nomics might be effectively replaced by its ecological counterpart (EE). Our conclusion is that, practically speaking, a transition of this sort is unlikely.

This leaves us in a quandary about how economic production and consumption can be reduced to levels compatible with environmental integrity. Persuasive as arguments against economic growth may be (like those offered in chapter 11, among others), the current tendency toward ever-increasing levels of economic activity will not be reversed by argument alone. As long as corporate executives value profit, and as long as individual consumers value the acquisition of goods, society will continue to heed the advice of mainstream economists.

Instead of talking about shifts in scientific paradigms, perhaps we should be talking about shifts in societal values. As matters stand, so-ciety is wedded to a set of values that encourages the proliferation of production and consumption. But there are other values that would have the opposite effect. Perhaps there is a workable solution to our en-vironmental crisis that relies on a shift away from the values that have fostered its onset. The remainder of the study is devoted to exploring this possibility.

PART 3

Ethics and Economics

13

Desire for Wealth in Free-Market Economies

13.1 Return to the Basic Problem

Our environmental crisis boils down to the fact that the biosphere is struggling under the impact of far more entropy than it can get rid of. One pervasive form of entropy is low-grade heat, which when trapped in the atmosphere leads to problems of global warming. Another is the decimated layer of ozone in the stratosphere, allowing the entry of ultraviolet radiation in amounts causing increasingly severe damage to the earth's essential food chains. Other damaging forms of entropy discussed in previous chapters include the accumulation of nonbiodegradable wastes, progressive desertification of once-fertile crop lands, and a loss of species diversity necessary for a healthy biosphere.

As already observed, a lion's share of this excessive entropy is due to human industry, contributed in amounts that have been steadily increasing since the beginning of the Industrial Revolution. Viewed in a positive light, the time since then has been a period of economic growth, resulting in steadily increasing amounts of consumer goods and services. The downside of this rising tide of affluence is that it has been generated by profligate consumption of fossil fuel. This extravagant consumption is a major cause of the entropy overload now afflicting the biosphere.

Economic growth is brought about by the expenditure of negentropy (energy and raw materials) taken from the biosphere. When this negentropy is expended, an equivalent amount of entropy is returned to the biosphere in such forms as waste material and low-grade heat. For every increase in GNP, there is a corresponding amount of additional entropy for the biosphere to deal with. In a biosphere already overloaded with entropy, every significant increase in GNP results in further deterioration of the ecological structures supporting human society.

The cumulative upshot is that the biosphere is becoming increasingly incapable of supporting human society at its current level of economic activity. Although no one knows all the ways society will be affected when its ecological support systems reach the point of collapse, it almost certainly will encounter increasingly widespread famine, spreading social conflict, and extensive failures of material infrastructure (e.g., loss of coastal cities).[1]

Nation by nation, and sector by sector, the world economy is running out of control, producing increasingly more entropy than the biosphere is able to discharge back into space. A question of utmost urgency at this point is how economic production worldwide can be scaled back to a level the biosphere can support.

A necessary first step toward answering this question is to isolate the factors that drive economic growth[2] and to gain an understanding of how they operate in a free-market context. The next step is to look for ways this effect could be neutralized. Given the likelihood that several factors contribute to this effect, we should look for those that might conceivably be altered in ways that would bring excessive economic growth under control.[3] This sets the task of the present chapter. Subsequent chapters will consider how needed changes might be brought about.

13.2 Economic Growth Driven by Desire for Wealth

To orient our search in the right direction, let us recall Adam Smith's metaphor of the invisible hand (described in section 10.3). As Smith himself put it, although an individual "intends only his own gain" and pursues only "his own interest," the resulting economic activity frequently promotes the interest of society at large. Subsequent commentators on this passage have glossed "pursuing self-interest" as advancing one's own good (Bishop Butler), as satisfying one's private appetites (Bernard Mandeville), and as maximizing personal profit.[4] Although Smith may not have said so explicitly, the passage suggests that he viewed pursuit of self-interest as the driving factor behind economic activity generally.

It seems likely that Smith thought of self-interest in this role as covering personal benefits of various sorts he left unspecified. But the concept became more specific in the thought of his followers. In discussing the classical notion that economic phenomena follow logically from a few basic principles, Georgescu-Roegen identifies as two of those principles (1) that a greater good is preferred to a smaller and (2) that everyone seeks the greatest quantity of wealth with the least labor and self-denial.[5] Behind this, he says, is the classical view that all economic phenomena are grounded in the desire for wealth. Slightly rephrased, this is the view that the desire for wealth is the driving force behind all economic activity.

This classical view that economic activity is driven by desire for wealth continues to apply in today's economic climate. Given our current economy's emphasis on the "bottom line," it seems adequately clear that people engage in economic activity for the income (added wealth) it provides.[6] But economic activity in a general sense is not equivalent to economic growth. Let us sketch out a line of reasoning relating desire for wealth to the phenomenon of economic growth specifically.

The argument begins with the commonplace observation that economic activity is a major *source* of wealth. Given a broad enough conception of economic activity, it may in fact be the only source.[7] This being so, the normal path to follow for someone desiring wealth is to engage in economic activity. Just as someone desiring knowledge will participate in educational activity, so someone desiring wealth will pursue activities by which wealth is acquired.

Once a person's economic activities begin to generate wealth, moreover, that person's desire for more wealth often remains unsatisfied. As thoughtful observers of this phenomenon have long realized,[8] desire for wealth is open-ended. Like desire for pleasure and desire for power, desire for wealth is not readily satiated. Once wealth has been accumulated in relatively small quantities, desire continues for yet further accumulations. By this dynamic, desire for wealth induces affected individuals into progressively higher levels of economic activity.

The next step in this line of reasoning is to observe that increased economic activity on the part of individuals who desire more wealth is tantamount to increased activity on the part of money-making organizations they participate in for that purpose. For example, individuals invest in corporations for additional income, enabling the corporations themselves to expand their business activity. Expanded activity on the part of particular business organizations, in turn, contributes to higher GNP on the part of the overall economy. In briefest summary, this is the manner in which desire for wealth on the individual level leads to growth on the level of the economy at large.

There undoubtedly are other forces propelling growth on the corporate level. One that might be suggested is a competitive market environment in which firms that fail to maintain a certain margin of profit soon go out of business. Keeping a business profitable requires investment, perhaps in more efficient technology, more extensive advertising, or business expansion. Rather than taking expenditures for such purposes out of previous profits (which typically go into salary increases and shareholder dividends instead), a firm may elect to fund them by loans at favorable rates of interest.[9] To pay back these loans without eroding future profits, however, the firm's gross income must continue to rise. And with growing incomes on the part of its constituent firms, the economy at large continues to grow. This indeed is one respect in which economic health depends on continuing growth, in accord with the dictates of mainstream economics (recall section 10.6).

Yet while some such dynamic presumably operates in contemporary free-market economies, it does not constitute an exception to the general thesis that growth is driven by individual desire for wealth. We may grant that an economy's growth results from an implicit need of its constituent firms to operate on a profitable basis. Nonetheless, this need is a consequence of arrangements by which executive salaries, bonuses, and dividends are siphoned off from corporate profits.[10] Without desire for wealth on the part of its individual managers, there would be no incentive to keep a firm afloat in the first place. Here is yet another avenue through which economic growth is driven by desire for wealth on the part of individual people.

This line of reasoning does not exclude other factors behind economic growth that have nothing to do with desire for wealth.[11] All it purports to show is that desire for wealth is the central factor. Without desire for wealth operating behind the scenes, modern economics would not be stimulated to grow incessantly.

13.3 Wealth a By-Product of Consumer Demand

Wealth takes many forms in modern economies. Individual wealth could take the form of stocks and bonds, real estate, or money deposited in a bank. The wealth of a corporation might consist in the market value of its shares, in its capital assets, or its holdings in other companies. The wealth of a nation, in turn, is the total wealth produced in it, commonly measured as GDP over a given accounting period.

Some forms of wealth can convert to others (e.g., cash and property), while others (e.g., GDP) cannot. What they have in common is that all trace back to the production of goods and services. This is why GDP is an appropriate measure of national wealth. A nation's composite wealth boils down to the goods and services produced by its individual and corporate citizens.

We must take care to note, however, that wealth does not derive from production alone. Defined generally, an economic good is something that can be sold in the marketplace.[12] To build up an inventory of things that no one will buy does not amount to the production of wealth. The production of things contributes to wealth only when people are willing to pay for them.

Wealth is created by producing goods that are ultimately purchased. This means that consumption is essential to the production of wealth. As far as economic growth is concerned, it means that growth entails increased production and increased consumption alike. To understand the dynamics of economic growth, we need to understand how production and consumption interact.

Goods and services flow to the consumer as a result of consumer demand. Needless to say, the goods that enter into that flow must be supplied by a producer. But the mere fact that goods have been produced does not set them in motion toward potential consumers. Without active consumer demand, goods would not leave the producer's warehouse. There is no way of *pushing* them into the consumer pipeline if no one is interested in exchanging money for them.[13]

In a manner of speaking, consumer demand sets up a suction that *pulls* goods through the pipeline. Think of a drinking straw as a simple analogy. By sucking on one end, a person creates a vacuum that is filled by juice entering the other end of the straw. Without suction, the liquid would stay in the glass. Similarly, in the case of the pipeline between producer and consumer, the flow of goods is set in motion by active consumer demand and not by a supply of goods sitting in the warehouse.

Given adequate availability of goods in the warehouse, moreover, the volume of goods flowing through the pipeline is directly correlated with the volume of consumer demand. The greater the demand, the more goods flow from producer to consumer. To the extent that the volume of goods flowing through the pipeline can be controlled, accordingly, a direct way of exercising control is by manipulating consumer demand. If some agent wants to increase economic activity in a given sector of the economy, one available means is to stimulate demand in that particular sector.

This is the modus operandi by which desire for wealth leads to economic growth. Wealth is a by-product of economic activity. Inasmuch as economic activity is keyed to consumer demand, managers desiring additional wealth will seek opportunities to stimulate additional demand. Driven by additional consumer demand, production of goods will increase as well. And increased production of consumer goods amounts to further growth of the economy in question. In brief, desire for wealth leads to economic growth by stimulating demand for relevant goods and services.

This emphasis on consumer demand as the key to economic growth follows the tradition of Adam Smith. In Smith's view, expansion of production (by division of labor, etc.) requires larger consumer markets to absorb the additional goods produced. A condition of this happening, he thinks, is that the class of consumers must be expanded to include ordinary workers, who must receive ample wages to buy more than bare necessities. Another condition is that capital (land, factory equipment) must be available to support higher levels of productivity. Once enabling conditions like these are in place, however, consumption is the key factor in an expanding economy.

As Adam Smith put it in *The Wealth of Nations,* production has consumption as its "sole end and purpose."[14] Consumption is the end of production not only in being the reason for which production is undertaken but also in supplying the impetus by which production is energized. The main departure of the present account from Adam Smith in this regard is its treatment of desire for wealth as a factor operating on its own, with no inherent connection (via the "invisible hand") to the common good.

13.4 Consumer Demand Stimulated by Marketing

Economic growth is keyed to increased consumption.[15] Consumption, moreover, is stimulated by advertising and marketing. We now undertake an analysis of certain marketing techniques that can be employed to induce consumers to buy more goods. These techniques gain added relevance when we turn in subsequent chapters to consider practical ways of curtailing economic growth.

Marketing techniques are aimed at manipulating consumer behavior, with varying degree of aggressiveness. Our discussion begins with commonplace forms of advertising and proceeds to more aggressive techniques.

(a) Product Presentation

In its simplest form, advertising is a matter of presenting a product to potential consumers. This might amount to nothing more sophisticated than simply making the product known. A product might be made known by word of mouth, by conspicuous labeling (e.g., a sign over a stall saying "Fresh Meat"), or by a simple description under "Classifieds" in a local newspaper. At this level, there is little concern to target specific groups of consumers. At the next level of sophistication, there is an effort to make the advertisement itself attractive to the consuming public. Examples in visual media include use of attention-getting formats (like bold type), bright colors, and illustrations. Auditory equivalents include special sound effects and commercial jingles. These techniques are aimed at attracting notice to the ad itself, still without regard to specific audiences.

Another notch upward in manipulation comes with ads designed to compel attention, as distinct from merely inviting notice. Techniques of this general sort in use today include increasing the volume when ads appear on television, beginning movies with commercials for local businesses, and designing ads that feature people in distorted (hence attention-getting) postures. As audiences become jaded by techniques like this, there is a tendency to raise their level of intensity (e.g., by adding more loudspeakers in movie theaters).

A natural extension of such techniques is adapting them to specific audiences.[16] Children in certain age groups are targeted by ads featuring cartoon characters. Older children are addressed by ads depicting peer-group attitudes (think of Joe Camel—"Joe Cool"—now banned for being too effective). And young adults are typically responsive to ads showing movie stars and professional athletes. Not surprisingly, senior citizens are attracted to pictures of healthy-looking people noticeably younger than themselves, like those found in ads for certain medications.

Just as ads for toys are found in maternity magazines, and ads for summer cruises in college alumni bulletins (but not vice versa), so there are preferred media for promoting endless varieties of other "special interest" products. A development along these lines has been the compilation of lists of mail-order customers with demonstrated interests in certain kinds of products. These lists can be sold not only to other retailers but also to fundraisers for various political and social causes.

(b) Preference Management

Advertising techniques like the above are basically ways of bringing products to the attention of potential buyers. A different dimension of manipulative skill comes into play when marketers attempt to shape the preferences of potential consumers to make them more receptive to particular products.

As might be expected, the transition from product presentation to preference management tends to blur on both sides. There is an element of preference management, for example, in the use of generational symbols ("Joe Cool," healthy-looking middle-aged people) to attract the attention of particular age groups. The effect of such techniques is not far different from that of designing packaging both to attract attention and to stimulate appetite for the product inside. Yet there is an important distinction between drawing attention to an advertisement and invigorating an appetite for the advertised product.

Preference management is a form of behavioral conditioning, and the procedures involved have been well-established by behavioral science. Recall the classic example of behavioral conditioning: the case of the dog trained to salivate at the sound of a bell by repeated experience of that sound in conjunction with appetizing food.[17] If the conditioning is effective, the dog will respond to the bell with basically the same behavior (salivation) as previously elicited by presenting the food.

Similarly, when a man sees an ad picturing a particular car in the presence of a receptive-looking woman, the response assumed to be elicited by the woman will become associated with the car as well. The reaction intended by the ad's designers is that the car itself will become a further object of desire. In like fashion, when a young woman sees an ad showing men and women of her age having fun together while drinking a certain brand of beer, her normal desire to be part of a group like that is extended to include the beverage as well.

A more aggressive form of preference management takes over when advertising capital is used to create specific "needs" that are conveniently satisfied by the product in question. A familiar example is the original offering of Coca-Cola, which contained an addictive substance (subsequently removed from the drink, although its name was retained). The natural appetite on which the soft-drink industry is based today is a craving for sweets, which humans share with bears, a few birds, and many insects (sweet substances are a common source of energy). But there is no natural desire for sugar-water of one or another particular taste. Now that soft drinks have been established as part of the human diet,[18] marketing efforts of the suppliers have been devoted to making their flavors more appealing than competing flavors. Here is a clear example of using advertising budgets to create "needs" with which consumers had not previously been burdened.

Another example of recent note is the marketing of video games and of various other ways of spending time with computers. The computer industry has spent billions of dollars convincing people they need computers and computer software for uses they had not dreamed of previously. The so-called electronic age might come to be recognized less as a landmark in the progress of human knowledge than as a testimony to the skills of preference managers.

(c) Neuromarketing

The next breakthrough in preference management may come with an application of neuroscience to marketing, aptly dubbed "neuromarketing." In experiments employing MRI (Magnetic Resonance Imaging) scanners, devices normally used to look for tumors, researchers have discovered specific regions of the brain associated with product preference. By studying correlations between brain responses and marketing stimuli, neuromarketers have been able to tell what aspect of an advertising presentation (e.g., color of box, brand name) consumers find attractive. Their intent is to apply information about brain activity to increase the effectiveness of brand-loyalty and other advertising campaigns.[19]

Although different brain sectors have been involved in different experiments, the key regions appear to be the nucleus accumbens, which activates with the anticipation of pleasure; the insula, associated with unpleasant experiences; and the medial prefrontal cortex, which is implicated in mediating the effects of pleasant and unpleasant stimuli. The general idea seems to be that when the anticipation of pleasure set up by a product display outweighs the associated displeasure (e.g., the monetary costs), the consumer is ready to buy the product. When displeasure dominates, the product is rejected.

One widely reported study at Baylor College of Medicine involved more than five dozen people, who underwent brain scans in a taste comparison between unlabeled Coca-Cola and Pepsi. Half the subjects chose Pepsi, which also registered a stronger response than Coke in the pleasure centers of those who chose it. But when the subjects were told which cola they were drinking as part of the test, about three-fourths chose Coke for its perceived better taste.[20] These results may reasonably be interpreted as showing that the Coca-Cola brand name elicited an anticipation of pleasure beyond that associated with its actual taste, for reasons having to do with a history of successful marketing.

In another study conducted for DaimlerChrysler at Ulm University, Germany, a dozen male subjects were shown pictures of sports cars, sedans, and smaller cars and asked to evaluate them for attractiveness. Not only did all subjects rate the sports cars higher, but the brain areas activated by the pictures of these cars were the same as those activated by other pleasure-inducing stimuli like sex and cocaine. A likely explanation is that these particular subjects had been previously exposed to numerous ads showing sports cars in the presence of sexy models, and that this association had been internalized in the pleasure centers of their brains.

Results like these seem to vindicate certain long-established techniques of advertising by disclosing some of the neuromechanisms that make them effective. Such results also show that successful marketing plays more to the emotions than to principles of rational choice, like those long dear to mainstream economists. Although it is yet unclear whether research of this sort will lead to radically new marketing techniques, it at least enables marketing firms to find out when their ads are arousing interest and to adjust them for maximum consumer response.

Despite its uncertain future, programs for neuromarketing research are currently under way in many major universities,[21] often with support by corporate sponsors. And several marketing agencies now offer neuromarketing services to their clients. Among major firms known to have taken advantage of such techniques are Ford, Daimler-Chrysler, Coca-Cola, Proctor and Gamble, and Motorola.

13.5 How Consumption Contributes to Ecological Destruction

Thus far in this chapter we have seen how economic production is driven by desire for wealth, how growth in production of goods is dependent on increased consumption, and how consumption is stimulated by various kinds of marketing. Earlier in this study we saw that production of economic goods takes in negentropy and puts out entropy in the form of various types of ecological degradation. It follows that increased consumption, by way of enabling increased production, results in greater amounts of entropy being discharged into the biosphere. This entropy is the nub of our environmental crisis.

To make this complex set of interactions more perspicuous, it will be helpful to represent them in the form of a diagram (see figure 13.1).[22]

Figure 13.1. Generation of Wealth in a Market Economy

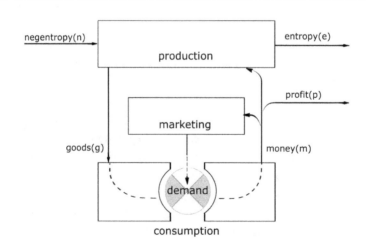

The process of production is represented at the top of the diagram in figure 13.1 as a "black box"[23] with two inputs and two outputs. One input is the negentropy (n) going into production of goods, and one output the entropy (e) resulting from that production. This part of the production process is primarily physical. The other output is the flow of goods (g) to the consumer, the other input the portion of money (m) from the sale of those goods that goes back into the production process (capital investment, etc.). In relevant respects, this part of the overall process is economic.

The configuration representing the consumption process at the bottom is left open in the middle to show a motorlike component providing the "sucking" action (as described in section 13.3) that draws goods from the producer and sends out a flow of money in the other direction. This "motor" represents consumer demand. The flow of money it sets up splits into three branches: (1) money reinvested in production (capital improvement, etc.), (2) money retained as profit, and (3) money diverted into advertising and marketing. This configuration should be viewed as a "black box" as well, leaving many features of the consumption process unrepresented. Among such are sources of money spent on consumer goods (e.g., wages), which may include profits (p) rejoining the money flow.

The marketing sector, also represented by a "black box," responds to the influx of funds with activity designed to stimulate consumer demand. This influence is represented by the line directed to the center of the "motor," signifying the tight coupling between demand and marketing. Among details unrepresented here are various marketing techniques (as described in section 13.4) that might be employed.

These basic components in place, let us extend our reading of the figure by assigning quantitative significance to the three lines n, e, and p at the periphery. A lengthening of p indicates increasing profits, and similarly for lines n and e. With this provision, we can symbolize lesser and greater amounts of these three factors.

Let us now imagine a "behind-the-scene" manager setting the whole system in motion with the intent of maximizing profits. While making sure that reinvestment is adequate to keep the factories running, the invisible manager adjusts expenditures on marketing to achieve maximum profits symbolized by extension of line p.[24] By maximizing p, the skillful manager enables a generous return on investment and secures more income for personal use.

While manipulating distribution of funds among reinvestment, marketing, and profit, the manager will be unconcerned with quantities n and e. Given its role in the system, however, increasing expenditure on marketing causes the "motor" that powers consumption to draw more goods from the production sector. This in turn increases both the quantity of negentropy (n) brought into the system and the quantity of entropy (e) discharged into the environment.

Taking all these interactions into account, we can visualize the effect of the manager's manipulations on factor e. The more profit the manager gets out of the system, the more entropy the system discharges into the environment. In terms of the symbolism of figure 13.1, extending line p brings about an extension of line e as well. In terms of real-world consequences, greater profits generated by a free-market economy result in more damage inflicted upon its supporting ecosystem.

The behind-the-scene manager in this thought experiment personifies the desire for wealth by which economic growth is driven. Motivated by this desire, the invisible manager sets in motion a series of causes resulting in ever more grievous ecological damage. Figure 13.1 thus illustrates how desire for wealth leads to increased consumption, how increased consumption leads to economic growth, and how economic growth leads to further degradation of the biosphere.

13.6 Limiting Marketing's Influence on Consumer Demand

One task set for this chapter was to identify the factors responsible for economic growth and to explain how they operate in a free-market economy. Figure 13.1 responds to this purpose. An ensuing task is to locate a factor that might conceivably be altered as a means of curtailing economic growth. The figure assists in this task as well.

Marketing, with its influence on consumer behavior, stands at the heart of the interactions portrayed in figure 13.1. Its influence is symbolized by the arrow directed downward from the marketing box. Without this connection between marketing and demand, marketing has no influence on other factors depicted in the diagram. Without this connection, the invisible manager (personifying desire for wealth) would have no way of manipulating consumer behavior to maximize profits. Thus disconnected from the causal sequence between consumption and production, desire for wealth lacks capacity to spur economic growth.

One way of curtailing economic growth, accordingly, is to restrict the effect of marketing on consumer demand. At first glance, this may sound like a far-fetched proposal. There are conceivable circumstances, nonetheless, under which restrictions like this might actually be imposed. During the height of Communism in Poland, for example, commercial advertising was entirely prohibited.[25] Advertising has also been restricted for moral (as distinct from political) reasons. Contemporary Albania prohibits advertisements "that are contrary to human dignity and moral principles," such as any that "encourage pornography and violence."[26] One can imagine a society in which commercial advertising is considered "evil"[27] and universally forbidden on ethical grounds.

Although banning marketing for ethical reasons is a conceivable course, the restrictions called for in present circumstances would not be so far-reaching. For our purposes, indeed, some kinds of advertising would remain in place as both needed and beneficial. This is so in particular with the less aggressive forms of product presentation discussed in section 13.4a. What should be restricted are ads aimed at creating new consumer "needs," notably ads under the category of preference management, as described in section 13.4b.

Let us return briefly to these two kinds of marketing, with an eye toward how they influence consumer demand. In its simplest forms, product presentation addresses needs that already exist. Regardless of their state in life, people need food, shelter, and clothing, along with tools and utensils for managing their daily affairs. These may comfortably be described as genuine needs.[28] When suppliers can offer products that meet such needs, presenting these by nonmanipulative advertising is a way of making them known to potential consumers. This kind of advertising thus benefits producers and consumers alike.

In their more pernicious forms, by contrast, techniques of preference management are aimed at creating new appetites for the consumer to satisfy. An obvious example mentioned previously is the highly flavored sweet water known as soda pop. Other examples are "brand-name" clothing and various electronic gadgets that marketers have made into signs of privilege, which people purchase to enhance their self-esteem. Apart from the influence of marketers in shaping our preferences, soda pop and designer clothing count among things that nobody really needs. The primary beneficiaries here are the marketers and the businesses that hire them. What the consumer gains is little more than temporary relief from an artificial "need" that is never completely satisfied.

Even worse, consumption of products designed to gratify artificially induced "needs" often are detrimental to the consumer's health. The corn syrup used to sweeten soda pop, for instance, is a major contributor to obesity, and the fried fat that makes potato chips tasty is a leading cause of clogged-up arteries.[29] Another negative consequence of manipulative marketing is that consumers are left with more things to want and are more subject to discontent when these wants go unsatisfied. All in all, techniques of preference management benefit the corporate producer at the expense of the consumer whose tastes are being manipulated.

Not coincidentally, preference management is the kind of marketing that contributes most substantially to economic growth. Product presentation helps satisfy the real needs of a populace by putting potential consumers in touch with suppliers who can meet those needs. But once this relatively modest set of needs has been met, increased economic growth would hinge on increased population. Growth from this source would not be enough to slake the boundless desire for wealth on the part of investors and corporate managers. The sort of marketing that swells dividends and corporate salaries is the sort that relies on the creation of new consumer "needs." By catering to the desire for corporate profit, preference management not only exploits the consuming public but also leads to higher levels of economic growth and hence to more extensive environmental damage.

Returning to the central box of figure 13.1, we now see that the restrictions on marketing required to curtail economic growth do not apply to all forms of marketing indiscriminately. What should be restricted is the kind of marketing that stimulates consumer demand for products that are not really needed, the kind of marketing that often works to the consumer's detriment. As just noted, this kind of marketing also feeds economic growth, with its resulting devastation of our natural environment.

The obvious question at this point is what anyone can do to help rein in marketing practices of the offending sort. This is not an easy question, and the answer will emerge only gradually through the remaining chapters of this study. By way of anticipation, it may be said that a key part of the answer has to do with making the consuming public *unresponsive* to the kind of marketing in question.

Think of it this way. Consumers have responded favorably to manipulative attempts to establish a market for sweetened water because they attach a positive value to pleasurable experiences. If they didn't value pleasure, they wouldn't buy pleasant-tasting liquid that is bad for their health. Similarly, people buy brand-name clothing because they value the favorable notice they attract by wearing it. If people didn't value being the center of attention, marketers could not sell designer clothing by appealing to the approval it will attract from their peers.

Preference management works only to the extent that the product being pushed offers something the consumer has learned to value. Without values of the relevant sort in place, the potential consumer will not respond to the marketer's stratagems.

As matters stand, economic growth depends on the success of manipulative marketing, which in turn depends on the presence of supporting consumer values. A shift to a less supportive set of values, it now appears, would contribute to the curtailment of economic growth. If this perception proves sound, the path toward a resolution of our environmental crisis may lie in the direction of value change. This line of thought is developed in subsequent chapters.

14

Environmental and Other Ethics

14.1 A Turn toward Values

Philosophers and social scientists (including economists) commonly distinguish value judgments from judgments of fact. It is a factual matter whether something happens or not, apart from any value attached to its happening. Value judgments, on the other hand, assign values to facts and are not limited to facts that actually happen.

For example, it is a fact that the United States attacked Iraq in 2003, quite apart from whether that action is judged to be a good or bad thing. A pacifist might hold, on the other hand, that all acts of aggression are wrong, including the attack in 2003 and any others that might or might not occur. To assess a thing as good or bad, right or wrong, or proper or improper (and the like) is to make a value judgment.

Philosophers and social scientists also debate whether any field of study can be entirely free from value judgments. Hairsplitting aside, most would agree that the first three chapters of this book deal with topics minimally involved with value considerations. Thermodynamics and biology, that is to say, are largely value-neutral.

Chapters 4 and 5, which deal primarily with matters of ecology, are similarly value-neutral. While ecology has ramifications that are heavily value-laden, the discipline itself is a branch of biology and hence relatively free of value considerations. Chapter 6, in turn, is primarily historical. As far as history is concerned, whereas values are involved in the choice of material, the events covered either happened or did not, regardless of any merit attached to their happening.

Concern for values begins to quicken in part 2. Chapter 7 treats the interaction between economies and the environment, a topic some economists evaluate as simply wrong-headed. Chapters 8 and 9 criticize common views about how the environmental crisis should be handled. And criticism of any sort involves evaluative discrimination between right and wrong.

Value considerations become increasingly prominent when we turn to the topic of economic growth. The discussion in chapter 10 is motivated by the fact that most mainstream economists consider growth to be a good thing, which conflicts with the finding in chapter 7 that growth exacerbates environmental damage. Chapter 11 takes up several reasons why mainstream economists think growth is a good thing and subjects these rationales to detailed criticism. And chapter 12 is concerned mainly with ecological economics, a discipline quite open in its treatment of value considerations.

Part 3 is occupied with values from start to finish. Desire for wealth, the central topic of chapter 13, is basically a matter of assigning a high value to possessions and to means of obtaining them. For reasons examined in chapter 13, this value attached to wealth is a primary source of our environmental crisis. Other values contributing to the crisis will be examined in subsequent chapters. The underlying theme to be developed in these remaining chapters is that the crisis cannot be resolved until certain values now operating in industrial society are replaced with others more conducive to environmental health.

Additional resources are needed for this final phase of the study. One reason is that the ecological problems we have been examining bring up issues of an explicitly ethical nature. Moral philosophy is a discipline that specializes in such issues. As we turn to the role of values in our environmental crisis, it is appropriate to consider what help moral philosophy might have to offer.

14.2 Moral Quandaries

At this point in the study we have good reason to believe that the biosphere is losing its ability to support human society, and that human society itself is largely responsible for this impending breakdown. There is nothing novel in this assessment, but lack of novelty does not make it any less disturbing. There seems to be something deeply wrong about a course of human activity that threatens the demise of the very society that pursues it. Human self-destruction of any sort seems morally problematic.

Moral problems of another sort arise with statistics regarding distribution of wealth like those cited in section 11.4. Approximately one-quarter of the world's population accounts for about four-fifths of its total energy consumption. The privileged one-quarter (mostly in industrialized nations) thus use about twelve times more energy per capita than people in poorer countries. Since energy consumption correlates directly with GNP (as described in section 7.2), and since GNP is commonly taken as a measure of a country's overall standard of living, the implication is that people in industrialized countries enjoy standards of living many times higher than those in poorer areas.

The disparity is even greater between particular countries at opposite ends of the scale. The per capita GNP of the United States in 2008, for instance, was more than $47,000, in contrast with $1,000 or less for nineteen poorer countries.[1] Given per capita GNP as a relevant measure, the standard of living in the United States is an astounding sixty times greater than that of Malawi (roughly $800). Within this broad pattern there is a tendency for most of the world's wealth to become concentrated in the hands of a few individuals. At the turn of the twenty-first century, assets of the world's wealthiest two hundred people added up to more than the incomes of the poorest 41 percent combined.[2] In 2006, the richest 2 percent owned more than 50 percent of the world's household wealth, while the poorest 50 percent of the world's total population owned barely 1 percent.[3]

This situation, to put it mildly, appears morally unjust. Our sense of moral injustice is exacerbated by our emerging awareness that the economic practices responsible for making a few people exceedingly wealthy are bringing humanity's ecological support system to the point of imminent collapse. This all adds up to a picture of the world's human population sharply divided into rich and poor as the result of industrial and economic processes that have brought humanity itself to a point of imminent decline. In addition to being unjust, this borders on insanity.

The situation is aggravated by the fact that no nation or group of people has what might be called a "natural right" to more of the earth's resources than any other group. Beyond access based on force, there is no exclusive ownership of the resources we appropriate and convert into wealth. Apart from military power securing its possession, the earth is not ours to buy and sell. A deep moral quandary of our era is what ought to be done when a few people commandeer massive amounts of the earth's nonrenewable resources in support of lifestyles from which the rest are excluded, and in the process cause such damage to the biosphere that human society as we know it becomes endangered. For purpose of discussion, this will be designated moral quandary 1.

The inequities noted above concern the distribution of wealth among currently living people. Moral issues of a different sort arise when the range of people involved is extended to include future generations. It is one thing to question whether it is morally right for rich people to consume far more of the earth's resources than their poorer contemporaries. It is quite another thing to ask whether our great-grandchildren (rich and poor alike) have a just claim to a stock of resources no less extensive than that available to present human enterprise. Is there some kind of moral injustice in our pursuing lives of luxury by using up resources our descendents might need just to stay alive? Is it morally right for us to leave behind toxic wastes (such as spent nuclear fuel) in forms that might well prove lethal to future human beings? These and similar questions constitute moral quandary 2.

Yet another type of moral concern emerges when we begin to think about our current treatment of nonhuman species. Do we have a morally defensible right to wipe out entire species of living organisms as an indirect consequence of our economic profusion? Or do our moral duties include a mandate to treat other living things compassionately, in keeping with the compassion we often feel toward the individual sentient creatures we keep as pets? In considering the consequences of our economic and social practices, does the welfare of other species count along with our own? Such questions are typical of moral quandary 3.

These quandaries provide common ground for our subsequent discussion of alternative ethical theories.

14.3 Ethical Theories

Questions of moral rights and duties fall within an area of moral philosophy known as normative ethics. Other areas of moral philosophy are metaethics, which deals with the meaning of moral discourse and the grounds of moral principles, and descriptive ethics, dealing with moral values in effect within actual societies. Roughly speaking, descriptive ethics tells us about the moral practices of particular societies, metaethics explains (or attempts to explain) why moral principles have normative status, and normative ethics identifies (or attempts to identify) principles possessing that status.

Put another way, the task of normative ethics is generally understood as one of developing and refining rational systems of morality by which people are said to be bound as rational agents. One purpose of such systems is to provide criticism of moral practices in effect within particular societies and, when these are found wanting, to show how they should be corrected. Rational systems of this sort are usually referred to as ethical theories.[4]

By standard accounts, ethical theory traces back to the time of Aristotle, and many varieties have developed during the subsequent twenty-five centuries. Ethical systems that have proved influential in the Western tradition can be classified in various ways, but any useful categorization will distinguish theories of at least four basic types. One type focuses on principles of obligation, such as Immanuel Kant's "categorical imperative" to the effect that one should act on a principle one wills to be universally binding on all rational agents. Another basic type focuses on the consequences of action, as in the case of J. S. Mill's maxim that one should act so as to bring about the most overall utility (happiness, pleasure, satisfaction). These two types are usually classified as "deontological" and "utilitarian" theories respectively.

A different type of ethical theory that has recently become prominent is exemplified by John Rawls's *A Theory of Justice*, focusing on ways of reaching consensus regarding the equitable distribution of goods. Yet another is so-called virtue ethics, which goes back to the ancient Greeks but recently has enjoyed a resurgence of interest. This theory focuses on traits of character that make for worthy membership in a well-ordered society.

Another general classification of ethical theories takes into account special fields of human endeavor to which they apply. Thus we can distinguish social ethics, political ethics, religious ethics, medical ethics, business ethics, and environmental ethics, along with various others. These two classificatory schemes are interactive. Someone interested in political or social issues, for example, might work primarily with theories of justice, while someone concerned with religious ethics might find deontological theories especially relevant. And as we shall see momentarily, there are historical as well as theoretical ties between environmental ethics and utilitarianism in particular.

14.4 Environmental Ethics

Despite theoretical affinities like those just noted, moral problems pertaining to the environment might be pursued in the context of any of the four basic types of ethical theory. Someone with Kantian leanings, for example, might try to come up with a "categorical imperative" to the effect that it is wrong to inflict distress on humans yet unborn. Or someone interested in virtue ethics might try to show that even-handedness is a virtue to be exercised with respect not only to humans but to other sentient creatures as well. And so forth for other types of ethical theory one might be inclined to pursue.

Given the broad range of interests involved in environmental issues, however, some moral philosophers think that environmental ethics requires its own special kind of theory. One example is a set of views known as deep ecology.[5] According to its advocates, deep ecology is distinct from what they call "shallow ecology," which focuses on issues like air pollution and resource depletion. Whereas shallow ecology is concerned primarily with human well-being, and hence is anthropocentric (human-centered), deep ecology stresses the importance of an ecocentric perspective. According to deep ecology, all life on earth is interrelated, and all living things have value on their own apart from their usefulness to human beings.

Deep ecology proposes essential changes both in the way we think of ourselves and in the way we think of the rest of nature. Instead of conceiving ourselves as isolated individuals, we should think of ourselves as integral parts of the ecosystems in which we are rooted. A consequence of this view would be that we stop thinking of the rest of nature as a source of resources to be exploited for human purposes and start thinking of it as a web of life in which we are inextricably connected. Thinking of the biosphere in this way, we would be no less concerned to maintain the health of environment than to preserve our own individual health.

Deep ecologists have been put on the defensive by violent forms of political activism (tree spiking, arson) aimed at protecting natural systems from human destruction. Some critics worry that deep ecology might lead to "ecofascism" in the form of a centralized world government that would suppress individual freedom as a means of preserving the integrity of the biosphere at large. But these fears seem misplaced. To the extent that deep ecology has a political agenda at all, it calls for a decentralization of the excessive power presently exercised in environmental matters by international corporations and heavy industry.

Another movement in moral philosophy inspired directly by ecological issues is known as ecofeminism. In its earliest version at least, feminism was a school of thought devoted to analyzing the oppression of women and to devising strategies for women's liberation. Ecofeminism in particular is premised on the perception that men dominate women in much the manner that humanity dominates nature, which suggests that the best strategy on both fronts is to attack these forms of dominance together. Ecofeminists are divided, however, on the extent to which these two forms of subjugation are causally related. Some maintain that male dominance is the cultural source of our environmental problems, while others think that the two forms of dominance are separate in origin. A result is that some ecofeminists are concerned more with gender than with ecological abuses, while others are involved with environmental issues independently of their bearing on matters of gender.

These are only some of the forms environmental ethics has taken since it emerged as an academic discipline in the 1970s. What all versions have in common is a rejection of an exclusively anthropocentric view of nature and a propensity to think of nonhuman forms of life as having value in themselves.

14.5 The Link between Utilitarianism and Moral Issues of Economic Origin

Each in its own way, the quandaries identified in section 14.2 invite formulation in economic terms. Quandary 1 concerns the inequitable distribution of economic goods. Quandary 2 centers on the diminished availability of resources needed by future generations for their economic well-being. And quandary 3 pertains to the damaging effects on the biosphere of our current economic practices. Given this common bearing, we might expect that utilitarianism, among the various ethical theories reviewed previously, has most to offer toward the resolution of these economically based quandaries. This is so for at least three distinct reasons.

One reason is that J. S. Mill, a founding father (with Jeremy Bentham) of utilitarianism, was also an important contributor to early economic thought. Mill's *Principles of Political Economy,* published in 1848, was the leading textbook in economics for the remainder of that century (see section 10.4).

Another is the close link between cost-benefit analysis as practiced by economists and the so-called hedonic calculus integral to utilitarian theory. While cost-benefit analysis can take many forms in economic decision making, it boils down to a matter of assessing the costs of one good thing (the benefit) in terms of other good things that must be given up to obtain it (the cost). As we shall see presently, a very similar procedure of weighing costs (disutilities) against benefits (utilities) is required to comply with the utilitarian maxim that all action should be aimed at bringing about the greatest overall utility.

A third reason is the prima facie connection between the view, advanced by some utilitarians, that people naturally act to maximize their own pleasures and Adam Smith's famous "invisible hand," by which the market supposedly channels the self-interested behavior of individual participants to serve the best interest of society overall (discussed in section 10.3). The view that people naturally act for their own greatest pleasure was put forward at the beginning of Bentham's *An Introduction to the Principles of Morals and Legislation* (1789). To someone who agrees with Bentham that the overall interest of a community is identical to the sum of its individual interests, Smith's "invisible hand" is nothing more than the summating process by which individual benefits are combined in the generalized form of the community's best interest. This reductionist interpretation of the common interest, found in Bentham's *Principles,* was also held by his friend James Mill, father of John Stuart.

If any ethical theory holds promise of helping us come to grips with the moral quandaries identified above, utilitarianism thus appears—at least initially—to be the most likely candidate.

14.6 Utilitarian Theory

For the rest of this chapter, it will be useful to characterize a given ethical theory under discussion with respect to: (1) a characteristic maxim or moral imperative, (2) its understanding of the good to be promoted by morally correct action, (3) its recommended manner of determining what actions are morally correct, and (4) the range of interests it takes into account in determining morally correct action. Following is a characterization of utilitarianism with respect to these points.

1. The characteristic maxim of utilitarianism is well-known. A standard formulation is that an action is right when it results in the greatest overall utility or least overall disutility—alternatively, when it

results in the greatest ratio of utility over disutility. In this context, the term "utility" means roughly what is beneficial or satisfying. The moral imperative of utilitarianism, accordingly, is that one ought to act so as to bring about the greatest overall benefit or satisfaction.

2. Utilitarians differ on what types of satisfaction ought to be taken into account. Bentham's theory centered on the maximization of pleasure and the minimization of pain, which he thought to be the two feelings primarily responsible for shaping human behavior. While not excluding satisfactions of this more hedonic sort, Mill extended the class of morally relevant benefits to include affections like contentment and aesthetic enjoyment as well. In any case, the good to be promoted by moral action boils down to happiness in the form of satisfying experiences.

3. A distinctive feature of utilitarian theory is the hedonic or utilitarian calculus needed to tally up satisfactions and dissatisfactions resulting from (or likely to result from) a given course of action. In contemplating the relevant consequences of possible future actions, the morally responsible agent would have to arrive at specific ratios (total utility/disutility) by which alternative courses of action could be compared. The agent's moral duty, of course, would be to undertake courses of action yielding the highest available hedonic ratio. And in evaluating the degree to which actions already taken comply with the principle, one would have to compare their hedonic ratios with those of other courses of action that might have been pursued instead.[6]

It is this purported use of the hedonic calculus, as already noted, that anticipates contemporary use of cost-benefit analysis in economic decision making. In lieu of comparing potential costs and benefits in monetary terms, however, utilitarianism theorists would have us compare courses of action in terms of their consequent satisfactions and dissatisfactions.

4. The fourth feature of utilitarianism to be considered is the range of individual interests that count in determining what actions are morally correct—that is, in calculating overall satisfaction and dissatisfaction resulting from a given course of action. In standard utilitarianism, this boils down to the interests of sentient creatures whose affective

states count in applying the hedonic calculus. Early versions of the theory seem to have taken for granted that the class of individuals whose interests count (i.e., who have moral standing) was limited to human beings with whom the agent has political, social, or business dealings (often including slaves).

In more recent utilitarian literature, however, the range of moral standing has been extended by some authors to include other species as well.[7] At one extreme, moral standing might be limited to the individual agent, which would amount to a form of egoistic hedonism. At the other, it might be stretched to include all sentient creatures without regard to species. The task then becomes one of determining which creatures are sentient, a daunting version of the "problem of other minds." (Dolphins, dogs, and cats, almost certainly; sparrows and mice, probably; worms and insects, probably not—but who can be sure where to draw the line?)

14.7 Utilitarianism's Response to the Quandaries

Thus conceived, utilitarianism appears to be especially well-suited as a context for working through the moral quandaries described above. In addressing the question whether it is morally unjust for people in the United States to enjoy a per capita income as much as sixty times greater than people in impoverished countries, for example, one's first step as a utilitarian would be to make sure that the range of morally qualified individuals includes all of the populations involved. The next step would be to prepare figures showing overall ratios of satisfaction over dissatisfaction accompanying various levels of difference in distribution of wealth. In the likely event that a sixtyfold difference does not provide the greatest overall ratio, the upshot would be that this difference in distribution is immoral. The very substantial question remaining would be what action should be taken to achieve the maximum ratio and thus to remedy the injustice.

Similarly, in dealing with the issue whether it is morally acceptable for wealthy people now living to use up resources that might be sorely needed by future generations, the utilitarian's tactic would be first to determine (presumably by theoretical argument) whether future people have the requisite moral standing. If indeed they do, the next step would be to figure out some way of representing them in the hedonic calculus. What corrective action is called for would depend on the results.

In response to moral problems regarding our treatment of other species, in turn, one would first consider what might be done to assure that they have moral standing. Whether or not certain aspects of this treatment (e.g., killing animals for meat) are morally correct would hang on the results of subsequent hedonic calculations.

14.8 Why the Utilitarian Response Is Unhelpful

Tidy as this may appear at first, utilitarianism's response to these quandaries is unsatisfactory in several respects. One is its notorious inability to deal satisfactorily with problems of justice. A stark example is its response to a hypothetical situation such as a small group of people being subjected to lives of unalleviated misery as a condition of extensive benefits enjoyed by the rest of society. By hypothesis, the overall ratio of pleasure over pain resulting from this arrangement exceeds that of any other situation in which the "sacrificial" group are allowed relief from their constant suffering. It thus becomes a moral duty to maintain this arrangement, despite the glaring injustice inflicted on the unfortunate few.

Our current situation of economic disparity among nations admits a comparably sadistic twist. Imagine that hedonic calculations have shown (not unexpectedly) that the current distribution of goods among rich and poor yields a relatively low ratio of overall utility to disutility, because of hardships borne by large populations in underdeveloped countries. Other calculations show that the ratio would peak if disutilities of the poor did not have to be taken into account. Since disutilities of dead people don't figure in calculations determining moral correctness of future actions, peak ratios could be achieved if all poor people were eliminated. The course of action to which more affluent people are morally bound, accordingly, is to aid and abet the (least painful) demise of these impoverished populations. Given this scenario, the developed world would enter a golden age of utilitarian morality without poor people around to depress the ratio.

Another type of problem besets any attempt to apply the hedonic calculus to future generations. By definition, pleasures and pains that might be experienced by people in the future do not yet exist. These experiences have no duration, no degree of intensity, and no other features by which pleasures and pains are supposed to be compared in calculating hedonic ratios. But since the hedonic values of such experiences cannot be assessed, they cannot be taken into account by the hedonic calculus.

It is of course possible to make speculative estimates ("educated guesses") about pleasures and pains future people are likely to experience. Suppose estimates were made assigning specific values to the hedonic experiences of nine billion people (another guess) that might occupy the planet when our great-grandchildren take over. A further problem then arises regarding how many further generations should be taken into account.

As far as anyone knows, human beings in some form will be around indefinitely (even if society as we know it collapses) and will continue to have experiences of pain and pleasure not unlike our own. In calculating the hedonic interest of any given future generation, it would be arbitrary to assume that this is the last generation worthy of moral consideration. A consequence is that, for purposes of hedonic calculations, there is no final generation to be considered and hence no limit to the accumulation of hedonic values on either side of the ratio. Both numerator (representing pleasures) and denominator (representing pains) "go on to infinity," rendering the ratio mathematically meaningless.

Problems like these add up when one tries to factor the interests of other species into the calculus. First the issue has to be settled of which species have moral standing. Then comes the problem of characterizing the (actual) experiences of the qualifying creatures in terms enabling comparison with our own. Having solved this, one would face the further question of how to factor the (currently unfelt) experiences of future animals into account. And then the problem would arise of how many generations of future animals deserve moral consideration. Generations beyond number would seem to qualify, threatening to render the calculation of hedonic ratios useless for moral decision making generally.

As anyone acquainted with moral philosophy knows full well, utilitarians are a resourceful lot, and there is no doubt that responses to problems like these can be and have been devised. Nonetheless, the existence of such problems shows that utilitarianism is less helpful than at first appears in resolving quandaries like the ones under discussion. We turn now to a more general problem in this regard, one affecting not just utilitarianism but other systems of normative theory as well.

14.9 Limitations of Normative Ethics in Solving Practical Moral Problems

A distinction has already been made between descriptive ethics and normative ethical theory. Descriptive ethics deals with moral values in force within actual societies. Given that incest is prohibited among the Trobriand Islanders, for example, this is a fact to be included in a description of the moral values operating in that society.

Normative ethics, by contrast, is concerned primarily with theories enjoining norms and principles to which moral agents ought to be held accountable. In arriving at these injunctions, the theorist typically pays little attention to moral values actually functioning in specific social contexts. In arguing for the maxim enjoining action bringing about the greatest good for the greatest number, for example, utilitarian theorists do not undertake surveys of socially immanent norms and values. Ethical theorists are not involved in descriptive ethics.

Unlike most theories in the physical science, an ethical theory has neither explanatory nor predictive powers. It cannot explain the presence of moral norms found in particular social circumstances, nor anticipate how those norms might change as circumstances vary. Someone seeking to understand why ritual suicide is morally approved in some societies but condemned in others, for example, might look to sociology or anthropology for help but probably not to theoretical ethics.

By the same token, an ethical theory is neither verified nor disconfirmed by empirical data like those reported in descriptive ethics. To the extent that a given theory is reliant on empirical data at all, these data are likely to be drawn from the "moral intuitions" of the theorists involved. An ethical theory might be attacked by bringing to bear intuitively plausible counterexamples it has difficulty accounting for. And it might be defended by showing how seeming counterexamples can be accommodated in a coherent manner. But in arguments for or against the theory, facts recorded by descriptive ethics are largely irrelevant.

In short, normative ethics is not a factually engaged discipline. Normative ethical theories thus are powerless to alter moral norms and values that happen to be operating in particular social contexts. To be sure, advocates of this or that ethical theory might be rhetorically effective in persuading a segment of the general public to take their views seriously. A case in point is the growing influence of the animal rights movement. But any change in social values will be due to the rhetoric, rather than to any inherent power of the theory itself to legislate values that are socially in place.

A consequence of this lack of factual engagement is that ethical theory has little to offer by way of resolving moral quandaries like those laid out previously. Inequities in distribution of wealth might be alleviated if there were effective values operating in the business community prohibiting profit made to the detriment of others. And hardships caused to future generations might be reduced if people in industrialized societies valued the interests of their more distant descendents no less than those of their children and grandchildren. Whatever the moral anomaly in question, it seems safe to say that one could specify values and norms of action that would mitigate the anomaly if they were effectively in place. The point is that ethical theory in itself cannot provide norms and values with the standing necessary to be practically effective.

To drive the point home, let us suppose that the quandaries under discussion would be resolved if there were a moral principle in force prohibiting anyone from consuming more natural resources than needed to lead a moderately comfortable life. Let us suppose further that all major ethical theories had converged on this principle, with the result that the entire profession of theoretical ethics is ready to affirm that it is morally wrong to consume more resources than needed for a comfortable existence. As far as moral norms actually in place are concerned, however, nothing changes when the ethical theorists arrive at their agreement. There was no such principle in effect before this result, and nothing changes in this regard when the agreement is reached. Ethical theory has no competence to bring moral values and norms of action into practical effect.

When our imaginary theorists proclaim that it is morally wrong to consume more resources than necessary, this pronouncement lacks descriptive force. It is not a description of what is actually the case. A less misleading way of putting the pronouncement would be to the effect that the moral principle in question *ought* to be in effect. Theoretical deliberations ("moral intuitions," counterexamples, and all) have indicated that the principle would have good results if actually in force. The pronouncement couched in terms of what is morally right or wrong is basically a recommendation about what *should* or *should not* be the case.

Along with competing ethical theories, utilitarianism offers advice about moral principles (e.g., the utilitarian maxim) and procedures (e.g., the hedonic calculus) its advocates think *should* be operative in actual society. Given the amount of time and rational effort these thinkers have devoted (and continue to devote) to such issues, their advice certainly should be taken seriously. In view of the general disinterest among ethical theorists in the social values that have led to our moral quandaries in the first place, however, it is unclear how helpful this advice might be in resolving these anomalies. More useful advice might be forthcoming from a moral perspective more directly in touch with the actual moral values involved. The rest of the chapter is occupied with such a perspective.

14.10 Leopold's Land Ethic

Although its author was not a professional moral philosopher, Aldo Leopold's *A Sand County Almanac* is probably the most-cited work of the twentieth century in the burgeoning field of environmental ethics. Of particular relevance is a relatively brief section entitled "The Land

Ethic." As befits Leopold's profession of wildlife conservation, the section is concerned with the proper relation between the human race and other species in the biosphere.

The characteristic maxim of the Land Ethic is succinctly stated in the following much-quoted passage: "A thing is right when it tends to preserve the integrity, stability, and beauty of the biotic community. It is wrong when it tends otherwise."[8] Each phrase in this maxim calls for careful consideration. In the process, we will have occasion to discuss other features of the Land Ethic, continuing in the format adopted for utilitarianism above: specifically, (2) its understanding of the good promoted by morally correct action, (3) its recipe for determining what actions are morally right in particular circumstances, and (4) the range of interests it considers relevant in making this determination.

The phrase "tends to preserve" stands at the center of Leopold's maxim. Whether a thing is right or wrong depends on the consequences it tends to produce. To qualify as right, a thing does not have to produce the relevant consequences invariably; it is enough if it *tends* to produce them—that is, produce them on most relevant occasions. In this respect, the Land Ethic is similar to rule-utilitarianism,[9] which prescribes acting in ways that as a rule produce the most useful consequences.

A departure from utilitarianism is that right action, according to the Land Ethic, is aimed not at bringing about a desired state of affairs (such as the greatest good for the greatest number) but rather at *preserving* something already present. The sense is that the biotic community is able to maintain its "integrity, stability, and beauty" quite well when left to itself and that our duty as humans is to interfere with that process as little as possible. (Leopold, as already noted, was a wildlife conservationist, and conservation is a matter of leaving things as they are.)

Of particular interest is the state we are exhorted to preserve. Integrity, stability, and beauty enter into the maxim as three conditions that stand or fall together. There is no hint that right action might favor stability over beauty on one occasion and beauty over integrity on another. The three rather are aspects of a healthy land community that are mutually enhancing. What is right, fundamentally, is what keeps the land healthy.

According to the Land Ethic, in brief, the good promoted by morally right action is ecological health, as manifested in the "integrity, stability, and beauty" of the biotic community at large. In contrast with utilitarianism, the consequences that render a thing right or wrong have little to do with the satisfaction of individual people. Like deep ecology, the Land Ethic is nonanthropocentric. While individual humans might tend to find some amount of satisfaction in their roles within the biotic community, such satisfaction is not a goal with which the Land Ethic is primarily concerned.

This brings us to the topic of how morally correct action is to be determined on particular occasions. Instead of a comparative tally of competing utilities and disutilities, the Land Ethic's recipe for discerning what is morally right involves the perceptual skills of ecology. As Leopold says explicitly, ecological awareness "does not necessarily originate in [academic] courses bearing ecological labels; it is quite as likely to be labeled geography, botany, agronomy, history, or economics." However labeled, ecology for Leopold is "the perception of the natural processes by which the land and the living things upon it . . . maintain their existence." At its best, ecological science works "a change in the mental eye," and it is the mental eye thus transformed that discerns what is right and wrong in particular circumstances.[10]

A further contrast with utilitarianism concerns the range of interests relevant in determining right action. Whereas the utilitarian maxim prescribes the greatest overall total of individual satisfactions, the maxim of the Land Ethic is aimed at the well-being of a collective entity. Thus while the utilitarian theorist must take a stand on the range of individuals whose interests count in hedonic calculations, the range of interests relevant to the Land Ethic is built into it from the beginning. In determining what is morally right, we are called on to protect the interests of the land community as a whole.

We should note finally that Leopold's conception of moral correctness applies not just to human actions or rules directing them, as does that of utilitarianism, but rather to anything that might affect the health of the biosphere overall. A thing is right, the maxim says, when it tends to maintain and to support land health, and this might apply to legal systems, social institutions, and corporate policies, as well as to individual actions and practices. It might even apply to actions withheld, as when one fails to take steps that ought to be taken to preserve the health of the biosphere that humanity shares with many other species.

14.11 The Land Ethic's Response to the Moral Quandaries

By way of reminder, the three moral quandaries described at the beginning of the chapter are: (1) whether it is morally acceptable for a few people to maintain lifestyles unavailable to the rest by taking over a lion's share of the earth's resources at enormous cost to the biosphere, (2) whether we are morally accountable for the wasteful use of resources that might be vital to future generations, and (3) whether our extensive destruction of other species can be morally justified. We turn now to consider what the Land Ethic can contribute to our attempt to think through these quandaries.

One thing to note immediately is that this ethic has nothing to say explicitly about the justice (or lack thereof) involved in the enormous disparities of wealth among economic systems. Not only is it silent on matters of social justice generally, but it is similarly uninformative about moral relations among individual persons. To Leopold, such observations would not count as criticisms. The Land Ethic he called for is not a comprehensive ethical theory. Put in his words, it should be conceived instead as "a mode of guidance" for dealing with problematic ecological situations, to be added to moral guidelines already in place dealing with "the relation between individuals" (the Ten Commandments are mentioned as an example) and "the relation between the individual and society" (e.g., the principles of democracy). The Land Ethic was to be an extension of these previous guidelines "to land and to the animals and plants which grow upon it."[11]

Despite its lack of an explicit response to quandary 1, however, the Land Ethic's verdict on massive economic disparity is loud and clear. Without systematic exploitation of the biosphere for human gain, large-scale economic inequities of this sort would be unlikely to occur. Judged by the Land Ethic, this exploitation is a basic moral transgression itself, in view of the ecological damage that inevitably results. By extension, economic and social inequities resulting from such exploitation are morally blameworthy in turn and should be eliminated along with their causes. If the exploitative practices of large industrial powers were curbed, as the Land Ethic prescribes, economic disparities among nations would tend to disappear as well.

The Land Ethic's response to the other quandaries is more direct. As far as future generations of humans are concerned, we can be quite certain that they would benefit from a healthy biosphere and would find living in an unhealthy biosphere harmful. Thus our general duty to preserve the health of the biosphere is in effect a duty to act in ways that serve the interests of future generations. In the context of the Land Ethic, we are morally accountable for profligate consumption of resources other people might need, either now or in the future, because profligate use of resources is morally wrong given the ecological damage it usually produces.

As far as our treatment of other species is concerned, we are morally obligated under the Land Ethic to deal with other members of the biotic community in ways conducive to its overall health. According to recent estimates (see section 5.9), human industry is responsible for an ongoing annihilation of biological species at thousands of times greater than the normal background rate. This massive destruction of species is a major factor in the dangerous erosion of ecological stability now under way. From the standpoint of the Land Ethic, the practices responsible for this destruction are morally unjustifiable and should be eliminated.

14.12 The Practical Significance of This Response

As already noted, Leopold's Land Ethic is not a comprehensive ethical theory, on a par with Kant's rationalistic deontology or the utilitarianism of Bentham or Mill. If it were, it would remain largely disengaged from our current environmental predicament. It would provide little insight into how this predicament arose, and it would have little advice to offer regarding appropriate responses. This is because our current environmental plight was brought about in large part by values currently operating in industrial societies, a matter with which ethical theory is not typically concerned.

Moral philosophers relegate the study of values operating in actual societies to the field of descriptive ethics, which they are usually content to leave to anthropology and sociology. Leopold's Land Ethic fits in somewhere between normative ethics and descriptive ethics. It is not normative because it does not constitute an ethical theory. Its orientation is empirical rather than deductively systematic. And it is not descriptive because the values it is concerned with are not typical of current societies. Whereas descriptive ethics is occupied with values playing active roles in actual societies, the Land Ethic is concerned with values that should be playing active roles but at present are not.

The Land Ethic, in short, is an assortment of ecologically oriented values that are being recommended as replacements of various environmentally damaging values currently operating in industrial society. Reasons recommending this replacement are based upon an empirical understanding of how ecosystems function rather than on abstract principles of ethical theory.

In a memorable sentence at the end of his book, Leopold characterizes human history as a series "of successive excursions from a single starting-point, to which man returns again and again to organize yet another search for a durable scale of values."[12] The "starting point" to which he refers is a stage in human affairs (unfortunately recurrent) when humankind begins haltingly to put together a viable civilization. As scholars of social history know all too well, many such attempts have foundered because of environmental degradation (e.g., the Inca civilization and that of the Easter Islanders). The "durable scale of values" to which Leopold refers curtails human actions that are environmentally destructive. The values advocated in his Land Ethic, if implemented, might be reasonably expected to have that effect.

What would it be for values of this sort to be socially implemented? And how might such values work to protect the environment? These questions are addressed in the following chapters, taking Leopold's work as a model.

15

Typology of Social Values

15.1 Social Values at the Root of the Crisis

According to UCLA's annual survey of first-year college students, approximately three-quarters of students canvassed in 2006 thought it was "essential or very important" to be well-off financially. Another recent survey by the Pew Research Center found that 80 percent of college-age U.S. citizens see "getting rich" as a top life goal for their generation. Previous sampling by the Gallup Poll in the 1990s found that 80 percent of Americans earning more than $75,000 a year said they would like to be richer.[1] These figures, even if only approximate, indicate that U.S. society places a high value on wealth. Put otherwise, they indicate that desire for wealth is a dominant motivation among the U.S. populace.

This ties in directly with the central thesis of chapter 13, to the effect that desire for wealth is a major contributor to our environmental crisis. In very brief summary, the argument leading to this thesis is that desire for wealth drives economic production (see section 13.2), that economic production is closely linked to ecological damage (see section 7.1), and hence that desire for wealth is largely responsible for the worsening condition of the biosphere. The complex interactions between the ecological and economic factors involved are laid out more fully in section 13.6.

In another news item from the Associated Press, it was noted that two-thirds of today's U.S. college students ranked high on a scale measuring self-esteem, more than double the number of twenty years ago.[2] This report also mentions the UCLA results above, suggesting that possession of wealth is a source of self-esteem. As proposed by Veblen in *The Theory of the Leisure Class* (chapters 1 and 3), one reason people desire wealth is because its possession enhances public esteem of the possessor. Possessing wealth not only makes one feel good about oneself but also attracts admiration from others.

Most people desirous of wealth presumably understand that they will never become billionaires. Billionaires result when people already wealthy desire to become wealthier. For most of us, however, great wealth is not essential to self-esteem. Most of us are content with our lot if we have adequate income to satisfy our desires as consumers. Most people are content to "keep up with the Joneses" in the quantity of things they bring back from the mall.

Generally speaking, for people of mid-range incomes, self-esteem comes with being able to acquire the things one wants. Placing high value on acquisition is a defining characteristic of consumer society. Valuing acquisition, people keep going back to the mall and coming home with more purchases to impress their neighbors (Veblen's "conspicuous consumption").

The value attached to acquisition thus joins that attached to wealth as factors enabling the operation of free-market economies. Valuing wealth, corporate managers and stockholders are motivated to seek increasing profit. This is likely to involve marketing initiatives to stimulate consumption of the products their corporations sell (see figure 13.1).

For potential consumers to respond to this stimulation, however, they must be antecedently inclined to spend their money on purchases (rather than saving it or giving it to charity). This is where the consumer value of acquisition comes in. Specialized marketing aside, what makes consumers ready to buy new products is the value they see in coming to possess them. Only when acquisition is established as a consumer value will people be ready to submit to the blandishments of targeted marketing.

For wealth or acquisition to be established values in a given society is for them actually to be valued by that social group. Their status in that case is not merely theoretical, like the values (e.g., universal justice) recommended by normative ethics. Values established in actual societies influence the behavior of actual people. Their status is such that they have substantial effect in the societies concerned. In what follows, we will refer to them as *operative* values.

Section 15.3 below attempts to clarify what it means for a value to be operative in a given society. The chapter then proceeds to a classification of different sorts of social value, with attention to how different values interact. This will be useful in subsequent chapters, dealing with the replacement of one set of operative values with another.

Before any of this can be done, however, we need to be clear about the nature of the values in question.

15.2 The Sense of the Term "Value" in the Present Context

The term "value" has various usages. Mathematicians and logicians speak of values taken by formal variables ("let x be the square root of 2"). Numismatists speak of the monetary values of ancient coins, card players speak of the relative values of the several suits (hearts rank higher than clubs), and economists speak of the market value of this or that commodity. Our discussion of values in this chapter is not concerned with any of these usages.

The meaning of the noun "value" in the present context is tied to a particular sense of the verb "to value." This is the sense it shares with "to esteem" and "to consider worthy." A social value is something valued within the society in question, meaning something thought of favorably or viewed in a positive light.

Let us contrast this sense of the noun with the other senses mentioned above. When logicians speak of the value taken by a variable, they are not talking about something they hold in esteem. When economists speak of the market value of a particular commodity, they are not talking about something considered especially worthy. And so on for the face value of an ancient coin and for the relative ranking of hearts and clubs. There is nothing in these other senses suggesting overt approval, whereas approval is essential to the sense now being explained.

A social value is something viewed in a favorable light within the society in which it operates. But what sort of thing becomes a social value when favorably viewed? It will be helpful to have a few examples at hand in dealing with this question.

One familiar example is the value of honesty. When the value of honesty is operative in a given social group, its members will tend to think of honesty in a favorable light. This means that they will generally approve of honest behavior in others and be guided by standards of honesty in their own behavior. What is viewed favorably is not the behavior as such, which might turn out to have unfavorable consequences. What is valued is the quality of honesty itself. Honesty is valued in this group wherever found in action, speech, or goal intended.

Another example is the value of moderation. Members of a society in which moderation is an operative value will generally look favorably on moderate behavior. They will encourage such behavior on the part of other people and will consider moderation a worthy standard in their own activities. The practice of moderation is looked upon as a commendable habit. What is valued is the exercise of moderation itself, as distinct from various forms of activity in which it might be evident.

Not all social values are morally edifying. A social value is constituted by the approval of the society in question, even though other groups might view it with disapproval. The value of vengeance provides an illustration. Most of us have probably heard of societies in which harm inflicted on one family or clan by members of another calls for retribution against the offending group. Vengeance in such a society is looked on favorably. It is considered worthy to harm other people when vengeance is the motive. Violent action against the offending group will be met with approval when it is viewed as satisfying a need for revenge. What society values in this case, however, is vengeance itself and not the violence committed in bringing it about.

In the first example, what is valued is the quality of honesty in its role as a standard of individual and group behavior. What is valued in the second example is the practice of moderation, which amounts to general esteem for a particular mode of conduct. In the third case the value illustrated is a motive for action, serving to authorize action taken under that motivation.

The definition of "social value" can be taken a step forward with these examples at hand. A social value is a guide of behavior that is (or might be) generally approved in one or another particular society. Such guides might take the form of standards, of practices, or of motives for action. Presumably they might take other forms as well. What is essential to a social value, at any rate, is that it is generally esteemed (is valued) within the group concerned and that it influences behavior within that group.

One manner of influence is that values of this sort contribute to the shaping of individual *preferences*. In a society valuing moderation, individual members typically will prefer temperate over dissolute modes of behavior. Such values also can establish *priorities*. When the value of vengeance is operative in a given group, members of the group might tend to rank honor over forbearance and forbearance in turn over acquiescence. Another role of social values is to serve as *criteria* for judging social acceptability within the group. In groups where honesty is an operative value, people will tend to judge dishonesty as socially unacceptable.

An attempt is made in subsequent sections of this chapter to characterize different roles values like these might play in one or another social context. By way of preparation, we turn next to a discussion of what it means for a value to be operative in a given society.

15.3 Operative Social Values

Vengeance serves as an operative value in some societies but not in others. The same may be said for the values of honesty and moderation. Let us attempt to identify the conditions that must be met for a given value to have operative status in a given social context.

As a beginning, we should note that a value's operative status is not directly tied in with character traits of the society's individual members. Take honesty again as an illustration. Personal honesty is a character trait, as is its opposite, personal dishonesty. Most societies will have a mix of honest and dishonest members. But for honesty to be operative in a given society is not a simple matter of a certain proportion of its members being personally honest.

To be sure, it is entirely possible that the value of honesty might have operative force even in societies where a majority of members are personally dishonest. Honesty might be publicly esteemed in a society because of social tradition or the influence of a conservative ruling class, while people in private go about their dishonest ways. This would be analogous to a society's endorsing marital fidelity in its public discourse while many individuals practice adultery in their private lives.

This is not to say that a society's operative values have *nothing* to do with individual character traits. A group of honest people is more likely than not to attach a high value to honesty on a social basis. And a society in which honesty is an operative value is more likely than not to be comprised of a majority of honest members. Such correlations notwithstanding, we should avoid any attempt to define operative social values in terms of personal values maintained by individual members. The values of a society are more than summations of values held by its membership.

Personal values are features of individuals, whereas operative values are characteristics of actual societies. Attempts to define the latter in terms of the former would be like trying to define a country's foreign policy in terms of the attitudes toward foreigners of its individual citizens. For present purposes, at least, let us proceed with a definition of operative social values that can be applied without reference to individual preferences.

We have already seen that a social value is something generally esteemed in the society concerned and that it serves in that society as a guide of behavior. At the very least, a given value can be socially operative only if it meets these two basic requirements.

But more needs to be said about what meeting those requirements amounts to. For a value to be socially operative is for it to exercise a certain force. In many cases, the operative status of a value is exhibited in the way it influences interactions among people in social contexts. Let us continue with honesty as an example in attempting to characterize this influence.

When honesty is operative as a social value, people generally will view honesty with approval wherever they encounter it. They will appreciate manifestations of honesty in the behavior of other people. And other things being equal, they will tend to maintain standards of honesty in their own behavior. One consequence of this general esteem for honesty is that people will tend to deal honestly with each other as a matter of course, unless under specific pressure to behave otherwise in particular circumstances.

Another consequence is that people will become accustomed to dealing honestly with other people. And just as they tend to treat others in a forthright manner, so they tend to expect honest treatment by others as their due return. While there will be exceptions to these tendencies, of course, social interactions within the group generally will be characterized by integrity and mutual trust.

A further mark of an operative social value concerns its role in the explanation of interpersonal behavior. As just noted, when the value of honesty is operative in a given social context, honest behavior is expected as a matter of course. Because of this, manifestations of honesty for the most part do not call for explanation. Dishonesty, on the other hand, is considered deviant and typically calls for explanation in egregious cases.

Besides being *self*-explanatory in this manner, when the value of honesty has force in a given social context, it can contribute to the explanation of *other* behavior that is initially puzzling. For example, if a hurried customer at an automatic teller receives an extra $20 bill and takes several minutes to notify a bank employee of the error, an intelligible explanation to an impatient companion might be "It was the only honest thing to do." If the customer were questioned about driving away with the extra money, on the other hand, "It was the dishonest thing to do" would not count as a plausible explanation.

The operative force of honesty shows up in its influence on interpersonal relations. There are other social values whose influence is exercised primarily in the behavior of persons acting individually. One such is the value of comfort, which will figure prominently in the discussion of subsequent chapters. Let us consider the influence of comfort as an operative social value.

As a matter of fact, the value of comfort is prominently operative in many industrialized societies. One sign of this is that people generally seek comfort in their daily circumstances. We generally dress in comfortable clothing, outfit our homes with comfortable furniture, and expect our work spaces to be maintained at comfortable temperatures. Lack of comfort, when we encounter it, is a condition to be remedied.

Given our own personal preference for comfort, moreover, we expect to find the same preference in other people. To facilitate sale of their products, furniture manufacturers try to design chairs and sofas their customers will find comfortable. To keep their employees from complaining, office managers try to keep their work areas at pleasant temperatures. And marketers of consumer products ranging from automobiles to bed covers realize that emphasis on comfort usually makes a good sales pitch.

The value of comfort also plays an explanatory role much like that of honesty examined previously. In contexts where comfort is an operative value, people engage in activities conducive to comfort as a matter of course. As far as the resulting comfort is concerned, accordingly, such activities do not call for explanation. If something were said to be done for the sake of discomfort, on the other hand, further explanation would be required to make it intelligible (e.g., masochism, penance, comic effect).

Moreover, comfort can serve in the explanation of behavior that is puzzling for other reasons. In the case of someone turning restlessly on an unfamiliar bed, for instance, the explanation might be that he or she is looking for a comfortable position. By contrast, it is hard to think of types of behavior for which "because it is uncomfortable" would serve as a plausible explanation.

A value is operative in a given social context if it exercises the kind of influence illustrated by honesty and comfort in these examples. Let us draw on the features shared by these two illustrations to arrive at a general characterization of a value's operative status.

For a value to be operative in a given social group requires (1) that it be held in general esteem by members of the group in question. Esteem of the requisite sort is shown by (a) a tendency on the part of its members to conduct their own affairs in accord with that value and (b) a tendency to rely on other people in the group to do the same.

For a value to have operative status requires further (2) that the value typically plays an explanatory role, both (a) insofar as action taken in accord with the value does not thereby call for explanation and (b) insofar as being in accord with the value can serve in an explanation of other puzzling behavior. When the value in question admits a direct contrary (as dishonesty to honesty and discomfort to comfort), moreover, the contrary lacks this explanatory ability.

Requirements 1 and 2 apply to social values generally. For a social value of any sort to have operative status requires meeting these two conditions. Other conditions come into play in distinguishing social values of different types. Our purpose calls for distinctions among approbatory, commendatory, and normative values. The remainder of the chapter is given over to these distinctions.

15.4 Approbatory Values

A social value encourages one or another form of behavior within a particular social context. Social values might operate (among other forms) as standards, practices, and motives for action (as described in section 15.2). The distinctions we shall be drawing among approbatory, commendatory, and normative values have to do with differences among such forms. The first two pertain to motives for action. We begin with approbatory values.

There are various impulses to action that generally are considered to be "hard-wired" or innate. Such motivating factors are established genetically and not by socialization. Among such impulses are the natural need for food and drink, the instinct for self-preservation, and the desire for sexual pleasure. To say that these impulses are not determined by socialization is to say that they hold more or less constant across all societies.

Although such impulses are present in all societies, there none-theless are vast differences in the kinds of behavior they motivate. For example, consider the wide variation in response to the need for food and drink, ranging from the gluttony typical of an ancient Roman banquet to the ritual fasting of an abstemious religious community. Other examples can be found in the ways different societies regulate the natural urge for procreation and in different mores regarding the choice of sexual partners. Whereas the underlying needs and desires are culturally invariant, the ways in which different groups respond is a matter of social conditioning.

Social conditioning of responses to inborn urges takes place largely through the formation of social values.[3] A major role is played by values of an approbatory sort. The primary effect of such values is to sanction certain types of response to natural promptings by implicit social approval.

Consider the approbatory value of pleasure as an example. Pleasant experiences are typically induced by eating and drinking, by sexual activity, by relief from discomfort (e.g., heat and cold), and so forth. When a particular activity is undertaken for the pleasure it provides, we say that pleasure is the motive of that activity. Gaining pleasure is the motivation or purpose of the behavior in question.

When operative, the social value of pleasure constitutes tacit approval of pleasure as a motive for action. By being operative, the value of pleasure conveys social endorsement of pleasure as an acceptable goal to be pursued. Entering into an activity because of the pleasure it provides is generally viewed in a favorable light. While things unavoidably may go wrong in pursuit of that goal, society generally approves of the goal itself.

As characterized above, a social value has operative status in a given group when (1) the value is esteemed by members of the group generally and (2) acting in accord with that value is sufficiently common not to call for explanation. In the particular example at hand, for pleasure to be operative as a social value is, first and foremost, for members of the group to view gratification of natural desires in a positive light. Individuals tend to seek gratification in their own activities and expect other people to do the same as a matter of course. When pleasure is an operative social value, moreover, gratification of natural desires generally does not require explanation itself, and otherwise puzzling behavior becomes intelligible with the realization that it was undertaken in pursuit of pleasure.

What makes pleasure an approbatory value is the nature of the warrant it provides. Insofar as the value is operative, it indicates general social approval of activity undertaken for the pleasure that follows. Its operative status has the effect of sanctioning such activity in the sense of tolerating or condoning it. Seeking pleasure is approved in the sense of being socially condoned, at least to the extent that countervailing values (like moderation) do not come into play.

But the approval in question does not extend to a *recommendation*. Nor does the operative force of pleasure as an approbatory value amount to a general *mandate* for pleasure-seeking activity. Recommendations and mandates are the business of commendatory and normative values, respectively. The approval built into pleasure as an approbatory value does little more than render pleasure seeking an acceptable activity, which is to say one worthy of being pursued, other things being equal.

When pleasure is in force as an approbatory value, members of the group concerned will tend to pursue pleasurable activity when an opportunity presents itself. Unless conflicting values come into play, they will do so expecting the forbearance of other people. Moreover, they will feel free to seek out occasions for pleasurable activity and even to devise opportunities on their own initiative. A society characterized by pleasure as an operative value is one in which pleasure seeking is generally tolerated.

Let us look at another value that often has approbatory force. According to Diogenes Laertius in his *Lives of Eminent Philosophers*, Epicurus defined pleasure as freedom from pain. Most theorists of bodily sensations today, however, would probably agree with Adam Smith in his criticism of Epicurus, holding that pleasure and absence of pain are qualitatively different.[4] At any rate, the instinct of pain avoidance seems distinct from the desire for bodily pleasure. Pain avoidance is another approbatory value, perhaps even more widely operative than pleasure in human society.

The value of pain avoidance is approbatory in that it sanctions evading pain as a motive for purposeful action. We wear shoes when walking on gravel because of the pain of going barefoot. And we choose anesthesia when having a tooth extracted because of the pain that would be experienced otherwise. On those rare occasions of deliberately submitting to pain (e.g., the prick of the anesthesia needle), most of us do so only to avoid greater pain in the future. Such activities are socially acceptable to the extent that pain avoidance is an operative social value.

Despite the prevalence of this value among human societies generally, however, there are various social contexts in which it is held in abeyance. One such is in athletic contests like boxing and rugby, where avoiding pain is likely to signify flagging commitment. In many cultures, moreover, learning to endure pain seems to be part of "growing up to be a man."[5] Countercases like these indicate that the value of pain avoidance might be operative under some conditions but not under others, and these conditions are subject to socialization.[6]

For purposes of general taxonomy, it should be noted that values both of pain avoidance and of pleasure come in more and less nuanced forms. Whereas some social groups might value pleasure seeking across the board, others might approve of seeking pleasure through eating but not through sexual activity, or vice versa. Similarly, pain avoidance might be approved without qualification in some social contexts, whereas approval is withheld in others under particular circumstances (e.g., athletic contests).

Drawing on these illustrations, we may attempt to formulate a general characterization of approbatory values. The key feature of such values is that they constitute group approval of one or another natural urge or instinct as a motive for behavior. The factor *motivating* the behavior is the urge or instinct itself. What the value in question contributes is a general endorsement of that particular motivation.

An approbatory value is operative to the extent that the motives it endorses are looked on favorably within the social group and as such are viewed as "normal" and not requiring explanation. In this respect, approbatory values share the features of social values generally. A mark of approbatory values specifically is that they give people leeway to act according to the value whenever an occasion arises. Given blanket clearance of this sort, the agent might seek actively to expand the range of circumstances in which the concerned motive can be exercised.

It should be noted at this juncture that lack of operative status on the part of a given approbatory value does not by itself amount to social disapproval. For example, there might be contexts in which pleasure seeking (e.g., eating chocolate) is neither approved nor disapproved by values that have operative status. In such cases, other types of social values would be needed to dissuade people from acting on their natural instincts and urges.

15.5 Commendatory Values

A further dimension of social approval comes into play when society holds up a particular manner of behavior as commendable or praiseworthy. In the case of approbatory values, the approval involved amounts to little more than treating a certain kind of behavior (such as pleasure seeking) as socially acceptable. With commendatory values, on the other hand, the endorsement involved has the force of a positive recommendation. The difference is analogous to that between an automotive magazine giving its "stamp of approval" to several models and singling out just one model it particularly recommends.

In the case of an approbatory value, furthermore, the behavior in question is motivated by a natural desire or instinct. But in the case of commendatory values, the relevant behavior is motivated by social approval itself. The driving force in the first case is an inborn urge, while in the second it is an inclination to do what is advocated by one's social group.

Once again, pleasure-seeking behavior takes forms dictated by the structure of the organism's nervous system. Being "hard-wired" in this fashion, such behavior tends to be culturally invariant. Behavior *motivated* by group approval, on the other hand, varies substantially from culture to culture. The reason is simply that different forms of behavior are encouraged by different societies.

In a straightforward sense of the term, behavior driven by group approval is shaped by social conditioning. There are various ways in which behavior can be affected by social conditioning, including operant conditioning, imitation, and cultural inheritance. Another way is through the influence of commendatory values. Let us see how this works with the help of a few examples.

Moderation was introduced in section 15.2 as an example of social values generally, without mention of any particular type. In fact, it is a typical instance of what we are calling commendatory values. In its commendatory role, moderation reflects society's general estimation of moderate behavior as especially praiseworthy. When the value of moderation is operative in a given social context, people will react favorably to moderate behavior whenever they encounter it.

As with all social values, moderation is operative when it is generally viewed with favor in the relevant social context and when moderate behavior is typically understood as not requiring explanation. A distinctive feature of moderation as a commendatory value is that people under its influence tend to look upon temperate behavior as deserving commendation. Such people tend to agree that such behavior is particularly admirable.

Moreover, the force of moderation as a commendatory value is both retrospective and forward-looking. As well as commending moderation in actions already accomplished, people will generally agree in recommending it as a feature of future activity as well. Save in exceptional circumstances (e.g., responding to imminent danger), moderation is deemed praiseworthy in actions both past and future.

Interestingly enough, moderation shows up (as temperance) in standard lists of both Aristotelian and Christian virtues.[7] And most other virtues on either list would probably correspond to commendatory values that might be operative in specific social contexts. As already noted, however, virtues are character traits of individuals, whereas social values are characteristics of social groups. What it is for a value to be operative in a given society cannot be defined in terms of character traits of its individual members. But this does not preclude consideration, in subsequent chapters, of the influence various virtues like temperance would have on individual behavior if adopted as operative values in a given social context.

It should come as no surprise to realize that not all commendatory values would meet with approval among ethical theorists. An example is the value of comfort, which seems to be operative as a hallmark of consumer society. Consumers look favorably upon the acquisition of commodities (such as air conditioners) that bring comfort to the user, and they consider comfort a major desideratum when relevant in future purchases. Advertisers are able to capitalize on our predilection for comfort by featuring the ease and contentment their products can be expected to bring.

Other commendatory values bound up with consumerism that are ethically suspect include convenience, self-interest, and acquisition, along with the key value of wealth itself. We return to these consumer values in chapter 16, in the course of considering their influence in current industrial societies.

Commendatory values like moderation, convenience, and comfort might be termed "positive," inasmuch as they *encourage* activities in which certain distinctive characteristics are notably present. But there are other values with commendatory force that encourage the *absence* of certain features and hence might be called "negative." Negative values of this sort *discourage* the presence of certain distinctive characteristics. An example would be a value discouraging pursuit of pleasure-seeking activity.

What a society would commend, if such a value were operative, is a general *inoperancy* of the pleasure motive. In effect, for such a value itself to be operative would amount to general endorsement of practices and modes of behavior that reject pleasure as a suitable goal of action. Abstaining from pleasure not uncommonly is advocated as a virtue among world religions[8] and is actively practiced in various religious communities. For a secular example, Victorian England came to be known for its commendation of practices that avoided sensual pleasure. In social contexts like these, pursuit of pleasure is actively discouraged.

These observation can be drawn together in a general characterization of commendatory values. Such values bring social esteem to bear in recommending certain practices or behaviors as particularly desirable. A sign of such values being in force is general agreement among members of the group that the relevant modes of behavior are worthy of encouragement. This agreement shows up both in evaluating past action and in shaping action yet to be undertaken. Motivation for acting in accord with the value comes from a desire for the social approval involved, rather than from an inborn urge for one or another sort of gratification.

Religious practices aside, the social encouragement extended by commendatory values does not amount to a mandate. In most societies where moderation is viewed as praiseworthy, immoderate behavior would be frowned upon but probably not prohibited. In societies commending comfort, for another example, accepting discomfort may raise eyebrows but is not deemed a transgression. Commendatory social values neither dictate nor prohibit. These effects come into play with normative values instead.

15.6 Normative Values

Just as there is an overlap of sorts between commendatory values and the traditional virtues (both encouraging certain practices considered praiseworthy), so normative values share features with the dictates of traditional ethical theory. In one version, Kant's categorical imperative prescribes treating other people not as means but as ends in themselves. Actions complying with that prescription are considered right, whereas noncompliance is considered wrong. Similarly, when the normative value of honesty is operative in a given social context, honest behavior is right and hence required, while dishonesty is wrong and hence forbidden.

The distinction between right and wrong becomes a factor with normative values. Like their commendatory relatives, normative values recommend certain forms of behavior as particularly desirable. What sets them apart is the force of this recommendation. The characteristic feature of normative values is the disjunction they set up between acceptable (right) and unacceptable (wrong) social behavior. For one type of behavior (e.g., honesty) to be prescribed is tantamount to its opposite (dishonesty) being unacceptable. And what is deemed unacceptable according to a given normative value is expressly singled out as requiring correction.

This distinction between right and wrong invites comparison with the difference between positive and negative commendatory values. An example of the positive sort is a value encouraging behavior conducive to comfort. Negative commendatory values are exemplified by a value encouraging the avoidance of uncomfortable behavior. While the two values of course might operate simultaneously, it should be noted that operancy of the former value (recommending comfort) does not require that the latter (recommending avoidance of discomfort) be operative as well. Avoiding discomfort is more than merely seeking comfort. For a society to encourage the pursuit of comfort (e.g., in purchasing consumer products) does not entail its taking a stand on behavior undertaken regardless of the discomfort that results (e.g., exercising).

With normative values, however, prescribing a certain type of behavior has the same force as proscribing its opposite. If society values honesty in a normative fashion, by the same token it values the absence of dishonesty. Right and wrong cover all options, in the same manner as a switch being on or off. If honesty is right, so is avoiding dishonesty, and if dishonesty is wrong, so is departing from honesty.

A normative value establishes a norm of acceptable behavior. In the relevant sense, a norm is a standard that must be met to qualify a performance as adequate for a particular purpose. To earn a medal in an athletic event, by way of analogy, the dominant contender might have to test free of drugs. Failure to meet that test would result in disqualification. Similarly, in order to avoid sanctions engaged by wrong behavior, a person must meet the standards defining right behavior within the relevant social context. Possible sanctions include public censure and penalties imposed by enacted laws.[9]

For a normative value like honesty to be operative in a given social context, accordingly, requires not only that it is deemed praiseworthy and beyond need for explanation. These conditions hold for all operative social values. For a normative value to be operative also requires that it serve as a society-wide norm distinguishing blameworthy from praiseworthy activity. Behavior judged blameworthy by that standard is socially unacceptable and considered to be in need of correction.

By way of generalization, we may say that a value has normative status in a given society if it is looked on as praiseworthy and beyond need of explanation, and if in addition it serves to distinguish socially prescribed from socially proscribed behavior. As a rule, this distinction will be marked by a public readiness to label prescribed and proscribed behavior "right" and "wrong," respectively. Wrong behavior, furthermore, will be met with sanctions tending to discourage repetition.

Other normative values to be discussed in subsequent chapters include equity, justice, and impartiality. Like most other social values, norms of this sort might come with built-in limitations. For instance, although societies conceivably might exist in which honesty is required without exception, the more common case is for it to be mandated under some circumstances but not under others (e.g., under hostile interrogation). Similarly, it is not hard to find cases where justice prevails in mainline society but has little influence along its fringes.

15.7 Interactions among Types of Value

The foregoing typology of social values is motivated by the analysis undertaken in previous chapters of the underlying cause of our environmental crisis. The crisis consists of the biosphere being loaded down with more entropy than it can get rid of. According to our analysis, this overload of entropy results primarily from the increasingly massive amounts of energy consumed by the world's industrialized economies. Increasing amounts of energy are required for economic growth, which is driven in turn by desire for wealth. Inasmuch as wealth is highly valued in most contemporary societies, the growth that it stimulates goes largely unchecked.

Boiled down to bare essentials, this means that contemporary human society at large is being undermined by its own prevailing values. If this line of analysis is even approximately correct, the most feasible way out of the predicament will be to rid society of the particular values that led to the crisis. Whatever else this amounts to, it involves rendering certain prevailing values inoperative and replacing them with others. This of course requires identifying the offending values (wealth is only one among several), a task undertaken in the following chapter. The remaining task of the present chapter is to gain a preliminary understanding of how change in social values might be brought about.

A first step is to observe that the three types of social values distinguished above are hierarchically ordered with respect to force. Approbatory values (such as pleasure) constitute social approval of certain natural instincts and urges as motives for action. Commendatory values (such as moderation) bring social pressure to bear in recommending certain practices and forms of behavior. Normative values (such as honesty), in turn, establish standards distinguishing acceptable from unacceptable activities, prescribing the former and forbidding the latter.

Approbatory values have the force of sanctioning behavior undertaken for certain preexisting motives. Commendatory values have the force of recommending certain forms of behavior and thus of motivating the behavior in question. Since approval goes along with such recommendation, of course, the force of recommendation includes that of approbation. In its effect, the force of recommendation is more extensive than that of mere approval.

Normative values require certain forms of behavior. But doing what is required is obviously recommended. Accordingly, the force of normative values is the most extensive of all. What is required is recommended, and what is recommended is approved. In addition to their own effect of mandating certain behaviors, normative values also have the force of both commending and deeming acceptable the activities in question.

In their negative mode, normative values proscribe behaviors that fall short of the relevant standard. What is forbidden, of course, is neither recommended nor approved. In this mode, accordingly, normative values have the force to countermand both commendatory and approbatory values.

Along with wealth and various others, the value of pleasure plays a major role in today's highly commercialized society. As such, the value of pleasure appears to be one that should be rendered inoperative in the interests of a healthy environment. With an eye on the interactions among types of values laid out above, it is interesting to consider how that might be brought about.

The value of pleasure is operative in consumer society to the extent that doing things for pleasure is generally approved. An obvious way of lessening the force of this value would be to diminish this general approval. Various tactics for accomplishing this are suggested in the final chapter.

Another step in this direction would be to set in place a value with opposing force. One such might be the social value of austerity, in the sense of renouncing bodily pleasures. Since there seems to be no inborn urge for austerity, this moves us into the range of commendatory values. If austerity gained the commendation of society at large, it would go a considerable distance toward rendering the value of pleasure inoperative. An effective combination would be simultaneously to diminish the approbatory force of pleasure and to increase the commendatory force of austerity.

Yet another possibility would be to establish some value like austerity (maybe asceticism) on a normative basis. For example, if asceticism were operative as a normative value, pleasure seeking as its opposite would be automatically proscribed. This not only would divest pleasure seeking of approbatory status but also would enlist a countervailing value. With the establishment of asceticism in a normative role, pleasure seeking would be considered unacceptable and be subject to penalties when it occurred.

This survey of possibilities gives us at least a tentative idea of how environmentally damaging values could be removed from social dominance. The next chapter attempts to single out prime candidates for removal.

16

Ecologically Destructive Values

16.1 Caveats and Procedures

Values entered the discussion in chapter 13, pursuant to the observations that economic production is impelled by consumer demand, that consumer demand is stimulated by marketing, and that corporate managers turn to marketing as a means of gratifying their desire for wealth (see figure 13.1). Acting through this chain of influences, desire for wealth is the driving force behind economic growth, which in turn leads to increasingly more extensive ecological destruction. The value attached to wealth by corporate managers and investors thus is a primary cause of our environmental crisis.

Values of another kind contributing to the crisis are those that make the consuming public responsive to the techniques employed by marketing specialists. These fall under the general category of consumer values. In conjunction with the value assigned to wealth, consumer values bear major responsibility for the damage human industry is inflicting upon the biosphere. As part of the strategy of alleviating this damage by curtailing economic growth, our best hope for social survival may be to replace consumer values now operative in industrialized countries with other values more congenial to environmental health.

In addition to their effect of making the consuming public responsive to advertising, consumer values often are responsible for particular kinds of environmental damage that can be traced back to them specifically. The tactic of the present chapter is to single out a few consumer values that are particularly detrimental and to examine some of their ecological consequences. Consideration of possible replacement values will be left for chapter 17.

Certain cautions are in order as we undertake this project. Since we all are members of societies where consumer values are operative, we shall be looking critically at values in which we might be personally involved. To remain objective, we should do our best to detach the values under consideration from our own feelings of personal involvement.

We need also to be on guard against possible charges that this undertaking is discriminatory or elitist. It is unavoidable that the results of our analysis will reflect unfavorably on the lifestyles of specific groups of people. One such case has already been encountered in chapter 13, where pursuits of the obsessively wealthy are shown to be contradictory to environmental health. Conclusions also will be reached in the present chapter that might be interpreted as contrary to the interests of ordinary people, of the working class, or of honest people just trying to make a living. Insofar as possible, we should refrain from judgments of this sort and do our best to proceed in an unbiased and impersonal manner.

In the case of each value treated, our procedure will be (1) to identify the value in question, (2) to describe its current role in some prominent practice, industry, or social institution, and (3) to look in detail at some of its deleterious environmental effects. The first value treated is that of pleasure.

16.2 Pleasure

Pleasure is designated by a variety of roughly synonymous expressions, including "gratification," "enjoyment," and "satisfaction." Although many different sorts of experience can provide pleasure (the gratification of applause, the enjoyment of a concert, the satisfaction of

meeting a goal), our present concern is with pleasures afforded by ful-
filling natural needs and desires. As a further restriction for ease of
example, we will stick with pleasures associated with food and drink.

Pleasure has already been identified as an *approbatory* value (see
section 15.4). When operative in status, it constitutes social approval
of sensory pleasure as a motive for action, which is to say approval of
pleasure-seeking behavior. The approbatory value of pleasure is operative in a given social group when members tend to look on pleasure
seeking in a favorable light and to take it for granted without further
explanation. A result is that members of the group generally will take
advantage of opportunities for pleasurable activity and will tend to seek
out opportunities under their own initiative.

Contemporary consumer society is premised on the value of pleasure, and more specifically of gratification. The role of pleasure in consumer behavior is well understood by marketing specialists. If a product
can be associated in the public eye with gratifying experiences, it is
almost assured of success in the consumer market.

Consumer gratification is the basis of the junk-food industry and
hence a contributing cause of the damage this industry inflicts upon
the environment. A case in point is the mammoth business enterprise
responsible for bottling and vending the wide array of beverages known
in some locales as soda pop. Soda pop recently has come under attack
by public-health advocates for its contribution to obesity and other
health problems. Leaving these to speak for themselves, our present
concern is limited to its environmental effects.

In 1998, people in the United States consumed roughly one gallon
of soda pop per person per week.[1] This compares with an average national drinking-water intake of slightly more than three gallons per
person per week. Carbonated beverages consumed in 2001 came packaged in some seventy billion cans, twenty-five billion plastic bottles,
and eight hundred million glass bottles. Of these, roughly one-half of
the cans and less than one-third of the bottles were recycled, leaving
over fifty billion containers to find their way into the nation's landfills.

The environmental impact of these discarded containers is not
limited to landfills and roadsides. Containers that are thrown away
have to be replaced. As far as aluminum cans are concerned, replacing

discarded cans with others made from virgin materials required the energy equivalent of over sixteen million barrels of crude oil and produced over three million tons of greenhouse gases. Although wastes involved in replacing containers could be sizably reduced by using bottles requiring deposits, the industry has been fighting laws to that effect for decades. Part of the reason, presumably, is that handling refunds on bottles cuts into vendor profits.

Another ecological hazard bound up with our taste for sweetened water is the many millions of vending machines worldwide that offer soda pop chilled to just the right temperature. An estimated three million soda-pop vending machines in the United States account for slightly less than one-tenth of 1 percent of the total electricity consumed commercially in the country today (i.e., around twelve to thirteen billion kilowatt hours).[2] In developing areas, the percentage could well be higher. A significant proportion of the ecological damage caused by electricity production across the globe thus can be blamed on the soda-pop industry and the consumer value of gratification that supports it.

Among more obvious effects of the junk-food industry are the containers and wrappings that make service at fast-food outlets so quick and easy. Despite recent efforts by certain chains to become more environmentally "friendly," most hamburger containers are still made of nonbiodegradable polystyrene. A small portion of these are recycled, but countless others (how many "billions sold"?) will remain in landfills indefinitely.

Other problems lie with the production of meat that goes into these delicacies. Raising beef cattle is an extremely inefficient means of producing protein for human consumption. It has been estimated that about 40 percent of grain grown worldwide (70 percent in the United States) is fed to livestock, despite the fact that a dozen or so people could be fed with the grain required to provide a beef diet for a single person. As matters stand, over one-half of U.S. cropland is used to feed cattle headed for slaughter.[3] By dominating the use of cropland in this fashion, the fast-food hamburger is a major contributor to land erosion, ecosystem poisoning, and general degradation of the nation's arable land.

An even more distressing aspect of the hamburger's environmental impact is that much of the industry's meat now comes from steers raised in Central and South America. Most of the land involved has been made available by destroying tropical rain forests. One study concluded that for every quarter-pound hamburger sold in fast-food restaurants, roughly 165 pounds of living matter has been destroyed at its place of origin.[4] This includes not only trees in the affected rain forests, but also birds and insects residing in them. Although estimates vary, more than a million acres of rain forest probably have been destroyed to serve the needs of the fast-food industry.[5] Such devastation would not have happened without the help of gratification as an operative consumer value.

16.3 Comfort

When comfort was introduced as a social value in the previous chapter (see section 15.3), it was identified as a value with *commendatory* force. Unlike pleasure, which it superficially resembles, the comfort we are concerned with here is not an approbatory value. The main reason is that comfort-seeking behavior seems not to be motivated by a natural desire. There appears to be no inborn urge for comfort to match our innate desires for food and sexual gratification.

The comfort we are presently concerned with is more specific than a general sense of pleasantness or lack of discomfort. It is a kind of comfort that borders on self-indulgence, often associated with what might be viewed as luxuriousness. For people with ample means, examples would be heated swimming pools, first-class seating on airplanes, and luxury hotel accommodations. Other examples, for people with ordinary resources, are overstuffed furniture, down-filled comforters, and central air-conditioning.

Whereas pleasure-seeking behavior is motivated by natural promptings, people are led to seek comfort primarily by social influences. The reason comfort is a commendatory value is that it provides motivation for behaviors and practices that society has come to consider desirable and worth recommending. Comfort-seeking behavior is motivated by the force of that general recommendation. To the extent that the value is operative, members of the group concerned will generally consider comfort something worth pursuing.

Through most of human history, pursuit of comfort in the relevant sense seems to have been limited to people with extraordinary means (kings, nobility, bishops). Since it was beyond the reach of ordinary people, there was little social incentive to look for comfort in the pursuit of daily routines. To the contrary, conditions we now consider distinctly uncomfortable were endured by our predecessors as a matter of course. Victorians put up with uncomfortable clothing, Shakers got along with uncomfortable furniture, and almost everyone (including castle dwellers) made do with uncomfortable housing. Only within roughly the last two hundred years have societies emerged with sufficient resources for comfort to be established as a general social value.

Once creature comforts of the relevant sort came within reach of ordinary people, however, they soon came to be perceived as generally desirable. Manufacturers were encouraged to make them widely available, and consumers were encouraged to buy them for personal use. People became increasingly aware of what they would miss by not purchasing the commodities in question. Previous luxuries became necessities, and lack of these commodities came to be viewed as a social liability. By the time this had happened, the commendatory value of comfort had achieved operative status.

In many cases, to be sure, products we are encouraged to purchase under the influence of this value (overstuffed furniture, heated automobile seats) may be no more damaging to the environment than less luxurious alternatives. A notable exception comes with contemporary society's increasing dependence on the comforts of air-conditioning. Let us look briefly at the deleterious effects of air-conditioning on the biosphere.

Around the turn of the twenty-first century, the United States consumed more electricity running home air conditioners than the overall national consumption of all but nineteen other countries.[6] At that point in time, 47 percent of dwellings in the United States had central air-conditioning, up from 23 percent twenty years previously. By now, that figure may be assumed to have extended past 50 percent.

The detrimental effects of coolants used in air-conditioning are well known. Although the United States and other industrialized nations have sharply reduced their use of ozone-depleting CFCs as a result of the Montreal Protocol (see section 5.6), use of CFCs in air-conditioning has been increasing in rapidly developing countries like China and India.[7] This makes air-conditioning a major contributor to ruptures in the ozone layer that threaten human health along with the viability of many chlorophyll-bearing organisms.

Other environmentally harmful side effects stem from the generation of electricity required to operate air-conditioning systems. The U.S. Department of Energy explicitly cites air-conditioning as a major factor leading to record high levels of carbon dioxide emission. Air-conditioning thereby is a major cause of global warming. Other environmentally harmful effects include smog, acid rain, and ecologically disruptive power lines.

Despite such hazards, air-conditioning has become a "basic need" in U.S. society. Regardless of income level, most Americans expect air-conditioning when looking for housing. In the southern part of the country, specifically, where people managed to get by for millennia without artificial cooling, more than nine out of ten dwellings now have some kind of air-conditioning. And other countries are beginning to follow suit. As of 2007, for example, over nine-tenths of urban households in China had air conditioners.[8] While China began to phase out CFCs for HCFCs in 2007, many of the units currently operating are older models still using ozone-depleting CFCs as coolants.

This widespread dependence on cooling equipment has multiple causes. Not to be overlooked are factors like improved technology and globalization, which have made the equipment less expensive and more readily available. (China is the world's biggest producer of domestic air conditioners.[9]) No less influential, however, have been changes in social attitudes shaping the way people think about personal comfort. Increased demand is a condition of increased production (recall section 13.4), and the growing prevalence of comfort as an operative social value is probably the main reason that air-conditioning is in growing demand. The prevailing value of comfort, in a manner of speaking, serves to lubricate the "conveyor belt" that moves cooling equipment from producers to consumers.

By reverse token, if the influence of this value were substantially weakened, there would be fewer air conditioners drawing electricity from polluting power plants. In most climates, at least, there are ways of keeping cool that are compatible with a healthy environment. One is to build houses that take advantage of natural circulation. Another is to build in the midst of trees, which cool the air not only by shade but by moisture evaporating from their leaves.

Contrary to what might at first appear to be true, renouncing self-indulgent comfort as a social value is not tantamount to resigning ourselves to discomfort in our daily routines. As far as air-conditioning is concerned specifically, it rather is to fall back on alternative (and often time-tested) ways of coping with hot weather that do not rely on energy-intensive technology.

16.4 Convenience

We live in a society devoted to labor-saving technology. We open cans, mix batter, and sharpen pencils in the least laborious way possible. Most doors in commercial buildings open without pushing, and once inside we avoid stairways when elevators are available. We microwave frozen meals for dinner or else head out to the nearest fast-food restaurant.

Our previous discussion of the fast-food industry focused on the gratifying taste of its products. Whereas trips to hamburger outlets are often motivated by pleasure, however, this should not obscure the fact that fast-food restaurants are eminently convenient as well. In many cases, it is both quicker and easier to buy a ready-made "meal" (hamburger, french fries, carbonated beverage) than to prepare something in one's own kitchen. Fast-food outlets are but one of many kinds of convenience store that have replaced the corner grocery in our commercial landscape.

Convenience emerged as a dominant social value in conjunction with the widespread availability of small electric motors in the late nineteenth century.[10] To attribute such a late date to the emerging value of convenience is not to suggest that people led lives burdened with inconvenience during earlier periods. Even hunter-gatherers presumably favored easier over harder ways of doing things. The point is that social groups had no occasion to develop a general predilection for labor-saving devices until electric motors appeared on the scene. Only then did a market develop catering to that general preference.

Once technology became available to produce labor-saving devices for the mass market, however, convenience took over as a major selling point in consumer society.[11] People bought things with convenience in mind, and by their example they encouraged other people to do the same. This set the stage for marketers to extol their products for the convenience they provide and for manufacturers to vie with one another in the design of convenience-providing products. As part of the whole process, the status of convenience became solidified as an operative social value.

Given that most labor-saving devices operate on electricity, expanding use of such devices causes corresponding increases in the amount of electricity we consume. And their use has been expanding rapidly in most parts of the world. Along with the proliferation of air-conditioning equipment goes an expansion of electricity-operated appliances like clothes washers and dryers, dishwashers, microwave ovens, vacuum cleaners, and food blenders. Most American households are likely to feature two or three dozen of such devices, consuming roughly as much electricity per annum as three or four window air conditioners.[12] It is obvious that labor-saving devices like these are responsible for a significant portion of the environmental damage stemming from the production, delivery, and consumption of electric power.

Among institutions and practices centered around our preference for doing things the easy way, however, the ecologically most damaging by far are those involving the private automobile. Once viewed as a rich man's plaything, the automobile has become part of the very fabric of all save the most underdeveloped of world societies. Although many factors contributed to its almost universal acceptance today, one of the more prominent is the sheer convenience it provides. The automobile has given ordinary people unprecedented mobility in daily travel routines, has enormously increased the range of places they can visit, and has enabled them to travel at almost any time of the day or night they choose. In addition, having access to a car enables a person to take advantage of auxiliary conveniences, such as banking by ATM; buying food, beverages, and medicines at pick-up windows; and returning library books and videotapes through after-hour depositories.

Book-length accounts have been written of the effects of automobiles on the environment.[13] Well over a hundred thousand more cars and passenger vehicles are being added daily to the total of about six hundred million on the planet's highways.[14] A substantial portion of the environmental damage caused by cars occurs before they are sold and put on the road. This includes damage resulting from extracting raw materials used in manufacture, damage as a result of energy consumed in converting these raw materials into finished products, and damage associated with conveying them to dealers. For example, it has been estimated that producing a single automobile requires around a hundred thousand megajoules of energy, generates about thirty tons of wastes, and results in over a billion cubic yards of polluted air.[15]

More frequently noted than figures like these are the amounts of fuel consumed by cars in operation and the pollution resulting from this fuel consumption. In 2002, more than eighty-five billion gallons of gasoline were consumed by transportation in the United States, about 90 percent of which was taken up by automobile travel. Today the United States consumes more oil for transportation than it produces, which makes our reliance on automobiles the primary reason for our country's dependence on foreign oil.

Before the Clean Air Amendments of 1977 regulating pollution from gasoline engines, a typical car discharged over 500 pounds of hydrocarbons, 1,700 pounds of carbon monoxide, and 90 pounds of nitrogen oxide for every ten thousand miles driven. In 1966, some 60 percent of the approximately 146 million tons of air pollutants discharged in the United States was due to motor vehicle traffic. While significant improvements in these conditions resulted from the 1977 amendments, automobiles remain a major source of greenhouse gases and ozone-depleting nitrogen oxide.

Once a car goes out of service, most of its metal and plastic parts can be recycled. But tires and fluids pose special problems of disposal. Although old tires often find other uses (e.g., in playgrounds and road construction), many are simply buried or left in large piles (where they sometimes catch fire). As far as fluids are concerned, every low-voltage automotive battery contains about a pound of sulfuric acid, which, along with oil, antifreeze, and transmission and brake fluids, must be sequestered indefinitely as toxic waste.

Ecological damage stemming directly from the manufacture, operation, and disposal of automobiles traces back directly to our active involvement with such machines. Given the systematic connections between the automotive industry and the rest of the economy,[16] however, there are indirect environmental costs we should be aware of as well. Some of these have to do with infrastructure, such as highways, parking lots, and city thoroughfares. In the United States specifically, there are more than 8.5 million miles of public roadway.[17] These covered a total of about 25,000 square miles (more than half the area of Mississippi).[18] Add to this an estimated 11,000 square miles of parking lots and driveways, and we have an area about the size of Indiana covered with asphalt, concrete, or other hard surfaces. Whereas much of this was once prime farmland, it is now ecologically dead. Apart from humans, the only living things that have anything to do with it are the countless millions of animals destined to become roadkill.

A revealing indication of the systematic interactions between the automotive industry and other sectors of its economy is the fact that, in Canada during the last half of the twentieth century, ownership of automobiles grew at almost exactly the same rate as the country's GDP.[19] In view of the close correlation between a country's productive output and its energy consumption (see section 7.2) and that between energy consumption and environmental degradation (see section 7.4), this suggests a comparably close correlation between environmental degradation and numbers of automobiles in service.

This summary account of the private automobile's transgressions against the environment has been intended to illustrate the adverse ecological effects of convenience as a social value. While convenience surely has been a factor in the general acceptance of the automobile as a means of transportation, however, it would be naive to suppose that other values are not involved as well. People are also drawn to cars for reasons having to do with freedom and autonomy. People with cars at their disposal do not have to rely on other people to take them where they want to go. Another value served by owning a car is that of individuality. In that regard, compare a Hummer with a Prius as a public manifestation of the driver's personality.

Yet another value bound up with car ownership is that of social status. Perhaps more than any possession other than our homes, an automobile can serve as a symbol of affluence and success. When we come to address the topic of conspicuous consumption in the next section, the automobile will continue to be available as an illustration.

16.5 Acquisition

By definition, a social value is something valued by society at large (see section 15.2). A given society might value a specific kind of experience (e.g., pleasure), a certain state or condition (e.g., comfort), or a particular manner of dealing with one's daily affairs (e.g., convenience). A distinctive feature of consumer society is the high value it ascribes to acquisition, which differs from the above in being a particular kind of activity.

Acquiring consumer goods is something people do. As such, it is distinct from the circumstance of possessing goods that have already been acquired. While the possession of goods can be a firmly entrenched social value in its own right, its role in consumer society is not much different from its role in previous ages.[20] Consumer society is premised on the continuing willingness of people to buy new goods, regardless of what they do with those goods subsequently (e.g., keep them, discard them, give them away).

As far as the three types of social value identified in chapter 15 are concerned, acquisition clearly is a commendatory value. It is not normative, because failure to acquire (although sometimes frowned upon—even considered unpatriotic) is seldom treated as categorically wrong. Nor is it approbatory, inasmuch as it does not result from satisfaction of a natural urge or desire that might be rendered socially acceptable by general approval.

Acquiring consumer goods rather is an activity that members of concerned societies urge upon each other, recommending it both by example and by group persuasion. Like other commendatory values, acquisition is a social construct. It is a value that evolved with the formation of consumer society, growing into its present role of abetting the process of consumption by which productive activity leads to corporate profit (see section 13.6).

The value consumer society attaches to acquisition derives from a social phenomenon that, following Thorstein Veblen, is commonly referred to as "conspicuous consumption." By Veblen's account, wealth and influence gain social standing for their possessors only when put on display before society at large. An effective way of making one's wealth evident is to spend lavish amounts on goods and services for which one has no obvious need. Consumption of this sort is conspicuous to the extent that it appears extravagant and unnecessary. Put briefly, conspicuous consumption is ostentatious expenditure on goods and services intended primarily as a display of wealth and social power.

In its traditional form, conspicuous consumption was a prerogative of rulers and noblemen, who would call attention to their status by maintaining large households, giving lavish banquets, and wearing clothes that ordinary people could not afford. With increasing economic productivity following the Industrial Revolution, however, goods and services once available only to the very rich came within reach of less affluent people as well. This had consequences for the wealthy and the not-so-wealthy alike. On one hand, opulent people had to rely on even more extravagant means to display their superior wealth. This means that their consumption habits became more and more wasteful.

As far as the less affluent were concerned, on the other hand, there was incentive to imitate the consumption practices of the privileged few. Included among these practices was the use of unnecessary consumption to establish status within one's social group. Whereas conspicuous expenditure once was confined to wealthy people demonstrating their superiority over the masses, accordingly, it now became a means by which ordinary people could prove their worth to each other. Although unlikely to catch up with the wealthy in quantity and quality of things acquired, the less wealthy came under social pressure to expand the tally of acquisitions they could flaunt before their peers.

Once this mentality became established, lavish expenditure of money (often beyond one's means) became a way of gaining the respect of other people, and accordingly of maintaining self-respect. Social status could be bought, so to speak, by an ongoing expenditure of one's limited resources in the acquisition of a never-ending stream of consumer goods. In the economic setting of consumerism, personal success is less a matter of what one does with possessions already acquired than of having demonstrated the ability to acquire them in the first place.

In both its original form described by Veblen and its current form enshrined in consumerism, conspicuous consumption is fraught with environmental hazards. For one thing, spending money on things that are conspicuously unnecessary is inherently wasteful. Producing such goods uses up natural resources, consumes energy, and poses problems of disposal, all to an unnecessary degree. Given a biosphere hard-pressed to meet the needs of its current population, it is harmful to stress it further by producing goods that are patently unneeded.

The harmful effects of unnecessary acquisition are amplified by the tendency of luxury items to transmute into "necessities." Air-conditioning has already been cited as a one-time amenity that has come to be considered a "basic need" in American society. Other examples that come readily to mind are designer clothing, stylish automobiles, and various kinds of electronic devices. While such things might be needed "to keep up with the Joneses," they nonetheless impose an added burden on the biosphere.

Yet another adverse concomitant of conspicuous acquisition is the incentive it provides for producers and marketers to find new ways for people to spend money wastefully.[21] Advertising a product as "new" or "improved" prompts consumers to think that it is yet another thing they can acquire without redundancy. And each new product adds to the inventory of superfluous items retailers can make available in shopping malls and specialty stores. A reasonable guess is that no shopping mall is without at least one store offering an array of items none of which anyone actually needs.

Acquisition thus joins gratification, comfort, and convenience as social values at the heart of consumer society. As a rough-and-ready definition, a consumer society might be characterized as one in which these particular values have dominant operative status. Drawing on this definition, we may think of the present chapter as an examination of the damage inflicted on the biosphere by consumer society.

16.6 Wealth

The final value to be discussed in this chapter is that of wealth. In chapter 13, desire for wealth was identified as the primary driving force behind the seemingly endless growth of free-market economies. But economic growth involves ever-increasing levels of energy consumption. Inasmuch as inordinate energy consumption was responsible for our current environmental crisis in the first place, the crisis was precipitated by desire for wealth. Moreover, given that wealth is desired because of the value we associate with it, the crisis is due ultimately to this particular value. It thus behooves us to look carefully at the value our society attaches to wealth.

To set our compass, we should note that wealth is not a consumer value. People purchase things for purposes of gratification (e.g., sweetened beverages), of gaining comfort and conveniences (e.g., air conditioners and automobiles), and of sheer acquisition. These are all consumer values. Barring financial investments and the like, however, people do not buy things for the purpose of becoming wealthy. Consumption tends rather to decrease one's wealth, which means that wealth is not a value by which consumption is motivated.

The connection between wealth and the consumer values we have been discussing is that the generation of wealth in a free-market economy depends on the latter values being operative in the consumer sector (see section 13.4). It is precisely because values like gratification and convenience are effective in present-day society that marketers can sell products by which these values are realized. And it is because such products can be sold that businesses are able to generate profits from which both corporate and personal wealth are derived.

With the three categories of social value in view (from chapter 15), let us see how wealth fits into the schema. It is clear that wealth is not a normative value. A value has normative status when it serves as a norm differentiating socially prescribed from socially proscribed behavior (see section 15.6). In typical cases, socially prescribed behavior is characterized as right, and hence required, while socially proscribed behavior is characterized as wrong and in need of correction. Even though some people in some circumstances may look upon gaining wealth as the right thing to do, however, there is no attendant intimation that lack of wealth is wrong and requires rectification. This is enough to show that wealth is not a value with normative status.

It is less clear, however, whether wealth is an approbatory or a commendatory value. Regarding the former, it is relevant to note that wealth is basically a matter of some people having control of substantially more resources than others. This is a social phenomenon that began to emerge only after people settled in fixed communities and began to accumulate agricultural surpluses (see section 6.3). Viewed this way, the value of wealth is a social construct and does not result (as do pleasure and gratification) from satisfying a natural urge or desire. These observations suggest that wealth is not an approbatory value.

Viewed from another perspective, nevertheless, wealth can be perceived as an offshoot of a natural instinct. Although a relative newcomer on the human scene, once wealth made an appearance, it soon became established as a form of power.[22] And humans are not different from other social animals in having a natural drive for power.

A characteristic of social groups generally is that they are arranged in hierarchies of relative influence. Following early research on birds, these hierarchies came to be known as "pecking orders." In human society, distinct examples are found in military organizations and in the hierarchical structures of large corporations. In both cases, the pecking order is a manifestation of relative power, having to do with who can issue orders and who has to obey them.

Given the prevalence of pecking orders within the animal kingdom generally, the tendency of people to arrange themselves in power structures of this sort appears to be innate. And since pecking orders are established by individuals vying for power, desire for power in some sense must be innate as well.[23] Insofar as wealth is a form of power, this makes desire for wealth an expression of a natural human instinct.

By this line of reasoning, wealth turns out to be an approbatory value after all. The natural urge behind it is not a desire for wealth as such, which as previously noted is a social construct. People rather desire wealth for the power it represents. In societies where this desire meets with blanket approval, wealth is a value with approbatory status. Contemporary society sanctions the pursuit of wealth in the manner that it sanctions pursuit of pleasure. Wealth thus joins pleasure as an operative approbatory value.

But wealth appears to be a commendatory value as well. The essential feature of commendatory values is that they are constituted by a general social recommendation of something as worthy of being pursued. As far as U.S. society is concerned in particular, there is abundant reason to think that pursuit of wealth enjoys widespread commendation. Recent evidence to this effect comes with the poll reported at the beginning of chapter 15, showing that 80 percent of young adults in the United States see "getting rich" as the main life goal of their generation. Comparable results undoubtedly would be forthcoming for other developed countries.

As far as commercial enterprise is concerned, we take it for granted that "the business of business is business" and that the purpose of business is to make a profit. When profitability is threatened by rising costs of employee benefits like pension plans and health insurance, as recently has become common, the public tends to acknowledge that those benefits must be sacrificed so that the business in question can preserve its profit margin.

In its dual status as an approbatory and a commendatory value, wealth receives the endorsement of society in two different respects. In its approbatory role, wealth is sanctioned as an acceptable goal to pursue. And in its commendatory role, it receives additional support as a goal that should be positively encouraged. This dual status gives wealth more persistence in its social setting than other approbatory and commendatory values taken individually. We take this additional staying power into account in the next chapter's discussion of how ecologically damaging values now operative might be replaced.

In concluding the present chapter, we should observe that there is nothing new in America's preoccupation with wealth. This country's zeal for making money goes back to the "Triangular Trade" of the late 1700s (see section 10.2), as part of which slaves were imported to boost its economy. Wealth gained by slave labor contributed significantly to the expansion of American capitalism in the nineteenth century. This made way for the era of the so-called robber barons (e.g., Carnegie in steel, Rockefeller in oil, Vanderbilt in railroads), who amassed enormous fortunes by sharp dealings on the market. The fact that these individuals often contributed large amounts to philanthropy does not affect their standing as symbols of American dedication to wealth.[24]

A more recent facet of our commitment to wealth is the additional means "captains of industry" have available to achieve it. Whereas the Carnegies and Rockefellers of the nineteenth century relied on unscrupulous business practices in taking advantage of expanding industrial frontiers, today's corporate managers also rely on techniques of marketing that are steadily growing in sophistication (see section 13.5). In the manner illustrated by figure 13.1, big business can generate profits by exploiting the prevailing values of the consuming public. These profits can then be converted into personal wealth by those in a position to do so.

Although wealth is not a consumer value itself, as already noted, it enlists the help of consumer values in working its influence. This means that it shares responsibility for their environmental consequences. When we turn to consider alternative values in the next chapter, we should bear wealth in mind as a social value urgently in need of replacement.

17

Values for Survival

17.1 Limitations of This Analysis

Viewed from the perspective of commerce and growth, the values reviewed in the last chapter call for endorsement rather than replacement. Desire for wealth keeps the economy humming, abetted by consumer values like gratification and convenience that induce consumers to part with their money. Judged from an environmental perspective, however, these values are deeply implicated in the ever-expanding economic activity that is destroying the very ecosystem on which such activity depends. From this latter perspective, it is a matter of utmost importance that these values be neutralized and replaced by others more conducive to a healthy environment. The present chapter is an attempt to identify a suitable set of replacement values.

There are several qualifications to make explicit before we begin. For one thing, it should be emphasized that the project at hand is not an exercise of ethical theory. Just as our criticism of consumer values in chapter 16 was not that they are morally wrong (although some may be), so the values discussed in the present chapter are not put forward as morally right. The factor recommending these latter values is the promise they show of being environmentally benign, in contrast with those that are environmentally harmful.

An example, by way of anticipation, is the value of moderation, which will be put forward as an antidote to the harmful value of gratification. Although moderation (and its close cousin, temperance) is a personal character trait recommended by moral philosophers going back at least to Plato and Aristotle, our concern here is not with personal virtues. Our concern is with moderation as a social value, which is to say an attribute valued by society at large. And the reason moderation should be preferred over gratification has to do with their respective ecological impacts, rather than one's superiority over the other from a moral perspective.

Another qualification concerns the dearth of evidence on which preferences like that of moderation over gratification can be based. The case against gratification is backed up by extensive data showing the adverse environmental effects of the fast-food industry (see section 16.2), to mention just one example among many. But there is not much evidence showing that the effects of moderation would be markedly better in that regard. The reason for lack of evidence is simply that social values have effects only when actually operative (see section 15.3), and there are no societies readily at hand in which moderation has operative status.

A consequence of this limitation is that the case for moderation as a replacement value must be largely hypothetical. *If* this value were operative in a given social context, I shall argue in effect, then its ecological impact would be far less damaging than that of the value (e.g., gratification) it is posited as replacing. While an inference of this sort falls short of certainty, at least it is subject to possible verification.

In order to keep inferences of this sort in focus, we shall consider potential replacement values in explicit contrast with harmful values they might replace. Thus moderation will be considered in contrast with gratification, and so forth. Association between opposites of this sort, however, will not always be unique. Moderation will also be considered as a possible replacement of acquisition, and values other than moderation (e.g., forbearance) will be considered as a potential replacement for gratification. Although values in the real world might not be grouped so neatly, this tactic will make our task more manageable.

17.2 Moderation

In the previous discussion of pleasure (see section 16.2), the junk-food industry was portrayed as a major beneficiary of this particular social value. The discussion focused mainly on the fare of fast-food restaurants, notably carbonated beverages and highly flavored hamburgers. Although fast-food establishments obviously are served by other values as well (especially convenience, as discussed in section 16.4), we shall continue to treat their products as sources of pleasure and gratification.

The value of pleasure is operative in social contexts in which gratification of natural urges is generally approved as a motive for action (section 15.4). A sign that contemporary consumer society answers to that description is that few questions are asked when children are raised on a diet of soda pop, hamburgers, and french fries. Designers of these foods make sure that most children find them enjoyable to eat. And given the general approbation of pleasure in consumer society, people find it normal that children (like everyone else) should eat what they find pleasureful and gratifying.

Given the operative status of this value, moreover, parents are not held in disapprobation when they take their kids regularly to fast-food restaurants. Teenagers feel free to indulge themselves daily on sweet drinks and fried fat because their peers can be counted on to do the same. And adults do not hesitate to "treat" themselves frequently because they have been brought up to think of it as normal behavior.

We have already noted that taste treats of this sort are a primary cause of obesity and its attendant health problems. Indeed, most of the recent coverage of junk food in the media has been preoccupied with these effects on health. Our present concern, however, is with their environmental effects. As detailed in section 16.2, both the production of junk food and its delivery to the consumer result in extensive damage to the biosphere. What can be done to curtail that damage?

One positive step in this direction is to impose limits on the marketing of the products involved. An example is the recent ban announced by the British Office of Communications on the advertising of junk food and drink in conjunction with TV shows that attract a large proportion of teenagers under the age of sixteen.[1] Another expedient would be to increase taxes on the offending products, in hopes of discouraging their consumption.[2] A potentially more effective (albeit less direct) approach, however, is to disengage the social values that encourage people to spend money on junk food in the first place. In the context of the present discussion, this amounts to rendering the value of gratification and associated values inoperative.

In effect, the value in question constitutes social approval of doing things expressly for the purpose of gaining pleasure. One thing needed to render the value of gratification inoperative is to eliminate, or at least to diminish, this approval. If something like this were to happen, people would no longer automatically assume that pleasure seeking on their part would go unnoticed by their peers. They would think twice before buying a carton of soda pop in the presence of strangers and would be less prone to take their kids to fast-food restaurants on a regular basis. Doing things just for pleasure would be a bit like scratching an itch in public; once in a while it might pass without notice, but doing it with abandon would make people stare.

Eliminating social approval is different from the onset of disapproval. A more positive tactic for discouraging pleasure as a social value would be the establishment of an opposing value that leads people to view pleasure seeking with some degree of disapprobation. One contrasting value of this sort mentioned previously is that of moderation. A close relative is the social value (not the character trait) of temperance.[3] Let us consider the effect of such values on pleasure-seeking behavior, supposing for the moment that they are operative in some relevant industrial society.

One thing to observe initially is that moderation and temperance are commendatory values. They are not approbatory because there is no natural desire to act in such a manner, and they are not normative because lack of moderation and the like is not proscribed as flatly wrong (so at least we will assume). Recalling the discussion of chapter 15, let us review what it is for a commendatory value of this sort to be operative.

A social value is operative if, for the most part, it is looked on favorably by members of the social group in question. This is indicated by a tendency of individuals to maintain the value in their personal behavior and by a general expectation that others will do the same. In the case of commendatory values specifically, the favorable outlook of society at large provides positive motivation for undertaking activity that exhibits the value in question (see section 15.5).

Given a social context in which the value of moderation is operative, accordingly, people in that context will practice moderation in their personal behavior. This means that they will avoid excesses of food and drink and generally will comport themselves without self-indulgence. In particular, it means that they will not overindulge in pleasurable activity. When presented with opportunities for experiencing pleasure, they typically will react with restraint rather than unbridled enthusiasm.

Moreover, they will expect other people to behave moderately as a matter of course. Self-indulgence of any sort will be considered unseemly, especially with regard to the pursuit of pleasure. Lack of moderation in particular cases will often call for explanation, insofar as it is perceived as a deviation from standard behavior. Avoidance of excess, on the other hand, will be considered normal and as such beyond need for explanation.

In contexts in which the value of moderation is fully operative, in brief, overindulgence in pleasure will strike most people as deviant and out of the ordinary. Seeking pleasure for its own sake will be generally discouraged. Contrary to the general approbation that marks pleasure itself as a social value, when the value of moderation is socially operative, seeking pleasure for pleasure's sake is generally held in disapprobation.

Hypothetical as such circumstances may be, if social approval of pleasure as a motive for action were withdrawn, and if a countervailing value such as moderation were to become operative instead, then the influence of pleasure on consumer behavior would be effectively nullified.[4] This would have significant consequences for the marketing of highly profitable products like soda pop, hamburgers, and french fries. While such consequences alone would not put fast-food chains out of business, their products would become less likely to sell just because of the pleasure derived from consuming them.

Other social values that might counteract pleasure in this general manner are those of restraint and forbearance. As far as its impact on the pursuit of pleasure is concerned, restraint from self-indulgence is tantamount to exercising moderation. Forbearance likewise is a form of abstaining from the immoderate pursuit of pleasureful activity.

It should be emphasized that the disengagement of pleasure as a social value is not a recipe for a dreary life. Even if values like forbearance and moderation were firmly established in the public arena, they would not eliminate our natural tendency to drink what we enjoy drinking, to eat what we find pleasant, or to engage in other activities we generally find gratifying (all pleasurable activities). Their effect instead would be to hold in check whatever inclination we have to indulge those tastes on every available occasion. Human life, after all, may be more enjoyable when it is not preoccupied with the constant pursuit of enjoyment.

17.3 Simplicity

In its primary sense, the term "simplicity" means absence of complexity. It can also mean lack of artifice and affectation, as well as freedom from ostentation. With respect to lifestyle in particular, however, it connotes the absence of luxury. This is the sense employed in the present section.

The present section is concerned with simplicity as a potential replacement for the social value of comfort. As may be recalled from chapter 16, comfort is a value largely responsible for the flourishing market in air conditioners, both in the United States and in urban areas of Europe and Asia. Alleviating the extensive ecological damage caused by air conditioners (see section 16.3) requires cutting back on their use. And this in turn involves dislodging comfort from its current status as a dominant value in consumer society.

Rendering the value of comfort inoperative is basically a matter of changing society's general endorsement of comfort as a state worth achieving. As with pleasure already mentioned, this change would occur in two phases. One would amount to a retraction of society's encouragement (i.e., its commendation) of comfort-oriented behavior. The other would be to establish a countervailing value that leads to comfort seeking being expressly discouraged.

Simplicity is one such value. Given that simplicity in this connection has to do with lifestyle rather than satisfaction of basic urges, it constitutes a commendatory rather than an approbatory value. The value change called for, accordingly, is the replacement of one commendatory value by another. For the change to take effect would be for society to stop recommending the pursuit of comfort and to begin recommending simplicity in personal pursuits instead.[5]

For society to encourage simplicity in the personal affairs of its members would be for simplicity to be operative as a social value. This would be tantamount to society encouraging its members to avoid luxury in their personal behavior. Insofar as comforts of the sort in question (e.g., air-conditioning) are forms of luxury, the effect would be to discourage individuals from the pursuit of luxurious comfort.

In contemplating this possibility, we should be assured that foregoing comforts like air-conditioning and cushy furniture does not entail a monastic existence. Abandoning one extreme does not require embracing the other. Simplicity is a matter of avoiding luxury, which is different from a Spartan self-denial.

Nor does a general avoidance of luxury require the withholding of amenities under special conditions. With regard to air-conditioning in particular, we should bear in mind that it serves many purposes other than personal comfort. We value air-conditioning for its role in shielding asthmatic children from allergens, for the climate control it provides in museums and libraries, for its making possible the installation of large computers in closed quarters, and so forth. There is no reason why such applications should be unavailable in societies rejecting the use of air-conditioning as a personal luxury.

Needless to say, the borderline between luxury and necessity in this regard remains elusive. There are certain uses of air-conditioning technology that at first appear essential but probably fall under the category of luxury instead. An example is its use in buildings without natural cooling capacities, such as glass-encased skyscrapers and high-rise apartment dwellings. In some structures of this sort, reliance on air-conditioning could be reduced by expedients as simple as operable windows. In cases where a building's design makes air-conditioning unavoidable, on the other hand, a society attuned to environmental impacts might view such a building itself as a dispensable luxury.

Another illustration of luxuries masquerading as necessities is the construction of retirement homes in sweltering climates for people who are not habituated to those conditions. Although air-conditioning might really be needed to make these structures habitable (even natives could not live in sealed-up buildings), their very existence might come to be seen as a luxury that society can ill afford in the first place. More will be said on this topic in the final chapter.

To recapitulate, if simplicity were to replace comfort as a dominant social value, many people would still be served by air-conditioning in their daily routines. This is true especially of people who spend time in museums, libraries, and children's hospitals, as well as those who are aged or otherwise infirm. As far as the general population is concerned, however, most of us would no longer consider it imperative to maintain our living spaces at a "comfortable" 68° to 72° Fahrenheit. Most of us would adapt to prevailing temperatures as a matter of course, like our ancestors did before air-conditioning was invented. This would be especially good news for a beleaguered environment struggling to cope with the impact of an invasive technology.

There are other social values that might replace comfort with similar effect, among them austerity and asceticism. Austerity is equivalent to extreme simplicity, connoting a lifestyle stripped of nonessentials. Asceticism is exemplified, in turn, by strict self-denial and by abstinence from creature comforts. Since both austerity and asceticism rule out indulgence in luxury, both would counteract the force of comfort as a consumer value.

It is not inconceivable that cultures might actually exist in which abstemious values such as these have operative force. Ancient Sparta comes to mind as a possible instance. But such values are often accompanied by authoritarianism and repression, which in themselves have little environmental merit. It is better to pursue simplicity with society's encouragement than to be deprived of luxuries by dictatorial authority.

17.4 Patience

Simplicity would also go a considerable way in counteracting the horrific effects of private transportation on the environment. A simple lifestyle would naturally lead one to rely, whenever possible, on simple means of transportation. Valuing simplicity in this regard, a person would choose to get around by walking or peddling rather than by motorized vehicles and, when motorized assistance was needed, would go by bus rather than by private automobile. Obviously enough, the more people there are making choices like these, the more the biosphere at large would benefit.

When the ecological damage resulting from automobiles was discussed in the previous chapter (section 16.4), however, it was taken to illustrate the adverse effects of the value of convenience. To some extent, the value of simplicity does indeed tend to counteract these adverse effects. For example, someone committed to a simple lifestyle would probably eschew conveniences like power cultivators and electric hair driers. But there are other values that seem to oppose convenience more directly. One such is the social value of patience.

It may at first seem counterintuitive to think of patience as a social value. We commonly consider patience to be a quality of character, thinking of a patient person (a person with that quality) as someone who can endure setbacks and inconvenience. The transition in thought from character trait to social value is enabled by the observation that society can either value or fail to value the quality of patience on the part of its individual members. Inasmuch as a social value is something valued by society generally, patience counts as a social value to the extent that it is held in esteem by society overall. A little reflection shows that it is a commendatory value, meaning that people are motivated to behave with patience by the positive encouragement of other people.

Given this meaning of patience as a social value, its opposition to convenience seems obvious. In contexts in which the value of convenience is operative, people will look for convenient ways of doing things as a matter of course. They will tend to find labor-saving devices attractive and be inclined to buy them when an occasion arises.

Manufacturers will find it profitable to develop labor-saving products, which can be effectively marketed on the basis of the convenience they provide.

In contexts in which the value of patience is operative, by contrast, avoiding inconvenience will not be taken as an end in itself. People will be inclined to spend the time and effort necessary to do things "naturally" and generally will not find labor-saving devices particularly attractive. Manufacturers will shy away from convenience-oriented products, and marketers will not emphasize convenience in presenting products to potential buyers. When patience is operative, in brief, it will tend to nullify the effect of convenience as a motivating value.

As with the consumer values of pleasure and comfort above, two steps are probably required to render the value of convenience ineffective. One would amount to withdrawing social encouragement of convenience-prompted behavior. If this were to happen, people would no longer consider it "cool" to own the latest snowblower and riding lawn mower (regardless of area to be shoveled or mowed). The other step would amount to the establishment of a countervailing social value, patience or some effective equivalent. If this were to happen, people would start looking at purchases motivated primarily by convenience with lack of sympathy or even disapproval.

As far as the curse of private transportation is concerned, we have already allowed that use of automobiles is motivated by various values other than convenience alone (in section 16.4). People own cars for reasons having to do with expressing individual personalities (e.g., Hummers versus hybrids), as well as with autonomy and social status. But convenience remains a major factor in their day-by-day use. We find it easier to drive than to walk when visiting a friend around the block. We would rather drive the kids to school ourselves than contend with the inconveniences of a car pool. And we find it convenient to drive our own cars to work even when public transportation is readily available.

It is with respect to mundane activities like these that the value of patience would probably make the most difference if it were socially operative. To be sure, dealing with a car pool can be a nuisance and a bother, and waiting for a bus can be annoying. But these are inconveniences that a patient person can typically endure. If a substantial number of current car owners were to show this kind of patience, the environmental benefits would be considerable.

There of course are other values that would tend to dampen the environmental effects associated with the use of private automobiles. In addition to simplicity, as already noted, there is the old-fashioned value of self-reliance. To the extent that a person is self-reliant, he or she is less dependent on automotive technology. Another is the value of independence, in the basic sense of managing things oneself without need of assistance.

Given that automobiles often count among our most prized possessions, moreover, other countervailing values would include those that act contrary to the value of acquisition. One such is the value of personal contentment, to which we turn next.

17.5 Contentment

A characteristic mark of consumer societies is the high value they place on acquisition. In such societies, people are motivated to go shopping both by social pressure (typically applied through advertising) and by the example of their peers. Acquisition thus is a commendatory value, encouraging the purchase of goods and services that are often unneeded.

The activities of supplying and consuming unneeded products impose unnecessary stress on an already overburdened biosphere. From an environmental perspective, it appears urgent that the motivation to buy such products be effectively neutralized. This amounts to neutralizing the influence of acquisition as a motivating value, which (as before) might take place in two stages. One is to render the value itself inoperative, thus eliminating social encouragement to make superfluous purchases. The other stage amounts to establishing a countervailing value by which needless purchases would be actively discouraged.

One value that would have this negative effect is that of contentment. The sense of contentment intended here is not a matter of uncaring complacency, as when one is content to leave other people to fend for themselves. The sense is rather that of being satisfied with what one has, when one already has enough for a decent existence. To be content is to be at ease with one's current circumstances and not to feel the need for anything more.

With respect to personal possessions in particular, to be content is to be uninterested in acquiring further private goods. Being content, of course, is a personal characteristic and hence distinct from contentment as a social value. Contentment becomes a social value when society at large comes to value that characteristic on the part of its individual members. In the manner of social values generally, the value of contentment is operative in a given social context when people are encouraged by their peers to show contentment in their personal affairs. Under the influence of that motivation, members of the social group will tend to feel content with circumstances that are already tolerable and will encourage others to do the same.

In societies where acquisition is a dominant social value, people will seek out occasions to make new purchases and will seldom be content with what they already have. In societies where contentment dominates, by contrast, people will tend to find what they have sufficient and will show little interest in new acquisitions. This is the manner in which the two values are opposed. Just as one is a mainstay of consumer society, so the other tends to stifle obsessive consumption.

As with other social values we have been considering, the value of contentment might be operative without applying to all aspects of social existence indiscriminately. Indeed, there may be certain contexts in which contentment is treated as something actually to be avoided. An example is the context of athletic competition, where competitors are exhorted to be content with nothing short of victory. Other examples could be found in the domains of health, education, and character development.

Our present concern, however, is restricted to contentment with respect to physical possessions. In contexts where contentment is operative as a social value, one can expect public support for remaining satisfied with material belongings adequate for a decent life. To the extent that this value becomes operative, it should tend to reverse at least some of the damage inflicted by consumerism upon our beleaguered environment.

Another value that might have a similar effect is moderation, considered above as an antidote to the environmentally pernicious value of gratification. Social pressure that moves people to avoid excessive pleasure seeking as unseemly might also motivate them to show like restraint in acquiring physical goods.

Character traits like moderation and contentment have long been recognized by moral philosophers and theologians as personal virtues. We now can see that they have environmental ramifications as well. When the exercise of these virtues is reinforced by public approval— i.e., when moderation and contentment are operative as social values— they also work to the advantage of the ecosystems that support human existence.

17.6 Equity

Our discussion of social values began in chapter 14 with a series of moral quandaries stemming from industrialized society's disproportionate use of the biosphere's nonrenewable resources. Prior to this, we had concluded that the current environmental crisis has been precipitated by excessive use of nonrenewable energy (chapters 5 and 6), that this excessive energy use is part of a pattern of continuing economic growth (chapter 7), and that economic growth is driven by desire for wealth (chapter 13). The upshot is that desire for wealth is largely to blame for our environmental crisis.

To all appearances, wealth is one of the most firmly entrenched values in modern industrial society. Moreover, desire for wealth appears to be among current society's strongest motivational influences. This is due, at least in part, to wealth's unusual status as both an approbatory and a commendatory value. Because of this dual role, pursuit of wealth receives social support in two different ways. Given its approbatory role, desire for wealth is sanctioned as an acceptable motive for action. And in its commendatory role, activity undertaken in pursuit of wealth receives positive social encouragement.

Because the value of wealth is so firmly ingrained in current society, it may be harder to dislodge than consumer values such as pleasure and convenience discussed previously. Whereas pleasure, for instance, presumably could be replaced by (1) withdrawing social approval of pleasure as a motive for action and (2) establishing as operative a countervailing (commendatory) value like moderation instead, the displacement of wealth as an operative social value appears to require stronger measures. This requirement brings normative values back into the picture.

In the classification of social values put forward in chapter 15, normative values were characterized as having more force than values of either approbatory or commendatory nature (see section 15.6). Like both of these others, normative values support certain kinds of behavior as socially acceptable. Beyond this, however, a normative value (like honesty) presents the behavior it supports as not merely acceptable but also obligatory. Normative values carry with them a distinction between right and wrong. What is right is deemed mandatory and hence prescribed, whereas what is wrong is proscribed and subject to sanctions. While sanctions vary from case to case, they must be strict enough to elicit compliance if the value is to maintain its operative status.

Put otherwise, normative values establish standards of acceptable behavior, and these standards are kept in place by adverse consequences that follow when they are not met. This is the kind of force that may be needed to counteract the effects of the value assigned to wealth in contemporary society. What normative value, or values, might be available for this purpose?

To prepare for an answer, let us return to the moral quandaries previously mentioned. The first (quandary 1 of section 14.2) is posed by the fact that a few people have accumulated enormous amounts of wealth in support of lifestyles unavailable to the vast majority and in the process have brought the biosphere to a state of near collapse. The lives of many are threatened by the wealthy lifestyles of a few. Whatever else one may think of it, this state of affairs involves a profound injustice.

Quandary 2 stems from the fact that our current opulent lifestyles are enabled by an extravagant expenditure of resources that future generations might need just to stay alive. This shows a callous selfishness, to say the least. In the view of anyone with moral sensibility, it is likely to appear inequitable to an extreme degree.

Quandary 3 arises with the observation that our practices of economic production result in the destruction of countless living creatures, often to the point of exterminating entire species. Although intuitions vary, this strikes many people as grossly unfair. Why should multitudes of other creatures have to die in order to sustain economic growth at a level profitable to human beings? If we are to be honest with ourselves, this question has no positive answer.

Our current economic practices result in an *unjust* distribution of wealth among people now alive, involve an *inequitable* appropriation of natural resources likely to be needed by future generations, and cause *unfair* injury to uncounted numbers of other living creatures. These grievous consequences are tolerable only in societies that permit injustice, inequity, and unfairness. Conversely, in societies where the normative values of justice, equity, and fairness are operative, consequences of this sort would not be tolerated.

In a way that might be instructive, these three quandaries epitomize our environmental crisis. *If* we weren't (1) engaged in lifestyles undermining the ecosystems on which human life depends, *if* we weren't (2) squandering resources likely to be needed by future generations, and *if* we weren't (3) killing off thousands of other species, *then* we probably wouldn't be confronted with an environmental crisis in the first place. Although the crisis extends beyond these particular moral issues, taken together they go to the heart of the problem. Any steps successfully taken to resolve these issues would go a long way toward resolving the overall problem as well.

Let us spell out why this is the case. A major portion of the present study has been given over to showing that the economic practices of industrialized society are largely responsible for our environmental crisis. These practices are also responsible for the moral quandaries we have been discussing in particular. Given the role desire for wealth plays in motivating these practices, it follows that both the general crisis and the particular quandaries are due largely to the value industrialized society ascribes to wealth. Both are due, that is to say, to wealth operating as a social value.

As just observed, furthermore, those particular quandaries would not arise in societies (unlike our own) where justice, equity, and fairness are fully operative social values. With regard to these quandaries specifically, the normative values of justice, equity, and fairness, if operative, would have the force needed to cancel out the morally pernicious effects of the value currently assigned to wealth. The reason is simply that the quandaries could arise only in social contexts where unjust, inequitable, and unfair practices are permitted, which is to say that they would not arise when such practices are not permitted. In brief, the quandaries would not arise in contexts where justice, equity, and fairness are in force as normative social values.

To the extent that these normative social values would cancel out the pernicious effects of wealth in the case of these quandaries, moreover, they would be effective as well in counteracting wealth as a primary cause of our general crisis. This is not to say that these values would be enough by themselves to set our environmental affairs aright. For reasons laid out earlier in this chapter, remedial values like moderation, simplicity, and patience would also be needed. As far as the value of wealth is concerned, however, its damaging effects would be largely nullified if these several normative values were to become operative.

As operative social values, it may be noted, justice, equity, and fairness amount to pretty much the same thing. While distinctions could still be drawn among them, in practical effect justice and fairness would be tantamount to equity. In social contexts in which equity is operative as a normative value, rich people would not accumulate enormous amounts of wealth at the expense of the poor, a privileged few would not engage in lifestyles that threaten the existence of future generations, and other species would not face extinction by our zealous economic practices. Thus it is that equity would serve as antidote to the destructive value of wealth.

But for equity to be established as an operative normative value, penalties must be forthcoming for noncompliance. What this means for practical purposes is that agents (individual or corporate) who exploit others in pursuit of wealth must be subject to social sanctions. Given the arena in which the value is to operate, such sanctions would be largely economic in nature. Examples might be progressive taxes on higher incomes and boycotts of businesses whose activities cause harm to other species. We return to the topic of remedial measures in the final chapter.

17.7 Living with Alternative Social Values

Before turning to possible remedial measures, let us reflect further on the kind of remedy required. In bare essentials, what is required is a worldwide shift in social values. Values built up over two or three centuries of consumerism got us into this mess in the first place. And the most promising way out requires eradicating those values in favor of others that are more environmentally friendly.

This obviously is a very tall order. A change in social values of such magnitude will carry with it fundamental alterations in the societies involved. It will also affect the individual lives of people within those societies, bringing about significant changes in their day-by-day existence.

For many people, the prospect of having to deal with significant changes in their daily lives can be a scary business. To make the prospect more palatable, we should realize that such changes for the most part would be for the better. For the most part, that is to say, individual lives would be more satisfactory in a society incorporating environmentally friendly values than in present-day society dominated by consumption and profiteering.

To help make this apparent, compare what we will call the "contentment ratio" of alternative future societies. A society's contentment ratio is the proportion of its members who are content with their lives to those who are chronically discontent. The two societies we shall compare in this regard are society P, which *perpetuates* the values of current society, and society S, in which these values have been *superceded* by values congenial to a healthy environment.

For the comparison to be realistic, it must take place far enough in the future to allow the changes in successor society S to take place. Let us assume a time lapse of two generations, or about fifty years. If no significant change in values takes place by then, society P will be in dire straits indeed. There will be widespread hunger from drought, massive dislocation of populations as a result of climate change, and few natural resources left by which human suffering might be alleviated. While a few pockets of wealth might remain, the vast majority of human beings will be living in poverty and destitution.

In the worst-case scenario, the biosphere will be degraded to a point where it can no longer support human society as we currently know it. In this case, the privileges of wealth will no longer be available, and what is left of society will be in a state of desperation. Worst scenario or not, the contentment ratio of society P in fifty years will be much lower than what would be found if the measure were applied today.

Society S, on the other hand, is the social order that will exist fifty years from now if the values of consumer society have been effectively replaced by the environmentally friendly values of the sort we have been discussing. During the interim, while these values were taking effect, the relationship between the biosphere and its human inhabitants will have been gradually improving. (If not, we have not located the right replacement values.) At the fifty-year mark, climate change will be abating, food supplies will be stabilizing, and gross disparities in wealth will be mostly eliminated. On balance, most inhabitants of S will be better off than the majority of people living today, and their overall contentment ratio will be significantly higher.

However this plays out in detail, the contentment ratio of S will be higher than today, and that of P will be considerably lower. This shows that a shift from present-day consumer values to a set of values favorable to environmental health almost certainly would be a change for the better. Whatever vicissitudes you and I undergo in the process, as reasonable people we have no better choice than committing ourselves to doing the best we can to help bring such a shift in values about.

Be that as it may, the hard reality of the matter is that any concerted effort to rid contemporary society of its consumer values is sure to be met with determined resistance. It will be resisted by marketing specialists who make their living inducing consumers to spend money, by corporate managers paid high salaries to maximize profits, and by mainstream economists dedicated to perpetual growth. It will be resisted most vigorously, perhaps, by people who use money as a means to political power and by the politicians who benefit from their largesse. How can such opposition be overcome?

In opposition to approaches advocating violent revolution,[6] it seems clearly desirable to work within the current economic system in trying to resolve our economically based problems. One possible strategy would be some form of peaceful resistance, in the tradition of Thoreau (in his essay "Civil Disobedience"), Mahatma Gandhi, and Martin Luther King Jr. If large numbers of consumers acting en masse were to make a point of not buying soda pop, for example, manufacturers would soon learn that products cannot always be sold on the basis of gratifying flavor alone. This would be a point in favor of the value of moderation.

Another approach might be modeled after the techniques of jujitsu, in which the force of an opponent's attack is turned to the attacker's disadvantage. An example would be using the power of advertising to instruct people on how advertising can be resisted. A precedent of sorts is the surgeon general's warning printed on cigarette packages. Effective use of this technique in the present case would advance the social value of contentment by making people less prone to the blandishments of advertising.

Approaches like the above are pursued in a public setting. There are also tactics that can be pursued by individuals acting privately. One example is unplugging the television,[7] thus making oneself unavailable to corporate sponsors. If enough people were to do this on an individual basis, the effect would be like a mass of tree roots breaking up a stone. The wiles of marketing cannot prevail against a grassroots refusal to pay attention to its images. Success in this venture would strike a blow for simplicity against all the clutter introduced by watching television.

These examples provide a prelude to the final chapter. One standard way of ending a study like this is to propose a series of policies by which needed reforms might be implemented. The trouble with policy statements, however, is that they stipulate what should be done but leave the doing to other people. Another way is pursued in the final chapter, which suggests practical remedies one might undertake on one's own or else in cooperation with like-minded people.

18

What Can Be Done? What Can One Do?

18.1 The Need Restated

Our environmental crisis boils down to the fact that human enterprise has generated more entropy than the biosphere can dispose of. This excessive burden of entropy results mostly from excessive amounts of energy consumed in the production of economic goods. Extravagant production of goods is enabled by extravagant consumption, which is motivated in turn by a robust set of consumer values.

Put the other way around, if these values were not in force, we would not consume such extravagant amounts of goods. If such amounts of goods were not consumed, they would not be produced in such extravagant quantities. If goods were not produced in such quantities, we would not use such prodigious amounts of energy. And if we did not use such amounts of energy, the biosphere would not be affected by excessive amounts of entropy.

This sequence of dependencies points to one way of alleviating our environmental crisis. If our current set of consumer values were replaced by others more conducive to environmental health, the biosphere might have an opportunity to heal itself. Abandoning these damaging values at least should keep the crisis from worsening. In any case, we have a clear and urgent need to free ourselves from the values of consumerism and to replace them with other values like those laid out in chapter 17. This final chapter is concerned with practical ways in which value transformations of this sort might be accomplished.

18.2 Pitting Marketing against Itself

Three general approaches to the needed value transformations were mentioned at the end of chapter 17. One focuses on action individuals can undertake on their own, like making oneself unavailable to commercial sponsors by unplugging the TV. Another approach emphasizes actions taken in cooperation with like-minded persons, such as boycotting products like soda pop that cause appreciable environmental damage. We shall return to these approaches momentarily. But first let us look more carefully at the approach previously likened to jujitsu, which comes down to using the power of advertising to diminish its own influence.

Marketing specialists have become adept at manipulating the economic values individual people attach to particular commodities. The marketer's basic task is to induce potential customers to ascribe a higher utility to the product than to the money required to buy it. Successful marketing leads people to act out those evaluations by actually exchanging money for the product in question.

Marketing can be used in other arenas as well. Campaign managers use it to "sell" candidates to potential voters, colleges use it to attract potential students, and governments use it to gain backing for policy initiatives. In these cases, the products "sold" are the candidates, the colleges, and the policies, respectively. Another possible application we want to consider is to employ marketing techniques in reshaping the system of values governing consumer behavior to make them more environmentally friendly. In this case, the product would be an altered system of social values.

From one perspective, there appears to be little new in this suggestion. In honing their skills over the years, marketers have learned that selling products is more than a matter of crafting seductive images. Successful marketing also requires a receptive audience of potential buyers who have been conditioned to respond to those images in the desired manner. Realizing this need, the advertising industry has undertaken to create a culture in which its ads will be effective.[1] The only novelty behind the present suggestion is the thought that advertising might help create an alternative culture in which its effectiveness is actually diminished.

One model for such an approach is the genre of informational ads purchased by nonprofit groups ostensibly operating in the public interest. Examples are ads in newspapers offering free medical screening and billboards showing pictures of missing persons. One can imagine attractively formatted ads saying, "Next time you're thirsty, drink water not soda pop," or "Hamburgers from rain forests cause global warming." Although it would be hard to convince commercial advertising agencies to take on such projects, for reasons we shall look at momentarily, it's a fair bet that shopping areas saturated by ads like these would sell less junk food.

A more effective tactic might be to use entertainment media to make environmentally responsible behavior look "cool." One paradigm that might be followed here is the use of the 1980s sitcom *Cheers* to promote the concept of "designated drivers." After a designated-driver poster appeared on the bar during some 160 episodes over a four-year period, drunken-driving fatalities in the viewing area fell 25 percent.[2] It would be interesting to speculate on the effects of a poster saying, "Beer tastes better without air-conditioning," appearing week after week in the midst of a group of bar customers wearing sweaty T-shirts.

In thinking through the effect of approaches like these, however, we should bear in mind that designated-driver posters and missing-people billboards do not cut into business profits. Other outcomes would be likely in the case of efforts to enlist the aid of advertising in reducing the influence of consumer values, the very point of which would be to steer customers away from environmentally harmful businesses. If this started to happen, the businesses affected surely would seek out countermeasures to reverse the effect.

One means of response would be to initiate ads directly opposing those they found undercutting their profits. For example, a fast-food chain might mount a public relations campaign proclaiming (truly or falsely) that since their beef is grown in the United States, it's patriotic to eat their products. Another tactic, potentially more effective, would be to threaten retaliation against media organizations running ads contrary to the interests of the businesses affected.

Instructive in this regard is the experience of the Canadian-based Adbusters Media Foundation, specializing in what it calls anti-consumerist "culture jamming."[3] Its method is to produce "anti-ads" that mock well-known commercials, with the intention of making viewers of real ads aware of how they are being manipulated. One example is an "uncommercial" in which Joe Camel becomes "Joe Chemo," dying of cancer in a hospital bed hooked up with an array of life-support equipment. Other instances target ads by Calvin Klein, Chevron, and Absolut Vodka. The more these "anti-ads" caught on, it turned out, the more difficulty the Adbusters group experienced persuading networks to air them. As one network executive explained candidly, while he thought personally that a given Adbuster spot contained an important message, "we will never air that spot because we would have Revlon and Maybelline and Calvin Klein coming down our throats the very next day, and that's where our bread and butter is."[4]

A prima facie more successful attempt to turn marketing against itself is the surgeon general's warning appearing on cigarette packages since the 1960s, an early version of which said, "Cigarette smoking may be hazardous to your health." This message obviously is at odds with other design features intended to make the packages attractive to potential buyers. In the year or two after the warning was mandated by Congress in 1965, tobacco companies spent hundreds of millions of dollars in an attempt to maintain sales at previous levels. Within the next few years, nonetheless, per capita cigarette consumption began falling about 3.5 percent annually, and new starts were off by about 12 percent.[5] Although expensive public relations campaigns have continued unabated, the tobacco industry has been on the defensive ever since.

The only reason the surgeon general's warning remains on cigarette packages, however, is that its presence is required by federal law. Viewed in context, this is not a case of marketing being used successfully against itself but rather one of an industry being subjected to enforceable laws of disclosure. As matters stand, there are no corresponding laws requiring networks to run ads designed expressly to work contrary to the interests of their most lucrative clients.

To be sure, there is a fair chance that laws with this effect could be enacted if a majority of the voting public became convinced that they are needed. But this would not be likely to happen until many of the values that would be supported by such laws had already become socially operative—values like contentment, simplicity, and moderation. The upshot is that marketing techniques probably would not be effective instruments by themselves for altering the values of consumer culture.[6] Other means must be relied on to bring about the value changes in question.

18.3 How Individuals Relate to Social Values

The next few sections discuss things we might do as individuals to help bring value changes of the desired sort about. A value change occurs when one value yields its operative status to another. To prepare for this discussion, we need a more nuanced understanding of how individuals figure in the circumstances that make a social value operative.

As initially characterized in chapter 15, being operative requires that a value (i.e., the thing valued) be held in general esteem by members of the social group in question. Signs of such esteem are that members tend to conduct their affairs in accord with that value and tend to rely on other members of the group to do the same. A change in values thus would be a matter of the group's general esteem being transferred from one value to another.

This characterization, however, leaves it unclear how individual members of the group in question are involved in value change. One manner of clarification falls back on statistics. By way of illustration, suppose gratification were replaced by moderation as an operative value in a given social group. Even before moderation takes over, there likely would be a subset of the population that already held moderation in high esteem personally. And after the shift there might be a subgroup of individuals who still attach a high value to pleasure seeking. If we think of a value as being operative when maintained by a simple majority of a society's members, then value change would amount to a shift from a majority that valued gratification but not moderation to a new majority that valued moderation but not gratification.

But this approach seems unrealistic, assuming as it does that individual members could value either moderation or gratification but not both together. Although the two values might turn out to be exclusive in isolated cases, most individuals will probably value both gratification and moderation but in differing degrees. Personal values for most people, that is to say, are arranged in hierarchies. In some hierarchies, gratification ranks higher than moderation, and in others vice versa.

Personal value hierarchies come into play when the person involved faces a choice between courses of action motivated by opposing values. Imagine the case of someone deciding whether to take a second helping of a very rich and tasty dessert. If gratification outranks moderation in this person's value hierarchy, the nod will go to the second helping. But if moderation is the dominant value, the person will decide that one helping is enough. While the person in the latter case does not reject pleasure seeking *tout court,* in this case the value of moderation holds sway.

Thinking of value change in this more nuanced way, moderation supercedes gratification as an operative social value when the personal value hierarchies of individual members shift from a majority ranking gratification over moderation to a majority with the opposite ranking. The value of gratification then would be socially operative when a preponderance of the population assigns more importance to gratification than to potentially competing values. And gratification yields to moderation when the demographic shifts to a preponderance ranking moderation higher instead.

This way of thinking adds clarity to the goal of achieving an environmentally friendly set of social values. As matters stand, consumer society is marked by a preponderance of people who value gratification over moderation, acquisition over contentment, and so forth. The value change we want to bring about amounts to a transition from the status quo to a situation in which moderation is valued over gratification and contentment over acquisition. Put generally, the value shift we want to take place is a change from the status quo to a situation in which environmentally friendly values hold sway over consumer values in the general estimation of the societies involved.

This way of thinking also provides a clear role for individuals like us to play. By way of analogy, consider the role played by individuals in a national election. No one individual, nor small group of individuals, can select a new president. But each of us can affect the tally by which the winner is chosen.

Likewise, while no individual can establish a set of environmentally friendly values, each of us can add to the plurality by which such values would be rendered operative. With each person who shifts priorities toward a set of environmentally responsible values, society is nudged a bit closer to a situation in which those values dominate. Let us look next at various forms in which our votes might be cast.

18.4 Taking Individual Action: Downshifting

Downshifting is a voluntary reduction of one's material standard of living. People downshift for various reasons, such as wanting to reduce stress, wanting a more balanced life, and wanting to spend more time with one's family. People can also downshift for the express purpose of reducing their consumption of material goods. A survey taken in 1995 of eight hundred American adults found that 28 percent of respondents had recently taken steps "to scale back their salaries and lifestyles to reflect a different set of priorities."[7] Extrapolation from the sample suggests that several million Americans have made changes in their lives that resulted in less income but enabled them to spend more time in rewarding activities. On the survey, 87 percent of the downshifters describe themselves as happy with the change. Two-thirds say they are happy with their current economic circumstances, only slightly fewer than the other respondents who did not make a change. About one-third say they not only are content with their circumstances but actually do not miss the income they gave up.

Given its emphasis on reducing consumption, downshifting is a clear-cut example of altering one's priorities to favor an environmentally friendly set of values. As such, it illustrates one kind of step an individual might take toward making values like simplicity and contentment operative in current society. Other steps, though less direct, might also be effective.

One such takes advantage of the fact that people learn by example. A standard marketing technique is to display people engaged in activities the viewer might be led to imitate, such as eating certain snack foods and drinking certain beverages. The power of advertising, of course, depends on its images reaching large numbers of people. Less powerful, but effective nonetheless, are examples individual people display to others they meet on an everyday basis.

The next several sections lay out a variety of practices individual people can engage in to help establish a set of environmentally sound social values. Engaging in these practices not only will set a good example for others but also will help fortify the values they represent in one's own value hierarchy. The practices in question will be discussed under three headings: things done currently we should simply stop doing, things we must continue doing but should do in a different way, and things not done currently we should begin doing. The next section discusses things we simply should stop doing.

18.5 Setting Individual Examples: Things to Stop Doing

(a) Stop Buying Things for Yourself that You Don't Really Need

Although needs of course vary from person to person, here are a few simple examples of what this restriction amounts to. Most people who have been around a few decades know that clothing styles change in a predictable manner. Most people also realize that changes in fashion are driven by needs of the manufacturers and retailers to keep merchandize moving, rather than by changing consumer needs. Unless required by external (e.g., work-related) circumstances to display the latest fashions, an ecologically minded person should make a point of wearing his or her clothing through several seasons. Another possibility is to wear secondhand clothing. Buy new clothing only when really needed.

In a similar vein, many people buy a new car every year or so, even when their older vehicles are serving adequately. People concerned with the environment should avoid this practice. Another example is the flood of electronic entertainment devices (mp3 players, etc.) currently on the market for which no previous need existed. Although some of these might turn out to be genuinely useful, one should have a genuine need in mind before making such a purchase.

The point here is not to avoid purchasing things that are really needed but rather to break out of the syndrome of recreational shopping. Instead of buying things for the sheer enjoyment of acquisition, one should learn to find satisfaction in remaining content with what one has already. To practice contentment in one's own consumption habits is a small step toward the replacement of acquisition by contentment as an operative social value.

(b) Stop Buying Things for Others that They Don't Really Need

The intended upshot of this recommendation emphatically is not that we stop showing love and appreciation for other people by giving them gifts. The intent is that, when we choose gifts that are purchased in stores, we do our best to make sure they will be genuinely useful.

Although statistics on such matters are hard to come by, a reasonable guess is that at least one-third of the routine purchases we make in stores (cars and major appliances excepted) end up as presents to other people. This guess is supported by such facts as that retailers tend to make most of their profits during the Christmas season, that there are several holidays besides Christmas that feature gift giving (Easter, Valentine's Day, Mother's Day, etc.), and that industries providing "ritual gifts" like flowers and greeting cards have become major players in many present-day economies.

Another reasonable guess is that stores specializing in gift items can be found in almost every shopping center, tourist site, and travel destination around the world, probably totaling in the millions. And most of the items they sell are of little practical use. A result is that a substantial portion of our natural resources go into the production and distribution of objects that do little more than take up storage space, until they are transferred into the trash and hence to the landfill.

From an environmental perspective, a preferable course surely would be to give fewer useless gifts in the first place. This option places a burden on the would-be donor to limit purchases to items the recipient would be likely to find valuable, and when such items cannot be found to show affection in some other manner. Many potential recipients, presumably, would prefer gifts made by the giver rather than those bought at stores.

(c) Stop Making Yourself Available to Mass-Media Solicitations

An overview of marketing techniques available to stimulate consumer spending was offered in section 13.4. Most of these techniques involve media like newspapers, radio, television, and the Internet. Although yielding place recently to the Internet, television held sway during recent decades as the most effective advertising medium of all. To view commercial TV ipso facto is to expose oneself to solicitations by commercial sponsors. One positive contribution individuals can make toward dismantling consumer society is to make themselves unavailable to this source of solicitation.

A beginning step is to tune out ads whenever they appear on the screen. Like its close relative "channel surfing," however, this tactic requires constant attention. Another shortcoming of this tactic is that it does not deal with ads incorporated into the featured programs themselves, as when a sponsor's product is written into the script or a company's logo is inscribed on a playing field.

A more definitive step, accordingly, would be to stop watching commercial TV entirely. Commercial TV is a control mechanism by which marketers manipulate potential consumers, and one effective way of disengaging from consumerism is to unplug the mechanism. This does not entail throwing your TV out with the trash or missing programs that are genuinely worthwhile. What it does entail is turning on the TV only on carefully chosen occasions when you have more to gain than some corporate sponsor.

As far as the Internet is concerned, marketers are becoming increasingly sophisticated in capturing the viewer's attention with unwanted solicitations. No matter what one calls up on the screen, it is likely to be accompanied by flashing lights, bouncing balls, or dancing animations calculated to lure the viewer into some commercial scheme. Sometimes it is even necessary to restart one's computer to rid the screen of these intrusions. A partial remedy is to train oneself not to look at unexpected distractions. As with learning to ignore TV ads, the point is to make oneself unavailable to intrusive manipulation.

Compared with TV and the Internet, avoiding solicitation by newspaper ads is relatively easy. Ads for local businesses often come in colored sections that can be thrown out unopened. The same tactic often works for unwanted mailings. Who knows how many trees would be saved if people simply declined to bring printed advertisements into their homes.

18.6 Setting Individual Examples: Things to Do Differently

The fundamental cause of the present environmental crisis is that people are consuming too much energy. People in industrial societies alone use far more energy than the biosphere can tolerate. Whatever else may be needed to resolve the crisis, it is imperative that we adopt individual lifestyles that are less energy-intensive. We must learn to do ordinary things differently to consume less energy.

Given that over 90 percent of the energy we consume currently has fossil origins (see section 6.8), this means we must learn to cut back substantially on our individual use of fossil fuel. Needless to say, there are various things we do presently with fossil fuel that are essential to civilized life, including cooking our food, heating our living spaces, and getting ourselves from place to place. The challenge is to do these things in less energy-intensive ways and to whatever extent possible in ways involving minimal use of machinery operated with fossil fuels.

(a) When Possible, Travel by Means Other than Car or Truck

The main public alternatives to private vehicles are buses, trains, and airplanes. Although there are wide variations within each class, average energy consumption per passenger mile for trains and buses is three to five times less than for airplanes and private conveyances.[8] In situations where one must be carried from place to place, going by bus or train is far more energy-efficient.

Hybrid cars and trucks, of course, are more fuel-efficient than standard models. Some hybrids have the additional advantage of close to zero tailpipe emissions while operating at low speeds.[9] But since hybrid buses are now in service with the same advantage, the advent of hybrid technology does not substantially affect the environmental benefits of buses over automobiles.[10]

When circumstances permit, however, the best alternative is to avoid motorized transportation entirely. The main options here are bicycling and walking. For short commutes, riding a bicycle can be even quicker than driving. On a bike path isolated from speeding traffic, a healthy person can travel fifteen miles portal to portal in about an hour. Roughly the same time on average is required for a commuter to leave the office, get to the parking garage, reach the highway, drive fifteen miles in rush-hour traffic, and finally arrive home to park the car. Over a ten-year period, moreover, a bicycle costs roughly fifty times less than an average car to own and operate, and bicycles can last for several decades if properly maintained.

Bicycles are preferable to automobiles for various health-related reasons as well. More deaths and disabilities follow from the use of automobiles today than from any other nonnatural cause, whereas with well-maintained bike paths (excluding motor vehicles) and proper equipment, the use of bicycles is not life-threatening. Given the exercise it provides, to the contrary, traveling by bicycle is downright healthy. And getting around by foot on a regular basis can be even healthier.

Whether by walking or biking, the main environmental benefit of moving about without motors is that it requires no direct expenditure of fossil fuel. According to one estimate, bicycle travel in the United States today saves about seven hundred million gallons of gasoline annually.[11] Individuals can contribute to favorable statistics like these by reducing their dependence on motorized transportation to the lowest level possible.

(b) Choose Forms of Recreation that Do Not Involve Engines

There is a continuing controversy between conservationists and snow-mobilers over access to public recreational areas. Snowmobiles are noisy, emit noxious fumes, kill vegetation, and disrupt wildlife, and in addition they are very dangerous to operate.[12] From the perspectives both of the environment and of public health, it seems clear that the conservationists should prevail. Individuals who like winter recreation should avoid snowmobiles and rely on skis instead.

In the domain of water sports, similar reasons favor use of sailboats over motorboats and so-called Jet Skis. As far as winged sports are concerned, preference should go to sailplanes and hang gliders over motorized aircraft. An additional advantage of genuine skis, sailboats, and hang gliders, from an environmental perspective, is that they enhance one's sense of dependency on and cooperation with nature.

Other recreational activities that tends to heighten one's awareness of nature include canoeing, trail hiking, horseback riding, and bird watching. Even watching pigeons in a park can be environmentally instructive if one thinks about how the birds have adapted to their human surroundings. A salutary effect of activities like these is that they help us to shift attention from ourselves to the world around us. By so doing, they help us break the hold of consumer values like comfort, convenience, gratification, and acquisition, all of which gain force when one is preoccupied with one's personal circumstances.

(c) Live in Spaces Designed to Conserve Energy

Dwellings can be energy-efficient in various respects. Among the more obvious, buildings can be constructed to save energy by judicious use of materials (e.g., insulation), by appropriate design of interfaces (e.g., minimizing exposure of outside walls to chilling winds), and by arranging living areas according to energy needs (e.g., placing bathrooms and kitchens near water heaters). Particularly effective is an energy-conscious selection of methods for heating and cooling.

There are three general methods of using solar energy for space heating: passive, active, and photovoltaic. Passive methods involve exposing heat-retentive materials like water or masonry to solar radiation and allowing the heat thus stored to be gradually released into the living space by radiation and convection. Active methods, so called because they require outside energy to operate, involve absorption of solar radiation by liquids (water) or gases (air), which then are pumped into living spaces where they release their stored heat. The photovoltaic method, in turn, employs arrays of collector cells made from semiconducting materials that convert sunlight directly into electricity.

Advantages of the photovoltaic method are that the solar arrays can be located away from the space to be heated and that their electricity can be used for purposes other than heating. A disadvantage is that the collector arrays are expensive to manufacture (although their cost has been declining as the technology improves). An advantage of the active over the passive method is that the collectors (being relatively light) can be placed at any level of the structure, whereas passive collectors (being heavy) must rest on massive foundations. Among advantages of the passive over the active, on the other hand, are that its components are maintenance-free and that the system requires no outside energy to operate. Given the flexibility these several methods provide, most dwellings in temperate climates can incorporate solar features that significantly reduce their dependence on fossil fuel.

An additional advantage of a passive installation is that the components used for solar collection during the winter can also provide cooling after a hot summer day. Heat from within the house is absorbed by the dense collector bodies and radiated outward by black-body radiation, the same process by which other low-grade heat is returned from Earth to space (see section 1.5). Because of the narrow range of room temperatures involved, however, this expedient usually will not provide adequate cooling just by itself.

Another alternative to air conditioners that is effective in dry climates is evaporative cooling of air before it enters the living space.[13] This is basically the same process as cooling the skin by perspiration. A disadvantage is that it can raise the humidity in a building and hence is not suitable for home use in humid climates. Yet another cooling method not subject to this limitation is building a substantial portion of a dwelling's living space underground, where ambient temperatures in the range of mid-50° Fahrenheit (except in permafrost) prevail yearlong.

The most straightforward natural alternative to air-conditioning, however, is use of shade and ventilation. In climates with large shade trees (where evaporative cooling often is not practicable), an effective design for cooling combines a shaded roof with operable (opening) windows placed to allow evening breezes to circulate through the living space. Openings at upper levels of a structure enable hot air to escape as the evening cools, a process that can be aided by energy-efficient ceiling fans.

18.7 Setting Individual Examples: Things to Begin Doing

The rallying point of environmental economists (see section 12.2) is their advocacy of market prices that take environmental costs of production into account. Until this starts happening within the market itself, we can do something like it on an individual basis. What this boils down to is training ourselves to think of these "external" costs as part of the total price we pay for the products involved. Products to be discussed in this section are those produced by agribusiness, those stemming from genetic engineering, and those transported long distances from point of production.

(a) Limit Consumption to Organically Grown Products

Major portions of the foods we eat today—meat, fruit, and vegetables— are produced by agribusiness on factory farms. Agribusiness uses mass-production techniques relying heavily on petroleum-based fertilizers and pesticides and on genetically modified seeds, all of which have environmentally harmful side effects (see sections 5.8, 8.3). Use of these techniques enables factory farms to produce food at lower financial cost (while maintaining large corporate profits) than can be managed by smaller operations (usually family farms) employing natural farming methods. In effect, factory-farming enterprises are appropriating environmental resources that they do not own to subsidize the products they put on the market, and in the process are putting small farmers out of business.

Individual grocery shoppers can respond to these exploitative practices by refusing to make purchases that contribute to the profit agribusiness makes at the expense of the environment. One way of proceeding is to add the environmental subsidy (mentally) to the sticker price of factory-farm produce and to make one's purchasing decisions accordingly. While in most cases it is hard to assess these environmental subsidies in precise dollar values, a workable rule-of-thumb is that the environmental contribution is greater than the difference in sticker prices between factory-farm and naturally grown products. This estimating procedure is reasonable because the environmental subsidy will underwrite substantial profits for the agribusiness concerned, in addition to the difference in sticker prices itself.

The predictable upshot of this approach is that the biggest bargains in food shopping, all things considered, are likely to be found in the organic food section. An unfortunate limitation of the approach is that people struggling to make ends meet might not be able to participate. To the extent that one's means allow, however, staying away from mass-produced food will have the salutary effect both of avoiding complicity in the environmental outrages perpetrated by agribusiness and of declining to contribute to its corporate profits.

(b) Exercise Caution in Buying Genetically Modified (GM) Products

Previous discussion of GM techniques in this study concerned their use of pesticides (section 8.3) and their contribution to loss of biodiversity (section 5.8). Such techniques also have implications for public health. Avoiding GM products thus has the potential of contributing to both environmental and personal well-being.

A disconcerting aspect of the human health issue is that adverse effects of GM strains have been indicated but not conclusively demonstrated. Genetic effects can take several generations to become manifest, and responsible scientific opinion varies on how detrimental potential effects will turn out to be in the long run. Current facts of the matter are that GM food has been on the market since the mid-1990s, that about 75 percent of processed food sold in the United States now contains a GM ingredient, and that the long-term effects of this technology on human health remain largely unknown.[14] Reasons for thinking that some of these effects might be harmful include the following data.

Safety testing of the Flavr Savr tomato (the first GM food) showed that it resulted in stomach lesions in laboratory rats. A similar result showed up with GM potatoes in 1998. These events tie in with results from an earlier experiment with a GM nutritional supplement that caused severe toxic reactions in more than one thousand people in Japan. Despite these warnings, the U.S. Food and Drug Administration (FDA) has no program for detecting or monitoring toxicity in GM food products.[15]

There are also risks having to do with allergic reactions. In 1993 a Brazil-nut gene was artificially introduced into soybeans to improve their nutritional quality. Eating animals fed with this product was found to produce immunological reactions in humans with Brazil-nut allergy. FDA scientists warned subsequently that new proteins produced in GM foods could also prove allergenic to human consumers,[16] but so far the agency has no testing program in place to guard against this possibility.

Another sort of health risk is suggested by the recently discovered cross-species transfer of a gene from GM rape to bacteria in the guts of honeybees. Seed from the rape plant is used to make canola oil (so called for its original production in Canada), and a gene introduced into the plant for pesticide resistance found its way into bee guts as part of the pollination process. The perceived danger is that bacteria in the human intestinal tract might be similarly affected, causing unpredictable changes in our ability to digest food. This prospect becomes more alarming in view of the fact that the honeybee population in the United States currently is in precipitous decline, which some think might be due to the interaction of bees with GM plants.[17]

It might be decades before we fully understand the health risks of GM food. In the meanwhile, we should have the discretion to avoid it if we see fit. Discretion in such matters is impeded by the fact that seed companies and food processors thus far have been able to forestall governmental requirements that products with GM ingredients be labeled. Until disclosure laws to that effect are enacted, an interim precaution might be to avoid products with suspicious ingredients (e.g., soybeans, rapeseed) as much as possible.[18]

(c) Buy Locally

With or without GM ingredients, most of the food products offered in supermarkets are transported long distances from point of production. In her provocative essay "Lily's Chickens,"[19] biologist and novelist Barbara Kingsolver figures that the average food item set before the typical U.S. consumer travels thirteen hundred miles before it reaches the table. At a conservatively estimated ten food items a day, this adds up to about five million miles per person per year, with all the toxic fumes, greenhouse gases, and other environmental degradation such mileage entails.

Another problem with shipping staple foods long distances is that large quantities must be involved to make the operation profitable for the shippers. In recent times, large-scale production generally has involved factory farms, which rely on distant (nonlocal) markets for their profitability in turn. This means that agribusiness and long-distance food transport go hand in hand. So add the deleterious effects of agribusiness to the environmental costs of food that travels thousands of miles to the ultimate consumer.

The obvious response on the part of an environmentally concerned individual is to buy food produced close to home whenever possible. Many towns and cities worldwide have farmers' markets where meat and produce can be bought directly from the producer. Buying locally grown products has other advantages as well. Most local produce is fresh when bought, which avoids need for refrigeration during conveyance. Furthermore, if you have doubts about locally grown food being organic or free from GM ingredients, you can ask the producer, enabling more control over the quality of food you put on the table.

18.8 Joint Action for Value Change: Public Action

Our concern in this final chapter is with practical steps toward replacing the ecologically destructive social values of consumerism with alternative values favorable to a sound environment. This means shifting the preponderance of public opinion from one value orientation to another. While individuals can influence the value hierarchies of others by means of personal example, large-scale shifts in value of the sort required cannot be brought about by individual exercises like those we have been discussing. What more can we do as individuals to promote a shift to an environmentally friendly set of social values?

The seemingly obvious answer is to join effort with like-minded individuals, hoping to amplify our efforts with coordinated group action. But what kind of group action? In this section we consider action under the category of public pressure. In the next section we turn to cooperative ventures, in which people pool their efforts for mutual benefit.

(a) Boycott Businesses that Exploit the Environment

Economic boycotts in recent memory include Martin Luther King's call to black and white Americans in 1955 to stay off buses in Montgomery, Alabama, and Cesar Chavez's boycotts in the 1960s of grapes and lettuce produced by large growers in Southern California.[20] An effective boycott of General Electric products in the 1980s forced this company to withdraw in large part from the nuclear weapons industry.[21] Occurring at approximately the same time, a boycott against tuna caught by purse seine nets (which destroyed millions of dolphins) led to agreement by the world's largest tuna producers to conditions requiring the display of the "dolphin-safe" label on their products.[22]

The time is ripe for organizing boycotts against corporations and industries causing massive damage to the biosphere. A pacesetter in this regard is a boycott against Exxon Mobil (Esso in Britain) launched recently by Greenpeace and Friends of the Earth.[23] Within a year or so after the boycott began, more than a million motorists in Britain were participating, and Esso sales had fallen by roughly 25 percent.[24] Boycotts are also ongoing (or recently have been) against Coca-Cola for causing critical water shortages in India and against Monsanto for the environmental effects of its herbicides (e.g., Roundup) and genetically engineered crops (soy and rape).[25]

With a few exceptions (like the Exxon Mobile case), boycotts for primarily environmental causes seem not to have been particularly effective. This may be in part because it is hard to focus sustained public perception on a corporation's environmental transgressions. The Internet has untapped potential in this regard. A well-designed website with details on major offenders would give organized groups of individuals considerable leverage over corporate offenders.[26]

(b) Promote Environmentally Sound Businesses

Supporting all businesses with good environmental records is not an appropriate goal, inasmuch as the goods some produce may be relatively useless (think of "pet rocks" in recycled packaging). Businesses especially worth supporting include those supplying things that people really need and that are designed for durability rather than corporate profit. Among commodities people really need today are household fixtures made from metal instead of plastic, appliances that are easy to repair, and word processing systems that remain serviceable without frequent "upgrades" (like old-fashioned typewriters).

An intriguing possibility is for groups of consumers to band together with the intent of influencing the kinds of goods that are brought to the market. If large numbers of low-tech computer users were to identify themselves explicitly as potential buyers of a word processing system guaranteed to remain usable as purchased for at least a decade, manufactures presumably would respond in kind. Profits not forthcoming from frequent "upgrades" could be offset by high-volume sales not requiring expensive advertising.

The Internet, as we know, has enabled the compilation of massive lists of like-minded voters who can exercise considerable influence in the political process. In like fashion, it could also enable the compilation of large lists of potential customers with shared interests in buying certain kinds of products. By taking the initiative away from marketers in making these interests known, like-minded people can cooperate in bringing environmentally friendly values to bear in market transactions. The more successful they are in such endeavors, the more prevalent these values will become in society at large.

(c) Mobilize by Internet

Chapter 17 gave reasons why replacing wealth as an operative value might require an alternative value with normative force. Equity was identified as a value that might have the desired effect. Like other normative values, equity brings distinctions of right and wrong to bear. If equity were in force as a social value, people would be prohibited from using resources for their own advantage that are needed for the livelihood of other creatures. This would translate into a prohibition against excessive wealth.

For equity to be operative as a social value requires that penalties be imposed for inequitable behavior. One way of penalizing offending businesses is to impose boycotts like those just considered. For such boycotts to have the intended effect, however, they must be conducted in ways making it clear that society at large is taking the action and not just a group of political activists. How might actions of this sort be organized to have this broader effect?

It was suggested above that boycotts could be organized on websites providing detailed information on major corporate offenders. Suppose these details included not only environmental offenses but also disclosure of attempts to divert public attention from these offenses. Suppose there was also information about specific policies and practices that lead to environmental damage, along with descriptions of what the organization in question does with profits resulting from such policies and practices. Suppose, in short, that the organization's environmental profiles were spelled out in relevant detail, in a form easily accessible to public view.

Full disclosure in details like this would constitute incriminating evidence of an organization's environmental transgressions. The bad publicity resulting could be enough of a penalty to motivate the business in question to abide by the norm of equity in its subsequent dealings. To the extent that such disclosure induces corrective changes in the way offending businesses operate, it would help establish equity as an operative social value.

One can even imagine a comprehensive environmental website devoted to these several purposes and also to providing up-to-date news on happenings of environmental interest worldwide. Establishing and operating a comprehensive website like this obviously would take a great deal of work. It would also require continued access to a politically neutral Internet.[27] Despite the potential pitfalls, however, organizing a website like this is one more thing a group of like-minded individuals might do to alleviate our ongoing environmental crisis.

18.9 Joint Action for Value Change: Correcting Inequities

The main focus of the preceding sections has been on replacing the social values of consumerism with other values conducive to environmental health. But return to environmental health requires diminishing society's high regard for wealth as well. As pointed out in section 16.6, this will involve both curtailing society's general endorsement of wealth as a personal goal and establishing countervailing values that forbid accumulation of excessive wealth. Section 17.6 identified equity as a normative value that might be effective in this countervailing role.

One way of countering the value of wealth and its influence, accordingly, is to promote equity as a normative social value. And one way to promote the value of equity is to actively oppose existing inequities. Returning to the analogy of voting in section 18.3, we can think of measures taken in opposition to inequity as votes against the dominant social value of wealth.

Let us consider this tactic in the context of the current recession. The recession has exacerbated a swarm of social inequities resulting in many people being unable to meet their basic needs. Paramount among these are needs for adequate housing, adequate food, and adequate health care.[28] But means are at hand for meeting such needs by exchange of labor rather than exchange of currency. By taking steps to make these means available to indigent people, we can help eliminate the inequities in question. In addition to improving the lot of needy people, this will erode the force of wealth as an operative social value.

This section addresses the inequities in question, with attention both to the wealth-enhancing practices that brought them about and to things that might be done to help eliminate them.

(a) Housing

Our culture's predilection for individual home ownership (the "American dream") provides opportunity for profit in various industries. Tract developers profit from building multiple dwellings with the same design. Real estate agents profit from commissions, closing costs, and related fees. Mortgage companies profit from interest payments, and so on. Such is the case at least as long as the housing market remains stable.

In point of fact, attempts by realtors and mortgage lenders to increase their profits was a major factor in the market collapse of 2008. Profits here are keyed to volumes of sales, which led (while housing values were increasing) to large increases in home sales to underqualified (subprime) buyers. Mortgage banks hedged against default by using mortgaged properties as collateral in insuring their loans with second-tier lenders. These latter maintained their credit ratings by passing the risk on to yet higher-tier lenders (either governmental, as with Fannie Mae and Freddie Mac, or private, as with AIG). One way or another, mortgages purchased by low-income buyers fueled the profit-making activities on higher levels of the system. In effect, the entire investment system was based on credit (in a Ponzi-like manner) rather than solid assets. When growing numbers of subprime mortgages went into foreclosure with the rapid decline of the housing market in 2007–08, the entire system collapsed like a house of cards.

The system collapsed because of excessive risks by bankers and investment firms in their attempts to increase profit. As a result, hundreds of thousands of low-income families lost their homes by default and the economy at large went into recession. Here is an example of blatant inequity. Attempts by investment bankers to become even richer caused millions of poor people to lose their homes.[29]

As many have observed retrospectively, the financial maneuvers leading to this disaster were motivated by greed—which is to say, by the desire for wealth. Another culprit is the economic practice of treating housing as a commodity to be bought and sold. Insofar as the collapse resulted from the financial sector of the housing industry becoming unglued, it resulted as well from our practice of making housing contingent on the exchange of money.

But there are ways of acquiring housing other than cash payments and mortgages. Consider the age-old alternatives of fabricating one's own living quarters. Inuits, for example, constructed igloos from blocks of snow, Navahos made hogans from logs and dirt, settlers of the American plains made soddies from turfs of prairie grass, and so forth. In such cases, access to living quarters is gained in exchange for labor instead of cash.

Although the practice is uncommon, there are circumstances today in which housing can be acquired in exchange for labor. Consider the Amish tradition of barn raising, in which a community comes together to assemble a barn for one of its households.[30] The basic idea is that groups of able-bodied volunteers cooperate in building shelters for each other.

Modern construction requires the services of specialists like electricians, plumbers, and heating contractors. Such craftspeople might be part of the cooperating group, combining their special skills with the labor of other members. Another possibility is to pay these specialists wages financed by nonprofit lending societies set up for this purpose.[31] Building supplies could be provided by similar cooperative ventures. Instead of purchasing materials and equipment from profit-making outlets, participating members could obtain these through purchasing cooperatives (perhaps comprised of several building associations) that buy supplies in bulk and pass the savings on to their members.

What we are describing here is basically a system of interacting cooperative ventures (co-ops). Able-bodied people pool their labor in building houses for each other. These are the primary co-ops of the system. Necessary supplies and equipment are obtained through secondary co-ops (co-ops of co-ops), which obtain them at discount prices. And necessary financing is handled through yet other co-ops, which manage money for their members on a nonprofit basis.

In upshot, whereas in our current economic system most people can obtain housing only in return for money, in this alternative system labor will sometimes do as well. The parts of this alternative system are already available in prototype—barn-raising communities, building societies, nonprofit banks (more on this to follow), and cooperative supply centers. What is needed to put them together is to escape from the grips of the profit motive in the provision of necessities like housing. Desire for wealth is the main cause of inequity in housing. And every step we can take to reduce the inequity will help diminish the force of wealth as a social value.

(b) Food

One anomaly of the recession in 2008–09 is that food prices went up while wages decreased. In view of the general interaction between supply and demand, we would expect food prices to rise as a result of decreased supply. But the reason food prices rose in the United States during this period was not a shortage of food, for in fact the country has more food than it needs. The reason food prices rose was that reduced cash flow was decreasing profits at each juncture of the food-supply chain and that the enterprises involved responded accordingly.

To get a sense of how this happens, let us compare the way food reaches the table today with the typical food-supply system of a century ago. The standard process for getting food from producer to consumer back then involved (1) local farmers who trucked their own produce to market, (2) local grocers who sold that produce for cash or locally arranged credit, and (3) local customers who could buy fresh meat and vegetables within a day or two after they left the farm. Although vestiges of this simple system remain (fruit stands, farmers' markets), food reaches the tables of most consumers today through a much more indirect process.

Today, the typical sequence begins with (1) chemical companies that sell genetically modified seeds, fertilizers, and pesticides to industrial farms, along with (2) factory farms themselves that use heavy equipment to grow a few standardized food products. These products are then transported to (3) processors that make them tasty and increase their shelf life, employing the services of (4) container corporations that provide packaging, (5) marketing specialists who make the packages attractive, and (6) advertising firms that bring the packaged product to public attention. Next in line come (7) wholesalers and warehouses that supply large food chains, (8) the food chains themselves, and (9) their local outlets, where the products are retailed. Providing transport along the way are (10) various trucking firms, railways, airlines, and shipping companies. At the end of the line, of course, is the hapless consumer, often equipped with credit cards to make the transaction more convenient.

One possible way of viewing the overall process is to see the consumer as *beneficiary* of all these auxiliary enterprises. For someone with ample means, to be sure, it is relatively easy to rely on the local supermarket to meet all one's food needs. From the perspective of an average wage earner, however, the food-supply chain is a long sequence of profit-making businesses each accounting for a portion of the price he or she pays for food. The typical consumer is more benefactor than beneficiary.

Given the competitive economic environment in which these businesses operate, moreover, each is trying to maximize its share of the money changing hands as part of the process. And in times of recession, when business lags and profits dwindle, each will do what it can to maintain a favorable profit margin. The cumulative result is that prices increase at each stage of the process, all adding to the final price paid by the consumer.

Beyond the inequity imposed on the helpless consumer, there are several levels of this supply system that are environmentally destructive on their own. Notable among them are the factory farms manufacturing food out of chemicals and the transportation networks moving products long distances to the point of consumption. Consumer equity and environment alike would benefit from a system that brings producer and consumer into closer contact. Let us consider a few ways of bringing that about.

One promising model is Community Supported Agriculture (CSA), also known as subscription farming. In a CSA venture, members team with a local farm in sharing both benefits and risks of food production.

The group pledges to cover the costs of the farm's operation, including the farmer's salary, and in return individual members are allotted designated portions of the farm's produce. Members benefit by receiving high quality food (often grown without chemicals) at lower-than-market prices. And farmers benefit by not losing income in years of crop failure, by receiving better return on their crops (with middlemen eliminated), and by avoiding the extra burden of finding markets for them.

Of particular interest for our purposes is that some CSA projects encourage members to work on the farm in exchange for a portion of their membership costs.[32] Here is a case in which someone can pay for food by labor instead of money, parallel to that discussed previously of exchanging labor for housing. In the context of the current recession, a person without regular income could still put wholesome food on the table as a working member of a CSA.

Flexibility in such transactions can be enhanced by integrating CSA farms into systems of so-called local currency. Local currencies (also referred to as community currencies) are issued and traded within circumscribed communities and are not backed by a national government. Such currency typically can be used in exchange for merchandise in participating stores and sometimes can be used to pay local taxes. Not uncommonly, local currencies are supported by local banks, where they can be exchanged for national currency and deposited in savings accounts.[33]

Another way of eliminating profit-taking middlemen from the food-supply chain, of course, is to grow one's own food. Ready illustrations are the victory gardens grown in the United States during World War II and the gardens many people still plant in their own backyards. Where backyards are small or nonexistent, there are other ways of making space available to the individual gardener. In communities featuring dwellings with large lawns, for example, occupants could be encouraged to replace lawns with well-tended gardens. This would entail a value shift away from vegetation that is merely decorative to plantings that serve a genuine human need.

Even in densely packed cities, space can often be found for growing fruits, vegetables, and other food crops.[34] Typical areas include vacant lots, road berms, building tops, and street planters. According to one recent estimate, Havana, Cuba, produces about 90 percent of its food supply in or near the city, relying primarily on municipally owned property.[35]

An interesting twist on the theme of noncommercial food production is known as "permaculture," for permanent agriculture. Briefly described, permaculture is the development and operation of agricultural systems modeled on the structure of natural ecologies.[36] Whereas industrial farming focuses on single crops grown with chemicals and harvested with heavy machinery, permaculture features a variety of foods grown together relying on natural energy flows. This orientation assures that its products are organic and that their production has minimum impact on surrounding ecosystems. Like CSAs and similar farming ventures, permaculture is a way of removing inequity in the provision of food.

(c) Health Care

In 2000, the World Health Organization ranked the U.S. health system thirty-seventh among those of 190 nations worldwide.[37] Ranking higher were France (1), Italy (2), Spain (7), the United Kingdom (18), Germany (25), Canada (30), Australia (32), and Costa Rica (36). A Commonwealth Fund ranking in 2004 showed that the United States, despite spending twice as much per capita ($6,102) than the other five countries covered in its study, ranked last in health care behind the United Kingdom, Germany, Australia, New Zealand, and Canada. Why is the United States getting such a poor return on its health-care dollar?

Part of the answer comes with another salient difference among the six countries involved. The United States is the only one without a universal health-care plan. Instead, it relies heavily on private health insurance coverage. In 2004, the nation's health insurers gained $10.2 billion in profit. During that year, their gain in net profit margin was 3.78 percent, up from 0.38 percent in 1999.[38] This trend continued up to the time of the recession in 2008–09.

Insurers are not the only parties making profit as part of the nation's enormous health-care industry. Another sector contributing substantially to our health-care costs are the tens of thousands of hospitals and clinics that operate on a profit-making basis. Although a majority of health-care facilities in the United States are legally nonprofit (and thus receive tax breaks), the actual combined profits of the fifty largest nonprofit U.S. hospitals in 2006 was more than $4 billion. This amounts to an average of about $800 million profit per hospital. These institutions operate like profit-making corporations in other respects as well. In 2006, the retiring CEO of Chicago's Northwestern Memorial Hospital (one of the top fifty) received a $16.4 million "golden parachute."[39]

In a 2001 report by the consumer health organization Families USA, the pharmaceutical industry was identified as the most profitable of all businesses in the United States.[40] With $289 billion in U.S. sales that year and a profit margin of 17 percent, it gleaned close to $50 billion in profit. The industry is also known for the exceedingly generous compensation of its top executives. In 2000, Pfizer's chair made $40.2 million exclusive of stock options, and Bristol-Myers Squibb's CEO held $227.9 million in unexercised options.

In response to public complaints about such profligacy, the industry's defensive mantra is that high drug prices are needed to underwrite research and development (R&D) of new drugs. Contrary to that rhetoric, however, a recent report by the consumer health organization Families USA documents the fact that drug companies are spending twice as much on marketing and advertising than on research and development. The high drug prices that have exacerbated the health-care crisis for Americans are due, not to R&D efforts, but to record-breaking profits and exorbitant compensation for top executives.[41]

Other factors contributing to the high cost of medical care are physician appointments, diagnostic technologies (including MRI equipment), prosthetic applications, laboratory testing, and nursing homes. With the occasional exception of unaffiliated physicians, all these services are supplied and delivered by profit-making enterprises. The involvement of so many businesses that operate for a profit goes a long way toward explaining the inequities infesting our profit-driven health-care system.

What would it be like to have a health-care system focused on the mission of keeping people healthy rather than on making large profits for its providers? While a detailed answer is beyond our present reach,[42] we may consider a few general principles that a detailed answer should exemplify. These principles are predicated on the status of equity as an operative social value.

Like food and shelter, preservation of health is a basic human need. When health care is a commodity for purchase in the market, as in the United States currently, this means that people must pay money to satisfy a basic need. If the going price is beyond their means, they forfeit the prospect of continued good health.

The first and most fundamental principle is that, in a society with a durable set of values, basic needs can be met without paying the prices demanded by profit-seeking providers. Paramount among human needs, as previously noted, are food, shelter, and preservation of health. In the case of shelter, one way of complying with this principle is to set up cooperative housing societies in which workers pool their labor in building dwellings for each other. In the case of food, similarly, the principle is satisfied by CSA-like arrangements in which people can acquire provisions in return for work instead of money. Arrangements along similar lines are possible in the case of health care.

Proof of this possibility comes with the Ithaca Health Alliance associated with the local currency entitled Ithaca Hours. Included as parts of the Health Alliance are the Ithaca Health Fund and the Ithaca Free Clinic. The fund offers its members interest-free loans for dental procedures and eye care, and the Free Clinic provides general health-care services to uninsured residents of the local community. The clinic is staffed on a voluntary basis by health professionals among its membership. In addition to free care by volunteers, the clinic also provides its members access to discounted health-care services by over a hundred providers throughout the state of New York.[43]

Whereas the first principle applies to human needs generally, the remaining two pertain to health care specifically. Second is the principle that prevention should be emphasized over corrective procedures. The point is not that correction is not important when necessary, but rather that the more prevention is exercised the less correction is needed. Since corrective care is usually more expensive than prevention,[44] people with low incomes will be less at risk in communities where prevention is emphasized.

The sense of preventive care here intended, it should be noted, is not limited to such matters as vitamins and regular check-ups. Broadly speaking, prevention means the avoidance of circumstances with unhealthy consequences. Increasingly evident causes of health problems in Western societies today are sedentary lifestyles and lack of exercise.[45] With this in view, preventive health care should be understood to include community-sponsored playing fields, bicycle paths, and jogging trails, along with shopping centers located to be accessible by walking.

The third principle is that health-care practitioners should not be limited to licensed doctors of medicine. Although this varies from country to country and region to region, medical doctors tend to have a monopoly in the industrialized West. As with monopolies generally, a consequence is that health care often tends to operate for profit, rather than being focused on the needs of ordinary people.

Needless to say, doctors of medicine (and their close allies, doctors of osteopathy) are indispensable within the ranks of health-care professionals. But there are several other preventive and healing practices that are gaining recognition for their contributions to human well-being. Among them are acupuncture, dietetics, herbalism, massage, and physical therapy. Since these practices generally employ neither patented medicines nor expensive equipment, their services are available at fractions of the costs of standard medical care. Their contributions to the inequities of our current health-care system accordingly are minimal.

Although health-care reform in the United States is a politically volatile topic, removing inequities like these is essential if we are to join other countries in providing equity the operative status it needs to nullify the force of wealth as a social value. Unless wealth can be dethroned as a dominant social value, there is little hope of escaping environmental catastrophe.

18.10 Looking Behind and Ahead

We have laid out the nature of our current environmental crisis. We have identified its origins in the excessive amounts of energy consumed by industrial society. And we have seen how those excesses go hand in hand with economic growth, driven by the desire for wealth of corporate managers and investors.

In this capacity, desire for wealth is abetted by the values defining consumer society. Curtailing growth requires curtailing the effect of these consumer values as well. Taken together, these observations indicate that our current environmental crisis can be resolved only by a massive shift in values. Survival requires a change from our present set of social values to a set compatible with a recovery of ecological health.

In this final chapter, we have considered a number of practical steps that might be taken to accomplish this massive change in values. Some are steps individuals can take on their own, hoping to influence the priorities of other people by personal example. The more people are influenced by our example, the sooner we can arrive at a set of values that are environmentally sustainable.

Others are steps to be taken in cooperation with like-minded people. Among these are group actions intended to dissuade businesses from profiteering at the expense of the environment. Also included are actions in support of cooperative ventures enabling people to meet their basic needs without adding to corporate profit. To the extent that these actions are successful, they will help bring about an economic order keyed to the needs of individuals rather than to corporate greed.

Individual or joint, these steps all involve significant changes in our daily routines. Some involve changes in personal lifestyle, and some a measure of individual sacrifice. Although society at large will benefit in the long run by a shift to environmentally friendly values, the process of transition will be painful for many people.

As of this writing, the world economy is in the grips of a bruising recession. Hardships imposed by this recession include several that might be expected to accompany a transition to an ecologically sound economic order. Paramount among these are widespread unemployment, loss of homes, and ruinous debts. Recession after all is curtailed growth, and curtailment of growth is a precondition of a sustainable economy.

Experts generally seem confident that the economy will eventually recover from the current recession. Of utmost importance, however, is the direction it will take on the way to recovery. One alternative is to reinvigorate the greed-based market economy that has just collapsed under its own weight. This is the effect of the massive government bail-outs of multibillion-dollar banks and of the cash infused to prop up a stumbling automobile industry.

Rejuvenating the old economy, however, will renew opportunities for key players to pursue personal wealth at the expense of others deprived of bare necessities. Since this alternative involves a resumption of economic growth, it moves us further along the path to ecological destruction.

The other alternative is to take advantage of the current recession as a first step in the direction of environmental sustainability. Toward that end, we would institute measures making it more difficult for a privileged few to amass excessive personal wealth. And we would make it easier for most people to meet their basic needs without contributing to profits taken by the privileged few.

No less importantly, we would encourage our economists and business leaders to devise an economic order not dependent on growth for its continued well-being. Given what we have learned from the present study, this would be an economic order also conducive to environmental health.

The question of how we emerge from the current recession is momentous. If the recession ends with a return to the pattern of continuing growth that mainstream economists hope for, our environmental crisis will continue to worsen. But if we treat the recession as an opportunity to find ways of doing business that respect the needs of ordinary people, the biosphere will benefit along with humanity at large.

The bottom line is this. If we can shake off old habits induced by unending growth, our ailing biosphere will have a fighting chance to heal itself. Self-healing is the only antidote to the self-destruction that currently threatens human tenure on the planet. It is the only way to save humanity from becoming unearthed.

Notes

1. I have benefited substantially from discussions with Professor Fernando del Rio Haza, former president of the Mexican Academy of Sciences, on matters of entropy and thermodynamics. Shortcomings in the treatment of these topics in the present study, however, are in no way his responsibility but mine alone.

2. I am indebted to my brother John E. Sayre, author of several distinguished texts in economics, for help in dealing with some of the intricacies of this discipline. He is not to be held responsible for the conclusions of this study, with more than a few of which he disagrees.

3. A free-market economy is one in which prices are controlled by supply and demand, in contrast with managed economies, in which prices are controlled by administrative authority. Since most actual economies today have elements of both, they are more accurately described as mixed. The term "mixed economy" has come into standard use accordingly. When the term "free-market economy" is used in this study, it should be understood as referring to the free-market component of economic systems that may be partially managed as well (e.g., those of Canada and the United States, as well as Russia).

Chapter 1. Two Laws of Thermodynamics

1. In this study, degraded energy is energy that has undergone a diminution in its ability to produce work. As further elaborated in section 1.3, entropy

is the work capacity expended in the process. With these provisions in mind, we can think of entropy as a measure of energy degradation—or as degraded energy for short. The concept of degraded order will be clarified in chapter 2.

2. The following characterization of a closed system will suffice for our purposes. Broadly conceived, a system is an assemblage of variables (physical or otherwise) that interact in a nonrandom fashion. Euclidean geometry is an example of a nonphysical system. A physical system is one with physical variables that undergo change with time. Physical systems are either open or closed. An open system is one that can influence or be influenced by factors outside itself. The defining feature of open systems thermodynamically is that they can import energy to accomplish work and can export the resulting residues. Such systems will also be referred to as operating systems. Living organisms are examples of operating systems; as elaborated in chapter 3, they must draw in energy to remain alive and must rid themselves of wastes produced by their metabolic activity. A closed (or fully isolated) system, by contrast, is one that cannot interact with anything outside itself. A thermodynamic consequence is that a closed system can neither import energy for work nor rid itself of the resulting by-products. All residues of work accomplished by a thermodynamically closed system are retained within the system.

3. In thermodynamics, energy capable of producing work is commonly referred to as "free energy."

4. In Clausius's original sense pertaining to heat exchange specifically, increase in entropy equals the amount of thermal energy exchanged divided by the absolute temperature (Kelvin) at which the exchange takes place.

5. An often-cited comment by John von Neumann (a pioneer in computing theory and the theory of games) to Claude Shannon (the founder of communication theory) is that Shannon ought to call his newly formulated measure of information "entropy" for the reasons (1) that the term was already used for a similar function in thermodynamics and (2) that "most people don't know what entropy really is, and if you use the term 'entropy' in an argument you will win every time." The quotation is from Myron Tribus, who heard it from Shannon. See Levine and Tribus, *The Maximum Entropy Formalism*, 3.

6. See Dyson, "Energy in the Universe."

7. This is radiation characteristic of a body that absorbs all energy it receives, heats up to a certain temperature, and then reradiates its energy with an energy/wavelength correlation determined by that temperature. This definition is elaborated in the notes of chapter 5.

8. It must be emphasized that this model is not a depiction of actual energy transformations. Nor is it a model of actual systems in which energy is used to produce work. A consequence is that it has no significance in terms of actual physical processes studied in thermodynamics. Its purpose is merely to help the nontechnical reader to conceptualize the transformation of energy

from one form to another and to understand in a general manner how work involves the production of entropy (i.e., expended work capacity).

9. As noted previously, energy leaves the earth's surface by black-body radiation. By current scientific consensus, this black-body radiation shares characteristic features with the nondirectional cosmic background radiation predicted in Big Bang theory and empirically discovered in the 1960s. The cosmic space occupied by this background radiation has a temperature of about 3° Kelvin (which is actually quite chilly). See Weinberg, *The First Three Minutes,* esp. 70.

Chapter 2. Entropy and Disorder

1. Treatments associating entropy with randomness or disorder are standard today not only in branches of mathematical and social science but also in the physical sciences, where the concept was first employed. For examples in physics, see Schrödinger, *What Is Life?,* and Bridgman, *The Nature of Thermodynamics.* For chemistry, see Daintith, ed., *A Dictionary of Chemistry;* for ecology, Odum, *Environment, Power, and Society;* and for economics, Georgescu-Roegen, *The Entropy Law.* For communication theory and ergodic theory (both mathematical disciplines), respectively, see Brillouin, *Science and Information Theory,* and Petersen, *Ergodic Theory.*

2. There is a movement among recent authors of chemistry textbooks to avoid illustrations that introduce entropy in terms of disorder, on the grounds that ordinary connotations of the term "'disorder'" tend to mislead beginning students (e.g., see http://www.entropysite.oxy.edu/cracked_crutch.html). A case in point is a bowl of cracked ice that subsequently melts into water. This illustration can be misleading insofar as the jagged ice appears visually to be less orderly than the homogeneous volume of clear water, suggesting erroneously that melting results in a change from disorder to relative order. As any advanced student of thermodynamics should realize, however, the molecules in a piece of ice are more highly ordered than the same molecules in the form of liquid water. For the instructor to point this out would seem to be an appropriate exercise in a standard first-year chemistry course. At any rate, the reader should be forewarned that the concepts of entropy as degraded energy and as disorder (degraded order) are interconnected only when disorder is understood in an appropriate sense. An appropriate sense introduced later in this chapter is disorder as randomness, correlative with order as departure from randomness.

3. By way of analogy with the second factor, consider a number of birds perched on a power line. One hundred birds evenly spaced (say one foot apart) constitute a higher degree of order (further departure from randomness) than

a mere five or six also evenly spaced but further (maybe twenty feet) apart. This factor is examined more fully in section 2.5.

4. The heading "Technical Addendum" is used for passages containing technical material relevant to the main text but not necessary for following the overall argument. These passages are included for the benefit of readers who, for whatever reason, might be interested in more background than the main text provides.

5. Significant work has already been done in measuring the flow of energy through ecosystems. See section 4.6 and associated endnotes. Measuring the objective structure brought about by those energy flows is another matter.

6. By this, Schrödinger said he meant entropy with a negation sign. See *What Is Life?* 73.

7. Prominent examples are Brillouin (*Science and Information Theory*, e.g., 153) and Georgescu-Roegen, (*The Entropy Law*, e.g., 8). I adopted the term in *Cybernetics and the Philosophy of Mind*. See also my more recent "Cybernetics."

Appendix

1. Important preliminary work had been done by H. Nyquist and R. V. L. Hartley in the 1920s. Relevant papers of Nyquist, Hartley, and Shannon were all published in the *Bell System Technical Journal*. "The Mathematical Theory of Communication" was published in the July and October issues of 1948. Various reprints have since become available.

2. A brief survey of philosophic applications can be found in my article "Information Theory."

3. It is customary to refer to this quantity as the "entropy" of the source. This terminology will not be used here to avoid taking a gratuitous stand on the relation between this quantity and the entropy of thermodynamics, on which experts disagree. A discussion of the relation between entropy in communication theory and that in thermodynamics may be found in my book *Cybernetics and the Philosophy of Mind*, chapter 3, section 3.

Chapter 3. Life, Negentropy, and Biological Feedback

1. The Sagan quotation is from his entry "Life" in the *New Encyclopaedia Britannica*. Lorenz's remark appears in his *Evolution and Modification of Behavior*, 32. Fisher's description was quoted by Julian Huxley (biologist) in Young, ed., *The Mystery of Matter*, 521.

2. Schrödinger, *What Is Life?* 79.

3. Ibid., 76.

4. An example of a thermodynamically reversible process is the freezing of water (slowly, at 0° Celsius and one atmosphere of pressure), since ice can melt under the same conditions without additional expenditure of energy. But energy is expended in all metabolic processes, which means that all life processes are irreversible.

5. As defined in chapter 1, an operating system is a physical configuration of interacting variables whose values change with time. Several examples are discussed in chapter 2.

6. An information-theoretic analysis of sentient and anticipatory feedback is put forward in chapter 4 of my book *Cybernetics and the Philosophy of Mind*. This analysis is extended to reason in chapter 12.

7. In the words of James Kay, in his article "Ecosystems, Science and Sustainability," ecosystems are self-organizing, meaning "that their dynamics are largely a function of positive and negative feedback loops." In his view, ecosystems have multiple operating states, depending on the feedback loops within them that happen to be dominant.

8. The term "biosphere" was coined in 1875 by the geologist Eduard Seuss. Study of the biosphere was established on a scientific footing by the biogeologist Vladimir Ivanovich Vernadsky in the first quarter of the twentieth century. Vernadsky stressed the extensive interaction between the living and the nonliving components of the biosphere (see his *The Biosphere*).

9. One of the first to focus on this was Howard T. Odum, in his influential *Environment, Power, and Society*, 11.

10. The diagram is intended to represent incoming solar radiation only in the wavelengths involved in photosynthesis, from roughly 250 to 1,500 nanometers (visible light ranges from 380 to 750 nanometers). The nonmetabolic branch of that incoming flow also is intended to include only solar energy that is converted into low-grade heat (e.g., that is absorbed by dark surfaces). The portion of sunlight reflected directly back into space (the albedo effect) thus is not represented.

11. See the articles by Oort and Woodwell in *The Biosphere*, and Gates, "Biosphere."

Chapter 4. Ecosystems and Top Consumers

1. The term "trophic" comes from *trophē*, a Greek term meaning food or nourishment.

2. A highly relevant account of the Easter Island civilization may be found in chapter 2 of Jared Diamond's *Collapse: How Societies Choose to Fail or Succeed*. Also see chapter 1 of Clive Ponting's *A Green History of the World*.

3. Numerical estimates in this paragraph come from the *Scientific American* publication *The Biosphere*, in particular the articles by Oort and Woodwell, and from Gates, "Biosphere."

4. This is the "fifth extinction" relative to the namesake sixth of *The Sixth Extinction: Patterns of Life and the Future of Humankind,* by Richard Leakey and Roger Lewin; see especially 52–56. For a less technical account, see Flannery, *The Eternal Frontier,* chapter 1.

5. This recycling process is dramatized by Aldo Leopold in his story of the "odyssey" of atom X in *A Sand County Almanac.* After being locked in sedimentary rock for billions of years, X was pulled into the world of living things by the root of an oak and began an uphill journey during which it passed through acorn, then deer, then hapless Indian, and again through buffalo chip, spiderwort, rabbit, and owl, until a fox caught a gopher and an eagle caught the fox, and X reached its zenith in an eagle's nest. After that, it was a quick trip downward, through cottonwood, beaver, coon, and crayfish, until X was finally caught in a freshet and returned to the sea. Without life, X might have remained under the surface forever, but by joining life it "soared on eagle's wings."

6. Ecologists more commonly think of stability as the ability of an ecosystem to return to equilibrium quickly after perturbation. In point of fact, a stable ecosystem typically would provide a firm support for its top consumers. The definition adopted in the text focuses on that point directly.

7. Eugene P. Odum, *Fundamentals of Ecology.* In his later *Ecological Vignettes,* Odum states categorically that, in addition to homeostasis, "redundancy—that is, more than one species or component capable of performing a given function—also enhances stability in ecosystems" (121).

8. An early attempt to quantify trophic flows in ecosystems in terms of information theory (see the appendix to chapter 2) was made by MacArthur in "Fluctuations of Animal Populations." Other attempts to apply information theory to the description of ecosystems may be found in Margalef, *Perspectives in Ecological Theory;* Wicken, *Evolution, Thermodynamics, and Information;* Pahl-Wostl, *The Dynamic Nature of Ecosystems;* and Ulanowicz, *Ecology.* An interesting analysis of the upshot of MacArthur's early attempt is offered by Ulanowicz, *Ecology,* 65–66.

9. For a summary overview, see Pimm, "The Complexity and Stability of Ecosystems." Also see Ulanowicz, *Growth and Development,* 117–18.

Chapter 5. Entropy Trapped within the Biosphere

1. Black-body radiation is radiation emitted from a totally absorbing (hence visually black) surface, peaking at a wavelength inversely proportional to the body's temperature. With a typical night temperature of about 10° Celsius, radiation from the earth's surface would peak at about 10^{-3} centimeters (in the infrared range). This is by way of elaborating the discussion in chapter 1.

2. Data in this paragraph come from Goudie, *The Human Impact on the Natural Environment*, 227–31.

3. Like all plants, algae produce oxygen in the presence of sunlight but consume oxygen at night. Microbes that decompose dead plant matter also consume oxygen. Under conditions of eutrophication, consumption of oxygen exceeds its production.

4. Material in this paragraph may be corroborated in the report from the European Commission, *Eutrophication and Health*.

5. Data are from Komin and Nikkola, "Ice Ages."

6. This paragraph derives from web reports of several NASA studies, most importantly Weier, "A Delicate Balance"; Herring, "Does the Earth Have an Iris Analog?"; and Gache, "Global Warming Triggers Cooling Mechanism" (citing the work of Roy Spencer).

7. Weier, "A Delicate Balance"; Herring, "Does the Earth Have an Iris Analog?"; and Gache, "Global Warming Triggers Cooling Mechanism."

8. Data to this approximate effect can be found in many places. See, for example, "Sea Level Rise."

9. See "Sea Level Projections."

10. Data from "Global Warming and Hurricanes."

11. Immediate sources for this and the following paragraph include "Shutting Down the Oceans" and "Coral Reef Biology." See also Kolbert, "The Darkening Sea." More recent reports on ocean acidity and coral reefs include Dean, "Coral Reefs and What Ruins Them," and Dean, "Rising Acidity Is Threatening Food Web."

12. See Hecht, "Global Warming."

13. Displacement of fifty million people is expected over the next decade if desertification continues at its present rate (Kahn, "As China Roars").

14. Estimates like these vary with sources and require regular updating. These figures are from Brown, *Eco-Economy*. Other sources used in this section include Hawken, Lovins, and Lovins, *Natural Capitalism,* and Ponting, *A Green History of the World*.

15. A much larger dead zone, caused by agriculture waste, is growing at the mouth of the Mississippi River. In 2010 it covered more than eight thousand square miles ("The Dead Zone"). Dead zones have appeared more recently in the Southern Hemisphere, particularly in South America, Africa, and parts of Asia. By 2007, scientists had identified more than four hundred dead zones around the world. See Adam, "Suffocating Dead Zones Spread Across World's Oceans."

16. The U.S. population in 2000 was about 290 million ("U.S. Population, 1790–2000").

17. See "Environmental Management."

18. Experts estimate that China's landfills will be full by 2020. Ross, "Management of Municipal Solid Waste in Hong Kong and Taiwan."

19. Data taken from "Plastics."

20. Moore et al., "A Comparison of Plastic and Plankton."

21. Facts and figures in this section come mostly from McNeill, *Something New Under the Sun;* Ponting, *A Green History of the World;* Hughes, *An Environmental History of the World;* and Simmons, *Changing the Face of the Earth.* Other sources are noted when relevant.

22. The chemistry of this effect has been studied extensively in recent decades. The creation of ozone in the stratosphere is initiated by the action of ultraviolet radiation (of less than two hundred nanometers) on ordinary oxygen molecules (O_2), splitting them into oxygen atoms (O). These atoms are then free to combine with other oxygen molecules to form ozone (O_3). Ozone in turn is broken down into O_2 and O by radiation of somewhat longer wavelengths (two hundred to three hundred nanometers). While the ozone layer remains intact, nearly all the sun's ultraviolet radiation is absorbed by these two complementary reactions.

The approximate balance between ozone formation and decomposition is disrupted by the introduction of halogen compounds (primarily bromides and chlorides) into the stratosphere. These compounds release halogen molecules that serve as catalysts in breaking up ozone molecules without themselves being rendered chemically inactive in the process. It has been estimated that a single chlorine molecule can catalyze the breakup of over one million molecules of ozone before it becomes part of an inactive compound (Sparling, "Basic Chemistry of Ozone Depletion"). The effect is a radical reduction of ozone (the so-called ozone hole) that correspondingly reduces the amount of ultraviolet radiation absorbed in the stratosphere. The basics of the chemical reactions are clearly described in NASA's Earth Observatory series "Ozone."

23. Brown, "Ozone Layer Most Fragile on Record."

24. A useful overview of the nitrogen cycle may be found in Delwiche, "The Nitrogen Cycle." The empirical data in this section come from Vitousek et al., "Human Alteration of the Global Nitrogen Cycle."

25. Vitousek et al., "Human Alteration of the Global Nitrogen Cycle," 7.

26. Penman, "The Water Cycle."

27. Main sources used for this section are McNeill, *Something New Under the Sun,* and Kimbrell, ed., *Fatal Harvest.* Other sources are indicated in relevant notes below.

28. Leahy, "New Studies Back Benefits of Organic Diet."

29. Pesticides are designed to kill living creatures. More often than not, they kill creatures other than those intended. It has been estimated that only about 1 percent of agricultural pesticides actually contact the pests targeted (Ponting, *A Green History of the World,* 370). A well-known case in point is the use of dichloro-diphenyl-tricholoroethane (DDT), which is highly carcinogenic. Its toxicity affects workers applying the chemical, other people and

animals caught in the spray draft, and yet others drinking the water that has collected from runoff. According to one source (Ponting, *A Green History of the World,* 370), about twenty thousand people die from pesticide poisoning each year, with three-quarters of a million others suffering serious health effects.

30. Kimbrell, ed., *Fatal Harvest,* 71.

31. "Genetically Modified Crops in the United States."

32. Data in this section come from Wilson, *The Diversity of Life;* Leakey and Lewin, *The Sixth Extinction;* Pimm, *The World According to Pimm;* Kimbrell, ed., *Fatal Harvest;* and articles cited in relevant notes below.

33. Recent estimates say about one-third of the world's amphibian species are threatened, about one-fourth of all mammals, and about one-eighth of all birds. See Benjamin, "Threatened Species Red List."

34. Snoeks, "*Lates Niloticus* (Fish)."

35. "Rainforest Facts."

36. Dean, "Scientists Warn Fewer Kinds of Fish Are Swimming the Ocean."

37. Statistics in this and the following paragraphs come from "Rainforest Facts."

Chapter 6. The Rising Tide of Human Energy Use

1. Counting the beginning of life (about 3.5 billion years ago) to the present as a single day, we can date the onset of industrialization as occurring within the last millisecond or two. This is a very short time for the ills of industrialization to become so dramatically evident as they are today.

2. Celluloid, the first commercially successful plastic, was introduced by its inventor, Alexander Parkes, in 1862.

3. Smil, *General Energetics,* 101.

4. Smil, *Energy,* 19.

5. Smil, *General Energetics,* 101. This estimate varies with geographic area. Semipermanent settlements in salmon and whale territories may have included several hundred people.

6. Smil, *General Energetics,* 200.

7. Ponting, *Green History,* 90; Goudie, *Human Impact,* 9.

8. Smil, *Energy,* 23; McNeill, *Something New,* 8; Simmons, *Changing the Face,* 89.

9. Ponting, *Green History,* 295.

10. Ponting, *Green History,* 90.

11. Simmons, *Changing the Face,* 24. Simmons estimates twelve million calories per diem for early agriculturists, which converts to about 19 billion

joules per annum (one calorie is equivalent to about 4.2 joules). This figure has been reduced by a factor of three in the text above, under the assumption that even after farming became prevalent in certain areas the people involved constituted a relatively small proportion of the human population worldwide. Although estimates like Simmons's are based in part on observation of presumably similar farming cultures existing presently (40, 47–49), they should be understood as approximations at best.

12. Ponting, *Green History*, 271.

13. Smil, *Energy*, 48.

14. Smil, *General Energetics*, 120–21.

15. Smil, *Energy*, 30–32, 42–47.

16. Ponting, *Green History*, 90.

17. Simmons, *Changing the Face*, 24.

18. Smil, *Energy*, 61, 70.

19. Smil, *Energy*, 49.

20. A major use of draft animals in the twentieth century was for purposes of military logistics. Ponting notes (*Green History*, 272) that the British army used 1.2 million horses during World War I, and that the Germans (despite their mechanized Panzer divisions) used 2.7 million horses during World War II.

21. Smil, *Energy*, 137.

22. Smil, *General Energetics*, 142, 301; Smil, *Energy*, 108.

23. Ponting, *Green History*, 274.

24. Smil, *Energy*, 103.

25. Ponting, *Green History*, 275.

26. Smil, *Energy*, 114.

27. See Baker, "Gunpowder."

28. Smil, *Energy*, 167.

29. Simmons, *Changing the Face*, 199.

30. Smil, *Energy*, 159.

31. Simmons, *Changing the Face*, 24, 104; Ponting, *Green History*, 92; Goudie, *Human Impact*, 10.

32. Smil, *Energy*, 168.

33. Smil, *Energy*, 170. This despite the construction of the first major hydroelectric plant at Niagara Falls in 1886 (Ponting, *Green History*, 286).

34. Simmons, *Changing the Face*, 201–2; Smil, *Energy*, 170.

35. Hughes, *Environmental History*, 120.

36. Smil, *Energy*, 187. Per capita energy use estimates for earlier periods are based on extrapolations from comparable present-day circumstances. Estimates from the industrial era, on the other hand, come largely from sources recorded during the period in question.

37. Goudie, *Human Impact*, 10; Ponting, *Green History*, 241.

38. Population data for recent centuries are available from many sources (for example, Aubuchon, "World Population Growth History"). Entries for earlier times are taken from estimates documented in relevant sections above. A graph showing basically the same progression can be found in Goudie, *Human Impact*, 10.

39. E.g., Ponting, *Green History*, 38, 42; Smil, *Energy*, 22; Hughes, *Environmental History*, 48.

40. Smil, *Energy*, 186.

41. Unless noted otherwise, numbers cited in this section are approximations from graphs in Smil, *Energy*, 186–87.

42. Ibid., 187. Smil's graphs here lump fossil fuels with primary electricity (wind, hydro, tidal, and geothermal), which affects the overall pattern only marginally (188). One metric ton of oil equivalent equals 45×10^9 joules. It may be noted that Lester Brown, using more recent data, reports in *Eco-Economy* (112) about eight thousand million oil-equivalent tons for the year 2000.

43. An ecosystem differs from a mere collection of organisms in that its members are related by distinct patterns of functional interaction. These patterns evolve by a process known as "ecological succession." Succession generally involves a gradual replacement of fast-growing organisms by more slowly growing organisms, along with a growth in complexity of the interactions among these organisms—i.e., growth in functional structure. Ecologists often use the term "assembly" in reference to this process, the sense being that succession amounts to "constructing" or "putting together" functioning ecosystems.

44. According to the "species-energy theory" currently debated among ecologists, increasing amounts of energy available to an ecosystem tend to reduce the breadth of its niches. This tends to increase the numbers of niches and thus to increase the variety of species the system can accommodate. More species within an ecosystem is tantamount to more contenders in the evolutionary process of succession by which the functional structure of an ecosystem is assembled. For an early exposition of this theory, see Wright, "Species-Energy Theory."

45. Ponting, *Green History*, chapter 16.

46. This is the topic of Bill McKibben's *The End of Nature*.

Chapter 7. Economic Production and Its Ecological Consequences

1. World population in 1900 (about 1.6 billion) was approximately one-quarter that in 2006 (about 6.5 billion). If population had held steady at the 1900 level during the twentieth century while per capita energy use increased at its actual rate, present total human energy consumption would be

one-quarter of its actual amount today. In 1900, total energy use was divided almost equally between biomass and fossil fuel (see section 6.8); count those amounts as 1 unit each. During the twentieth century, use of biomass energy approximately doubled, while use of fossil energy increased roughly twenty-fold (from about five hundred to ten thousand million tons of oil equivalent; see Smil, *Energy,* 187). This makes 22 units total human energy consumption in 2000. One-quarter of this amount is 5.5 units. With only 2 units contributed by biomass, the remaining 3.5 units would have to come from fossil sources.

2. See http://dictionary.babylon.com/Underdeveloped_country (accessed January 2010).

3. The study from which these data derived was made in 1961. Similar treatments of these data can be found in Odum, *Environment, Power and Society,* 184; Cook, "The Flow of Energy"; and Meadows et al., *The Limits to Growth,* 70. These data were chosen for the present purpose because of both their familiarity and their easy accessibility.

4. Other terms designating comparable measures in economic literature are "economic productivity," "energy efficiency," and "energy intensity." The higher a country's energy conversion ratio, the lower its energy efficiency. In subsequent discussion of this measure, the per capita qualifications of figure 7.1 cancel out and will be dropped.

5. This is borne out in a study by John Schmitz published on May 17, 2007, based on data from the *2006 Report of the International Energy Agency.* See Schmitz, "Energy 'Consumption' and GDP."

6. A substantially lower ratio for the U.S. economy in 1960 is recorded in Taylor and Van Doren, "Energy." A somewhat higher value (adjusted for inflation) is reported by the Energy Information Administration in its *Annual Energy Review 2008,* Table 1.5, p. 12. Different values also result when GDP replaces GNP in the statistics.

7. See Wolf, "Energy Revolution."

8. See Stern, "Energy and Economic Growth." As already noted in section 6.8, during the twentieth century global use of coal increased by a factor of 5, compared with a 150-fold increase in the use of natural gas. During the same century, use of electricity increased about a thousand times (Smil, *Energy,* 186).

9. Statistics in this paragraph come from "GDP Growth in China 1952–2009" and Schmitt et al., "The Chinese Service Industry." China's economic growth has continued at about the same rate (9 percent) up through 2006 ("GDP Growth in China"). Partially as a result of its very large population, this growth has been achieved at a fraction of U.S. per capita energy consumption ("China Energy Data").

10. An account of the controversy is given in Darmstadter, Dunkerly, and Alterman, *How Industrial Societies Use Energy.*

11. See Gever et al., *Beyond Oil*, 82.

12. Stern, "Energy and Economic Growth," 31. Differences between ecological and mainline economics are discussed in chapter 12.

13. Khatib, *Financial and Economic Evaluation of Projects*, 10.

14. Data presented in section 6.8 show that use of renewable energy (mostly biomass) doubled during the twentieth century while use of fossil fuel increased twentyfold. With a small portion of nuclear energy factored into the mix, this means that fossil sources account for close to 90 percent of worldwide energy use today. Whether industrial or domestic, most of this use contributes to the gross global product and would be classified "economic."

Chapter 8. Technological Solutions to Ecological Problems

1. In the years following the initial agreement, China became the world's largest producer of CFCs. By 2007, it had shut down most of its CFC plants, and joined the United States in using HCFCs (hydrochlorofluorcarbons) for air-conditioning instead (Bradsher, "Push to Fix Ozone Layer"). The environmental effects of HCFCs are discussed elsewhere in the text.

2. "Desalination."

3. "Water Desalination Plants."

4. Lang, "Seven-Year Glitch."

5. "Genetically Engineered Crops and Pesticide Use."

6. Benbrook, "Genetically Engineered Crops."

7. "How Coal Works."

8. "Water Desalination Plants."

9. Ibid.

10. A revision of the previous estimate of 2050 to 2065 as a likely date for recovery of the stratospheric barrier was widely reported in December 2005.

11. Bradsher, "As Asia Keeps Cool."

12. "Brief Questions and Answers on Ozone Depletion."

13. The feedback interaction between photosynthesis and the chemistry of ozone production is elusive. Ingmar Grenthe gives an evocative description of this interaction in his presentation speech for the 1995 Nobel Prize in chemistry.

14. "Montreal Protocol on Substances that Deplete the Ozone Layer."

15. Lapper, "Market Makers."

16. Davis, "Technofix-ation."

17. Bayon, "Trading Futures." Comparison with the $300 per ton cost (in 2000) of reduction by emission abatement equipment cited previously illustrates why companies often find it cheaper to continue polluting. These figures of course are in flux.

18. Offsets have also been purchased by large corporations with unfavorable reputations for greenhouse gas emission. A recent example, according to Fahrenthold, "Cost of Saving the Climate," is American Electric Power, which reportedly purchased offsets for about 4.6 million tons of carbon dioxide. At a rate presumably close to $10 per ton, this amounts to an investment of about $50 million for a more positive public image.

19. See, for example, "Steaz Leadz the Way" and "Iron Seeding."

20. During the period from 1995 to 2000, there was an approximately 25 percent reduction in acid rain falling in the northeast region of the United States. This reduction has been attributed to the trading in SO_2 emission credits that got under way in 1995. See "Acid Rain."

21. As reported in Milloy, "Carbon Offsets—Buyer Beware."

Chapter 9. Replacing Fossil Fuel with Clean Energy

1. Fuel cells also are sometimes described as clean sources of energy. Fuel cells produce electricity by recombining hydrogen with oxygen after it has been extracted from water (i.e., separated from oxygen) at a previous stage in the process. Although recombining hydrogen and oxygen produces only "clean" H_2O as a by-product, most technologies for extracting hydrogen from water are run by electricity generated by other technologies that are polluting in their own right. Strictly speaking, moreover, a fuel cell is not an energy source in the first place. It rather functions as part of a larger system that operates in the manner of a storage battery, being "charged" by the electricity used to produce free hydrogen and subsequently "discharged" with the production of electricity by recombining hydrogen and oxygen.

2. According to an EPA study conducted in the early stages of the ethanol boom, factories making ethanol from corn are releasing carbon monoxide, methanol, and various carcinogens at rates far greater than initially expected. Among carcinogens involved are acetic acid and formaldehyde. See Cosgrove-Mather, "Ethanol Pollution Surprise."

3. It should be borne in mind that ethanol can also be produced from other biomass sources, such as perennial switchgrass, algae, and organic wastes. Switchgrass can be grown without commercial fertilizers or pesticides, and the fuel it yields burns with few if any carbon emissions. (See Glasgow and Hansen, "Setting the Record Straight on Ethanol.") Commercial efforts are under way at ExxonMobil to produce transportation fuels from algae (LeVine, "ExxonMobil's Algae Exploration"), and General Motors is experimenting with a plant to make ethanol from garbage and crop wastes (Wald, "G.M. Buys Stake in Ethanol").

4. See Tverberg, "Corn-Based Ethanol" for a recent study showing that, when fossil fuels used in growing corn and producing ethanol are taken into

account, using ethanol instead of gasoline for fuel decreases emission of green-house gases (CO_2 and N_2O) by only 13 percent.

5. "Nuclear Energy and Greenhouse Gas Emissions."

6. It has been estimated that 6.29 quadrillion BTUs of fossil energy consumption went to nonfuel use per year in the United States between 1990 and 1998 ("Emissions of Greenhouse Gases in the United States 1998"). Average total fossil energy consumption in the United States during that period was about 90 quadrillion BTUs (O'Brien, "Trends in U.S. Energy Use"). Thus about 7 percent of total fossil energy was put to nonfuel use in the United States during that period.

7. Specific numbers in this regard are hard to come by. According to Vaclav Smil (*Energy*, 187), one-quarter of the world's fossil fuel output was converted into electricity in 1990, up from one-tenth in 1945. A rough extrapolation from these figures makes it unlikely that the electricity portion will reach one-third within the next two or three decades.

8. People and goods would have to be moved in smaller vehicles operating at slower speeds, filling stations would have to be replaced by recharging stations, substantially larger power grids would have to be constructed to supply these stations, and so forth. Moreover, given that something like one out of ten jobs in developed countries depend on the manufacture, sale, and maintenance of petroleum-powered vehicles (Arabe, "Automotive Industry"), major disruptions of a social nature could be expected as well.

9. The U.S. estimate of 7 percent (see note 6) is rounded up to 10 percent for the present comparison, which does not require exact numbers.

10. This is a mid-range estimate based on Smil's 25 percent for 1990 and the projection of one-third for two or three decades from now (see note 7).

11. No estimate like this for hypothetical (currently nonexistent) circumstances could be more than a guess. The 10 percent suggested in the text may be overly optimistic.

12. An exception is advanced solar technology that concentrates sunlight with mirrors and requires cooling to operate at maximum efficiency.

13. Concern among scientists that positive feedback of this sort has already taken over was widely reported in late 2006, leading to the conclusion by one group that "the issue of global warming is no longer a question of just cutting down on the burning of fossil fuels to decrease carbon dioxide levels" (Maxwell, "Methane Menace"). A related effect, intensifying this concern, is that global warming might hasten the release of vast amounts of CH_4 trapped beneath open oceans ("Gas Escaping from Ocean Floor"). Natural feedback processes like these leave little hope that global warming at this point can be controlled by minor changes in human energy-use patterns.

14. See "Frequently Asked Questions."

15. Data in this paragraph comes from DeLiberty, "Energy Interactions with the Atmosphere and at the Surface." Specific surfaces reflect specific

percentages of solar radiation falling on them, a property referred to as their albedo. Fresh snow has an albedo of about 90 percent, thick clouds about 75 percent, and water (perpendicular to the sun's radiation) about 4 percent. Accordingly, ocean water absorbs a much greater percentage of radiation than dense clouds and snow.

16. Studies are under way investigating means of converting petrochemically derived plastic waste into biodegradable form ("Microbes Convert 'Styrofoam'"). Ways also have been discovered of producing plastics from vegetable starches that are biodegradable in their own right ("Bioplastic"), called "bioplastics" to distinguish them from petrochemical plastics that can be degraded by biological technology. These developments are ecologically beneficial but have nothing to do directly with clean energy.

17. According to a fact sheet produced by the United Nations Environment Programme, most commercial wind turbines operate today with average wind speeds of more than twenty-two kilometers (about fourteen miles) per hour ("Wind Power").

18. See "Wind Energy." Wind power capacity in the United States grew about 45 percent in 2007 and currently supplies about 1 percent of the country's electricity (Krauss, "Move Over, Oil").

19. "Hydro Electric Power."

20. Power grid operators have estimated that it would cost between $50 billion and $80 billion to construct transmission lines from the middle of the United States to the Northeast (Galbraith, "Up to $80 Billion to Transmit Wind Power").

21. Electricity transmission lines require areas to be stripped of vegetation for supporting structures and motorized access, and they cause the death of hundreds of millions of birds and wild animals a year (Sagrillo, "Putting Wind Power's Effect on Birds").

22. "Reference: Wind Energy."

23. "Al Gore Is Optimistic."

24. Estimates of twenty or thirty years' lifetime expectancy for photovoltaic collectors seem fairly common. For a fifty-year estimate, see Heckeroth, "Solar Building Design."

25. Bowyer, "Weighing Up the Benefits."

Chapter 10. History and Theory of Economic Growth

1. The addiction, of course, is behavioral rather than chemical. In a consumer society like ours, high levels of economic production provide goods and services that consumers tend to find attractive. To the extent that society fosters lifestyles to which such goods and services are considered essential, consumers tend to rely on them as necessary ingredients of their perceived well-being. This reliance tends to be habit-forming. When it sets in, the bene-

fits of high economic production become addictive to the individual in that without them the individual feels a loss of personal welfare. The high levels of energy use an economy requires to produce these benefits become addictive by extension. Inasmuch as a perceived loss in well-being on the part of individual consumers is likely to follow a decline in available goods and services, high levels of productivity are required to maintain a cumulative sense of welfare on the part of consumer society at large. This converts into a requirement for high levels of energy use. Consumer society is addicted to excessive consumption of energy in the sense that large amounts of energy are necessary to maintain consumer satisfaction.

2. Xenophon, *Oeconomicus,* chaps. 2, 3, passim.

3. "Ancient Roman Economy."

4. "Social Origins of Democracy."

5. See Rempel, "Mercantilism."

6. Williams, *Capitalism and Slavery.*

7. *An Inquiry into the Nature and Cause of the Wealth of Nations* was published in 1776, eleven years after Anders Chydenius's *The National Gain,* which espoused some of the same principles. Smith did not use the term "economics" in reference to his own work, speaking instead of "political economy."

8. *The Wealth of Nations,* book 1, chap. 11, in the digression on the value of silver.

9. This is a generalization from Smith's discussion of the origins of the division of labor in book 1, chap. 2.

10. This expression occurs three times in Smith's writing—twice in the following citations and again in his posthumously published "History of Astronomy." In the latter, the invisible hand is that of a deistic God who regulates events in nature.

11. Smith, *The Wealth of Nations,* book 4, chap. 2.

12. Smith, *The Theory of Moral Sentiments,* book 4, chap. 1, 10.

13. Smith, *The Wealth of Nations,* book 4, chap. 8.

14. Ibid., book 1, chap. 1.

15. Ibid., book 1, chap. 3.

16. Evans, "The Mathematics of Desire."

17. Smith, *The Wealth of Nations,* book 1, chap. 9.

18. Ibid., book 5, chap. 3. Economists disagree on whether this was the case. As far as the contents of *The Wealth of Nations* are concerned, it seems that he never addressed the issue explicitly.

19. Fonseca, "David Ricardo."

20. As laid out in his *Introduction to the Principles of Morals and Legislation* (1789), Bentham's utilitarianism maintains that one ought always to act in a manner resulting in the greatest overall pleasure. The hedonic calculus is a theoretical algorithm for maximizing pleasures and minimizing pains.

The internal optimizing mechanism assumed by neoclassical economics is supposed to maximize pleasures for the individual affected rather than for society at large.

21. Thorstein Veblen was an early critic of the neoclassical school. For the quote, see "Why Is Economics Not an Evolutionary Science?" 373.

22. See the introduction to the 1962 edition of Keynes's *A Treatise on Probability* by Norwood Russell Hanson, a prominent philosopher of science. Keynes's *The General Theory* was preceded by his *Treatise on Money* (1930).

23. By the time of his Nobel award in 1976, Friedman had recanted his New Deal leanings in the 1930s in favor of the libertarian view that the role of government in guiding the economy should be severely restricted. For Friedman's involvement in President Roosevelt's New Deal, see Ebenstein, *Milton Friedman,* chap. 15. In 1988, Friedman received both the Presidential Medal of Freedom and the National Medal of Science from President Ronald Reagan.

24. Solow, "Technical Change."

25. This brief account of the dynamics of Solow's model is a simplification of (already simplified) accounts in Stern, "Energy and Economic Growth," and McCain, "The Neoclassical Model."

Chapter 11. Why Economic Growth Is Considered a Good Thing

1. In the standard economics textbook *Macroeconomics: Private Markets and Public Choice* (now in its seventh edition) by Ekelund, Ressler, and Tollison, for instance, the authors claim that the "overall goal of macroeconomic policy is . . . to attain maximum economic growth in the present and future" (147). For another instance, according to the online journal *Ecommerce,* the supposition that "unending economic growth is both possible and desirable" is built into mainline economic theories (Shekhar, "Environment Does Not Allow Further Economic Growth in the World?").

2. "Population Growth Rate."

3. "Country Comparison GDP per Capita."

4. According to a World Bank study reported by Schurman and Lappé (*The Missing Piece of the Population Puzzle*), in countries where income of the poor increases by 1 percent, fertility rates drop by 3 percent.

5. The reason many of us tend to focus on population reduction in addressing environmental problems may be that we want to find a resolution that does not involve our sacrificing the personal benefits that helped generate these problems in the first place. It is convenient to assume (dubiously) that the problems can be overcome without lowering our own living standards by curtailing the number of people who share in the benefits.

6. Consider the cases of China and India, countries in which policies of population control have been applied most stringently in recent decades. De-

spite their use of techniques such as abortion and infanticide (Jones, "Case Study"), population in these countries continues to grow ("Population in China"). Not coincidentally, these countries recently have joined the ranks of the world's heaviest industrial polluters.

7. "Country Comparison GDP per Capita."

8. Smil, *General Energetics,* 201.

9. Figures in this paragraph are from Ponting, *A Green History of the World,* 342–45. In the case of Zaire, corruption on the part of state leaders was a major part of the problem.

10. According to Ponting, *A Green History of the World,* 341, for instance, three-fourths of Britain's bilateral aid in the 1980s was tied to the purchase of British goods, and another 14 percent was used to subsidize prices for British firms vying for Third World contracts.

11. Friedman, *The Moral Consequences of Economic Growth.* The morality Friedman is primarily concerned with focuses on social attitudes and political institutions, such as tolerance of diversity, social mobility, and dedication to democracy. Friedman is a professor of economics at Harvard University.

12. Economists have been frank in pointing out that there are qualitative dimensions of human life that elude numerical measurement, and from time to time they have proposed alternative modes of assessment. One alternative has been to distinguish between standard of living (measured economically) and quality of life, measuring the latter in terms of health or of lifestyle choices ("Health-Related Quality of Life"). Another has been to measure well-being by what is called a "Human Development Index," covering matters of life expectancy and education along with economic purchasing power (Wolfers, "What Does the Human Development Index Measure?"). Economists are far from unanimous regarding how quality of life should be measured.

13. This disproportionality has been evident in rich and poor countries alike (Indonesia, Thailand, Chile, Argentina, etc.). In the case of the United States specifically, a recent report based on data from the Economic Policy Institute issued the following summary statement: "GDP is up, but virtually all the growth has gone into corporate profits and the incomes of the highest economic brackets" (Jackson, "Good News!"). At mid-2008, the average salary for a CEO of a major U.S. company was estimated at nearly five hundred times the salary of the average worker (Lyon, "CEO Salaries"). Even in the midst of the "Great Recession," banks that were receiving multibillion bailouts from the U.S. government were still paying their top executives billions of dollars in bonuses (Story and Dash, "Bankers Reaped Lavish Bonuses During Bailouts").

14. This is not to deny that job loss goes hand in hand with depression. The point in the text is that conditions fostering growth can lead to loss of jobs as well.

15. Trends and data cited in this paragraph come from Sachs, "The Social Welfare State"; and "Economy Statistics." Luxembourg, a small country with the world's highest per capita GDP, is left out of these statistics. See also Power, *The Economic Value of the Quality of Life*.

16. See Leonardi, ed., "Economic Growth versus Genuine Progress."

17. Weehuizen, "Mental Capital."

18. As reported by Michael Mandel, chief economist for *Business Week,* in his review of Friedman's *The Moral Consequences of Economic Growth* ("What's So Good").

19. Government involvement of a different sort comes with the massive bailout program pushed through Congress in late 2008. When the Federal Reserve agreed to an $85 billion bailout of the troubled insurance giant AIG (American International Group) in September 2008, the move was described in the media as "the most radical intervention in private business in the central bank's history" (Andrews et al., "Fed's $85 Billion Loan"). In response to the takeover of Fannie Mae and Freddie Mac a few days earlier, the dean of the University of Virginia's Business School remarked, "If anybody thought we had a pure free-market financial system, they should think again" (Schwartz, "A History of Public Aid During Crises").

20. To repeat a quotation attributed to Edward Abbey: "Growth for the sake of growth is the ideology of the cancer cell."

Chapter 12. Economics without Continuing Growth

1. One account of the growing dominance of neoclassical economics cites Chicago, Columbia, Harvard, MIT, Princeton, Stanford, and Yale as among universities responsible for the spread of mainstream doctrine ("The Strange History of Economics"). According to this account, the U.S. Air Force, through the RAND corporation, set up a scholarship program in the 1960s for economics graduate students at these institutions, thinking that certain analytic techniques taught in these programs (e.g., game theory and linear programming) had potential use in national defense. This was seen as a continuation of the Defense Department's funding of mathematical economics beginning shortly after the end of World War II, which eventually led to the worldwide dominance of neoclassical economics in non-Communist countries.

2. Ecological economist Herman E. Daly, in the introduction of *Beyond Growth,* tells about an encounter with the former chief economist of the World Bank, Lawrence Summers (later U.S. secretary of the Treasury, subsequently president of Harvard University, and presently President Obama's chief eco-

nomic advisor). Summers had just come back from a conference on *The Limits of Growth,* which he found worthless, and was asked by Daly what he thought about a certain diagram in the book showing the economy in relation to the more inclusive ecosystem, which Daly considered similar to one he had put forward himself. As Daly relates it, Summer's response was a curt "That's not the right way to look at it" (6).

3. "Market Failure."

4. Thus the title, for instance, of Kahn's *The Economic Approach to Environmental and Natural Resources.*

5. See The McIlvaine Company, "Air Pollution Market Soars."

6. "Costing the Earth."

7. The relevance of the Second Law to economics had been pointed out earlier by Frederick Soddy, Nobel laureate in chemistry (1921), in his *Wealth, Virtual Wealth, and Debt.*

8. Edwards, "Global Comprehensive Models in Politics and Policy-making."

9. In his *History of Economic Analysis,* Joseph Schumpeter refers to the "preanalytic cognitive act" in which researchers "visualize a distinct set of coherent phenomena as a worthwhile object of . . . analytic effort," which then "supplies the raw material for the analytic effort" (41).

10. In *Beyond Growth,* Herman Daly attributes to G-R the image of an irreversible hourglass representing the unidirectional flow of high-grade solar energy from the upper chamber into the lower, where it becomes unusable entropy. On the upper arch of the lower chamber is a repository of materials, including fossil fuels, which are not yet fully degraded and hence are still usable in the production of economic goods. For G-R, the economic process is equivalent to converting resources from this finite repository into products of economic value.

11. Georgescu-Roegen, *Entropy Law,* 284.

12. G-R wraps up a substantial discussion of the nature of time with the remark that one of the main objectives of his book is to prove "that the economic process as a whole is not a mechanical phenomenon" (ibid., 139).

13. See Cleveland, "Biophysical Economics."

14. Daly, *Beyond Growth,* 197.

15. For example, Costanza et al., *An Introduction to Ecological Economics,* and Common and Stagl, *Ecological Economics.*

16. Also called biophysical economics, which has been defined as the use of ecological and thermodynamic principles to analyze the economic process (see Cleveland, "Biophysical Economics").

17. See Norgaard's "Ecological Economics."

18. See, for example, Daly, *Beyond Growth,* 14–15, 69, 166–67, passim.

19. This is a paraphrase of Daly's definition in *Beyond Growth,* 9. It derives from the report of the World Commission on Environment and Development, *Our Common Future,* which calls development sustainable if it "meets the needs of the present without compromising the ability of future generations to meet their own needs."

20. Daly has little to say about how long qualitative improvements of this sort could continue without additional quantitative throughput. In case the earth's carrying capacity has already been exceeded, moreover, drastic cuts in throughput might be necessary, in which case even limited qualitative improvement would be problematic.

21. In its focus on actual preferences instead of what *ought* to be preferred, conventional economics is able for the most part to eschew value-laden statements of principle and policy. But values surely are involved in its choice to present the economy as a context in which people seek personal benefits with monetary prices, as distinct from well-being of a nonquantitative sort (e.g., happiness, companionship, fulfillment).

22. E.g., Norgaard, "Ecological Economics."

23. E.g., Daly, *Beyond Growth,* chaps. 14, 15.

24. Norgaard, "Ecological Economics."

25. For one example, see Herman Daly's preface to *Economics, Ecology, Ethics,* entitled "Introduction to the Steady-State Economy." The terminology of scientific paradigms and paradigm shifts comes from Thomas Kuhn's much quoted *The Structure of Scientific Revolutions* (1970). Kuhn's scientific paradigm seems closely related to Schumpeter's notion of a preanalytic vision.

26. As Daly has stressed in private correspondence with me, EE is far more accurate than its mainstream counterpart in its description of how the economic process relates to the physical world at large. The descriptive shortcomings alluded to in the text have to do with the internal working of actual free-market economics.

27. Daly, "Five Policy Recommendations for a Sustainable Economy."

Chapter 13. Desire for Wealth in Free-Market Economies

1. A 2009 news report warns that climate change will pose profound strategic challenges to the U.S. military, citing "violent storms, drought, mass migration and pandemics," along with "food shortages, water crises and catastrophic flooding," which "could topple governments, feed terrorists movements or destabilize entire regions" (Broder, "Climate Change Seen as Threat to U.S. Security").

2. We are not looking for one or another particular area of a given economy (defense spending, the private sector, etc.) that happens to be "driving"

that economy at a particular moment in history. Our concern is with the driving principle of economic expansion generally, rather than with sectors that are particularly active from period to period.

3. This chapter was being completed during a period in which the U.S. economy registered successive quarters of negative growth, with corresponding retrenchment in other major economies worldwide. The immediate causes of this recession are well-known and have nothing to do with efforts to restore environmental integrity. In point of fact, economic crises like this tend to aggravate ecological damage with a loosening of environmental restrictions to hasten economic recovery. The changes needed to restore the environment should enable reduction of economic production without collateral erosion of living conditions, in the spirit of Daly's call for reduced throughput while sustaining the quality of economic services (see section 12.5).

4. Ariely, "Irrationality Is the Real Invisible Hand."

5. Georgescu-Roegen, *Entropy Law,* 323. G-R mentions J. S. Mill as one economist maintaining this position.

6. The term "wealth" in the following discussion is used in the general sense of resources that have exchange value and are capable of satisfying needs, as distinct from the sense of an abundance of riches. Desire for wealth, accordingly, is desire for (additional) income, not desire to become very wealthy.

7. An individual can become wealthy by inheritance or thievery, but these are not sources of wealth in the sense intended. Another source that might be suggested is the mining of precious metals. But precious metals do not constitute wealth until they enter the economic process (e.g., are made available for sale), which means that mining by itself (apart from economic activity) is not an independent source of wealth. Military conquest likewise is not a source of wealth, being rather the forceful acquisition of wealth that already exists.

8. In *The Wealth of Nations,* Adam Smith observes that, in contrast with the desire for food, desire for conveniences and most amenities requiring purchase "seems to have no limit or certain boundary" (book 1, chap. 11, part 2). In chapter 2 of *The Theory of the Leisure Class,* Thorstein Veblen states that "the desire for wealth can scarcely be satiated in any individual instance," and certainly cannot be satiated in humanity taken generally. John Maynard Keynes, in *Essays in Persuasion,* observes that needs associated with an individual's desire to feel superior (i.e., those served by wealth) "may indeed be insatiable." It may be noted that George Reisman makes a virtue of the limitless desire for wealth in his recent *Capitalism.*

9. In certain types of business, this tactic can take the form of leveraging—borrowing money to invest with the aim of making a profit beyond what is needed to repay the loan. Leveraging is financially successful only when the value of investments increases. One cause of the recent recession was a collapse in the value of highly leveraged investments in credit derivatives

(contracts transferring credit risk from one party to another) backed by sub-prime mortgages.

10. Key traders in today's financial firms often have bonuses written into their contracts. This means that bonuses are paid even when the firms involved are losing money. Among nine banks receiving government bailouts during the recent crisis, $32.6 billion was reserved for bonuses while those banks lost $81 billion (Story and Dash, "Bankers Reaped Lavish Bonuses During Bailouts").

11. Other possible factors are national commitment to military might, which requires a strong economic basis, and ideologically based development programs like the Soviet Union's Five Year Plan in the late 1920s. It should be understood that the present discussion of the desire for wealth is not focused on free-market economies to the exclusion of so-called managed economies. Not only are most economies today a mixture of both, but even managed economies are influenced by individual desire for wealth. It would be naïve to assume that the substantial increases in GNP recently achieved by Communist China, for example, have not been motivated by a desire for self-aggrandizement on the part of its economic planners.

12. For one source, see Moffatt, "Definition of Economic Good."

13. There is a substantial body of literature on what is often called "Say's Law" (after Jean Baptiste Say, 1767–1832), sometimes paraphrased "supply constitutes demand." This makes sense only under the gloss that supplying goods already in demand creates demand for additional goods. See Best, "Say's Law and Economic Growth."

14. Smith, *Wealth of Nations,* book 4, chap. 8.

15. As noted previously, factors other than consumer spending also contribute to economic growth, in which advertising accordingly is not so directly involved. Prominent among these are expenditures for government programs and military equipment. In these domains, advocacy of special interests tends to take the form of lobbying rather than advertising in the media. Consumer spending, nonetheless, remains the dominant factor and has been increasing proportionately over the past several decades. Although estimates vary, consumer spending in the United States accounted for about 63 percent of GDP in the 1960s and now has risen to about 72 percent (Nutting, "Consumers Unable"). Figures like these (presumably similar for other industrial nations) make it highly relevant to study ways in which consumer spending is influenced by advertising.

16. A step closer to the "advertiser's dream of sending a particular commercial to a specific consumer" is announced in the *New York Times* of March 4, 2009 (see Clifford, "Cable Companies Target Commercials to Audience").

17. This is known as respondent (Pavlovian) conditioning, as distinct from operant (Skinnerian) conditioning. In this example, the conditioned stimulus is the bell's sound, the unconditioned stimulus the food, and the conditioned response the dog's salivation.

18. Soft-drink machines can be found today in almost every country with electrical power to run them. According to Coca-Cola (www.coca-cola .com), Coke is sold in more than two hundred countries.

19. Information on neuromarketing was derived from Cross, "Neuromarketing"; Dooley, "Brain Scans"; Tierney, "The Voices in My Head"; and Kayne, "What Is Neuromarketing?"

20. "Coke versus Pepsi."

21. In addition to Baylor and Ulm mentioned previously, universities with research programs in neuromarketing include Emory, Duke, Yale, Carnegie Mellon, MIT, Stanford, Johns Hopkins, Pittsburgh, and University College London, among others. Research in neuromarketing began in the late 1990s in the Marketing Department of Harvard University (according to Kayne, "What Is Neuromarketing?").

22. The diagram in figure 13.1 represents only that part of an economy dominated by consumer spending, to the exclusion of other sectors mentioned in note 15 (above). Needless to say, there are ways other than advertising by which corporate managers can stimulate consumer demand (e.g., designing better products), and ways other than buying consumer goods by which desire for wealth can affect the economy (e.g., investment). The figure thus does not depict an entire economy. Its purpose is only to show how consumer spending (the major sector of a market economy) is affected by advertising (its main determinant) and how desire for wealth thereby leads to ecological destruction.

23. A so-called black box is a representation showing only inputs and outputs with no indication of how they are connected in the system represented.

24. This amount can be substantial. In 1910, 4 percent of U.S. national income was spent on advertising. By the end of the century, the $600 million spent on advertising in 1910 had climbed to more than $120 billion, out of a total of $250 billion worldwide. See Robbins, *Global Problems and the Culture of Capitalism*, 15–16.

25. Rabikowska, "Commercial Consumer TV Advertising in Poland"; "Penetrating Eastern Europe."

26. See Koci, "An Overview of the Regulatory Framework for Media in Albania."

27. Well-known people who have denounced advertising as evil (with various qualifications) include C. P. Snow, Arnold Toynbee, Nye Bevan (a Welsh Socialist), Malcolm Muggeridge (an English conservative), and Pope John Paul II. See Clark, *The Want Makers*, 371, 393.

28. Genuine needs unproblematically include food, shelter, and other provisions necessary for bare survival. By extension, accoutrements necessary to meet such needs, such as tools, weapons, and basic conveyances, also count as genuine human needs. More debatable are various wants that must be satisfied for someone to flourish in a given society, like comfortable quarters, healthy food, and some level of education, along with other provisions required for a satisfying lifestyle. Clear examples of "mere wants" (not genuine needs), on the other hand, are desires that could go unsatisfied without marked decrease in quality of life, such as tastes for unusual foods, designer clothing, and an unlimited variety of automobile styles.

29. For a sober view of these effects, see Pollan, *The Omnivore's Dilemma*, 88, and Pollan, *In Defense of Food*, 47.

Chapter 14. Environmental and Other Ethics

1. "Country Comparison GDP per Capita."
2. "Internet Increases Global Inequality."
3. Walker, "Richest 2% Own 'Half the Wealth.'"
4. Unlike scientific theories, ethical theories are not held accountable to experimental (or other empirical) data. They are neither confirmed nor disconfirmed by empirical circumstances. This is one reason why ethical theorists are generally disinterested in the findings of descriptive ethics. Ethical theories make no predictions about empirical states of affairs on the basis of which the theories could be established as warranted or otherwise. Whether a given ethical theory is right or wrong is established primarily by rational argument, often supported by "moral intuitions" brought to bear by individual theorists. The "moral intuitions" exercised by ethical theorists are comparable to what economists, following Joseph Schumpeter, sometimes call "preanalytic visions" (see section 12.4).
5. Most work to date on deep ecology and ecofeminism (mentioned subsequently in this chapter) has appeared in articles and anthologies. A useful anthology covering both is *The Deep Ecology Movement,* edited by Drengson and Inoue.
6. An obvious assumption here is that affective states like pleasure and pain are subject to quantitative comparison. Bentham thought that affective states could be compared with respect to such features as intensity, purity, and duration, as well as to relational properties such as likelihood of being followed by similar affections. Features such as these can be scaled in terms of more and less, if not compared by quantitative measures as the calculus seems to require.

7. For example, Singer in his *Animal Liberation.*

8. Leopold, *A Sand County Almanac,* 262. *A Sand County Almanac* has sold well over a million copies in its many editions.

9. In contrast with act-utilitarianism, which applies the maxim of greatest utility to individual acts, rule-utilitarianism applies the maxim to general rules of action, which are then used to determine what ought to be done in particular circumstances.

10. Leopold, *A Sand County Almanac,* 262, 290, 291.

11. Ibid., 238, 239.

12. Ibid., 279.

Chapter 15. Typology of Social Values

1. These figures have been widely reported. See, for example, Wyer, "Political Engagement among College Freshmen Hits 40-Year High"; Irvine, "Wealth Top Priority for Today's Youth"; and Myers, "The Secret to Happiness."

2. De Vries, "Are Today's Students Too Self-Centered?"

3. This is not to say that social conditioning shows up only in the formation of social values. According to some social theorists (e.g., Berger and Luckmann, in *The Social Construction of Reality*), social conditioning can also affect our perceptions of reality and of the world at large. Neither is it to say that when the socialization in question involves social values, it takes place through the formation of such values exclusively. Social conditioning of values very likely involves neurological adaptations as well.

4. Smith, *Theory of Moral Sentiments,* part 7, sec. 2, chap. 2.

5. Pain endurance is integral to the spirituality of certain Native American cultures. An example familiar to moviegoers of the 1970s is the extremely painful ritual of vertical chest suspension, shown in the movie *A Man Called Horse.* This borders on other religious practices in which self-inflicted pain is considered valuable in itself, such as the self-flagellation practiced by certain Christian and Shi'ite Muslim sects (illustrated in Dan Brown's novel *The Da Vinci Code*).

6. Pain response may be innate, but not social approval of pain avoidance.

7. A representative list of Aristotelian virtues would include courage, temperance, fortitude, munificence (an attitude toward money), magnanimity (well-founded self-esteem), and patience. Christian virtues include, first, the theological (spiritual) virtues of faith, hope, and charity, and next the cardinal virtues of wisdom (prudence), temperance, justice, and fortitude. Other Christian virtues often listed are humility, chastity, patience, and diligence.

8. In his reply to objection 4, article 3, question 85, part 3 of the *Summa Theologica*, St. Thomas Aquinas affirms that abstaining from pleasure is a virtue pertaining to temperance. Similar teachings can be found in Buddhism and Islam.

9. In his *Introduction to the Principles of Morals and Legislation,* Jeremy Bentham suggested that the pain accompanying social and moral sanctions contributes "binding force" to laws and rules of conduct.

Chapter 16. Ecologically Destructive Values

1. See Elert, "Volume of US Soft Drink Consumption." By 2005, the rest of the world caught up (see "World Consumption of Soft Drinks on Rise"). Other sources of data for this section on soda pop are Hindley, "Orion-list Water Consumption"; "More Aluminum Cans Trashed Last Year Than Recycled"; and Container Recycling Institute, "Plastic Bottle Waste." All figures in this chapter are approximate, and different estimates often can be found in different sources.

2. For the estimate of three million machines, see Wolff and Dansinger, "Soft Drinks and Weight Gain." Given an average consumption of about 400 kilowatt hours per machine (see "Delamp Vending Machines"), this adds up to 1.2 billion or more kilowatt hours and growing. U.S. commercial electricity consumption in 2002 was about 3.5 trillion kilowatt hours (see "United States Electricity" and Hutchison, "About Electricity").

3. Figures in this paragraph come from Schwartz, "Abolishing Intensive Livestock Agriculture"; Carlson, "15 Reasons to Stop Eating Meat"; and "Considerations for a Vegetarian Diet."

4. See Tilford, "Biodiversity to Go."

5. See Schwartz, "Abolishing Intensive Livestock Agriculture."

6. The U.S. Energy Information Administration reported 0.42×10^{15} BTUs (123×10^9 kilowatt hours) electricity consumption for home air-conditioning in 1997 (Battles, "Trends in Residential Air-Conditioning Usage"). Unless otherwise noted, this report is the source of other data in this section. Regarding total consumption in other countries, twentieth-ranked Poland consumed 119×10^9 kilowatt hours during the same period ("Electricity Consumption"). Such figures obviously vary year by year.

7. However, see chapter 8, note 1.

8. See "Gree Electric Appliances Holding 2009."

9. See "World Air Conditioner (ACU) Market."

10. According to Smil, electric motors of various sorts became available for industrial use during the 1890s. Smaller motors suitable for household use were common by 1900 (*Energy in World History,* 170).

11. The following remark appeared recently in a typical small-city newspaper (Smalls, "Redbox Becomes Red Hot"): "As do-it-yourself machines satisfy our society's increasing push for customer convenience, another phenomenon is finding its way onto the list—automated DVD rentals."

12. For purposes of comparison, five window units consume roughly the same amount of electricity as one 20,000-BTU central air conditioner ("Common Household Appliance Energy Use").

13. Among the better known are Freund and Martin, *The Ecology of the Automobile,* and Kay, *Asphalt Nation.* Sources on the automobile include Melosi, "The Automobile and the Environment," and Smith, "Environmental Implications of the Automobile."

14. Sevin, "Car Facts," reports that 137,000 new cars are manufactured each day. Hardy, "Cutting Traffic with Driverless Cars," estimates that there are 600 million cars worldwide on the road today.

15. Data cited here and in the next three paragraphs are from Melosi, "The Automobile and the Environment."

16. In Canada, a country with one of the highest numbers of automobiles per capita, one in twenty working people in 1990 had jobs linked to motor vehicles (Smith, "Environmental Implications of the Automobile").

17. Ezstreet, "Potholes Are Gone Forever."

18. Burkart, "Could Solar Roadways Power the U.S.?"

19. Smith, "Environmental Implications of the Automobile."

20. In chapter 11 of *The Theory of the Leisure Class,* Veblen writes of the possession of goods as evidence of the prepotence of their possessor over other members of society. An added dimension of dominance comes into play, he says, with the conspicuous acquisition of goods one does not actually need.

21. According to an industry source (Marketing Intelligence Service), 33,678 new packaged products (food, beverage, health, household, pet, etc.) were introduced in 2003 in the United States alone, up from 31,785 in 2002. See Marino, "From Cereal to Juice." Although packaged goods comprise only a few product niches among many, this gives a rough idea of the magnitude of new lines that appear annually.

22. Domhoff, "Wealth, Income, and Power," gives a nuanced and up-to-date discussion of the relation between wealth and power in the United States.

23. Think of the role of *machtgelust* (desire for power) in Nietzsche's *Human, All-Too-Human.*

24. A recurrent theme of Alexis de Tocqueville's perceptive *Democracy in America,* published in 1835, is the country's obsession with money making and the amassing of vast fortunes in the private sector. As Tocqueville put it, he knew of "no other country where love of money has such a grip on men's hearts" (47).

Chapter 17. Values for Survival

1. According to Sweney, "Supermarkets Feel Junk Food Ad Ban Bite," this ban has affected supermarkets particularly. An indication of the potential impact of such a ban can be gathered from a recent report that 82 percent of all promotions by Australian food companies are aimed at bringing junk-food products to the attention of young people. See "Junk Food Tactics Hit the Supermarket Shelves."

2. A tax of this sort (sometimes referred to as the "Twinkie tax") has occasionally been proposed to combat the problem of obesity (Murphy, "Should There Be a 'Twinkie Tax'?"). For various reasons, such initiatives have generally proved politically unpopular.

3. Moderation and temperance overlap in that each involves avoidance of extremes. The term "temperance" may be less suitable for our purposes because of its association with the temperance movement (of the early twentieth century), which opposed all consumption of alcohol, which of course can be used in moderation.

4. In *Global Problems and the Culture of Capitalism*, Richard Robbins remarks that the creation of consumer economy required a change from "such values as thrift, modesty, and moderation, toward a value system that encouraged spending and ostentatious display" (21). What is needed now is a reversal of that change.

5. A few days after this section was drafted, the Vatican's ambassador to the United Nations announced that Catholics "should do something about climate change by adopting a life of voluntary simplicity" (Sadowski, "Nuncio Says by Living Simply Catholics Can Help Protect the Earth").

6. In the last paragraph of *The Manifesto of the Communist Party*, Marx declared that the ends he and Engels advocated could be attained "only by the forceful overthrow of all existing social conditions" (44). Whether this is true of the ends Marx advocated, it is not true of the aims motivating the present study. This book is aimed at bringing society into harmony with its supporting environment, without essential regard to how one or another social class ("bourgeoisie" or "proletariat") is affected in the process.

7. This should be understood as a figurative reference to all electronic forms of advertising, including those accessed wirelessly, which of course are not simply "unplugged."

Chapter 18. What Can Be Done? What Can One Do?

1. As described in *Affluenza: The All-Consuming Epidemic*, by de Graaf, Wann, and Naylor, marketing is a "public relations industry that creates and

perpetuates our commercial culture" (156). A similar observation is made in the similarly titled book, *Affluenza: When Too Much Is Never Enough,* by Hamilton and Dennis, who say that for advertising to be effective it "must sell not only products but also a very particular kind of world view—one where happiness can be bought, where problems can be solved by a product, and where having more things is the measure of success" (40). In his *Global Problems and the Culture of Capitalism,* Robbins summarizes the transformation of advertising during the twentieth century by saying that the goal of advertisers "was to aggressively shape consumer desires and create value in commodities by imbuing them with the power to transform the consumer into a more desirable person" (15).

2. See Kluger, "Hollywood's Smoke Alarm." This article discusses an initiative by the Harvard School of Public Health to get Hollywood executives to eliminate smoking from movies. Harvard has been joined in this effort by the American Medical Association and by the attorneys general of forty-one states.

3. See "Adbusters—Culture Jamming."

4. De Graaf et al., *Affluenza*, 211. When the group brought this lack of media access to the Canadian courts, the ruling was that its "anti-ads" are political and that the only political commercials the networks must accept are those of candidates in political campaigns.

5. Tursi, White, and McQuilkin, "Chapter 7: The Ax Falls."

6. This despite the fact that the advertising industry's treatment of environmental issues has shifted radically during the past few years. With a major boost from Al Gore's celebrated documentary *An Inconvenient Truth* (2006), public awareness of climate change has brought about a "palpable" increase in demand for products with low environmental impact (Pfanner, "Gore to Bring Talk of Green to Ad Festival"). Companies with environmentally problematic records (including energy producers and automobile manufacturers) have been scrambling to establish "green" images, and the advertising industry has been quick in developing skills enabling it to lead the way. A sign of how far the pendulum has swung was the appearance of Al Gore himself as featured speaker at the advertising industry's annual international conference in late June 2007, at Cannes (site of the international film festival in which Gore's documentary had starred a few weeks earlier). Advertising's interest in things green is not to sell fewer products (quite the contrary), but rather to sell more products with green labels. A concurrent manifestation of "green going mainstream" was the launching of MTV's "public service advertising campaign" Switch, aimed at promoting "environmentally friendly lifestyle choices among youth . . . to reduce the carbon emissions that contribute to climate change" (see "MTV Networks International Launches First Youth-Focused, Global, Multi-Platform Climate Change Campaign"). According to reporter

Eric Pfanner, the tenor of the campaign is to ask young people "to make little changes in the way you consume" (like using energy-efficient light bulbs and unplugging mobile chargers), changes so small "you won't even notice them" (Pfanner, "Gore to Bring Talk of Green to Ad Festival"). MTV will benefit from this campaign, but the environment itself probably will not. As an industry in its own right, advertising has no interest at all in reducing the influence of standard consumer values.

7. Unless otherwise noted, material in this section comes from the 1995 study "Yearning for Balance: Views of Americans on Consumption, Materialism, and the Environment," conducted by the Harwood Group (as summarized in the *YES!* article "Yearning for Balance").

8. See Bradley, "Comparison of Energy Use & CO_2 Emissions from Different Transportation Modes."

9. Most hybrids shut off their internal combustion engines when stopped, thus avoiding emissions at stoplights. But several models cannot run on an electric mode alone, which means that they emit pollutants when starting up. Another compromising factor is that most hybrid trucks and SUVs use their electric motors to boost power without substantial improvement in gas mileage. In this case the only distinct environmental benefit is low (or nonexistent) emissions at stoplights.

10. Hybrid buses operating in Seattle each use about three thousand gallons of fuel less per year than the standard buses they replaced. It has been estimated that if the nine largest U.S. cities used hybrid buses exclusively, they would save forty million gallons of fuel per year (Meredith, "First GM Hybrid Transit Buses Go to Work"). This is a greater fuel savings than that accomplished by one-half million smaller hybrid vehicles.

11. See Ryan, "City Hopes New Bicycle Lanes Will Get More People Pedaling."

12. During 2003–04, snowmobiling accounted for 41 percent of all winter sports–related accidents in Canada. See "Snowmobile Accidents, on the Rise in Canada."

13. Evaporative coolers (also called "swamp coolers") were used in ancient Egypt (see Digshotep, "Daily Life—Ancient Egypt") and are still used extensively in low-humidity areas like Iran ("Evaporative Swamp Coolers"). They are also used under more humid conditions in industrial plants and in buildings containing large numbers of livestock. Wherever used, they require ample supplies of fresh water.

14. The first commercially grown GM food product was the Flavr Savr tomato, which was released in the United States without special labeling in 1996. By 2007, GM strains comprised 91 percent of the U.S. soybean crop, 87 percent of cotton (a major source of cooking oil), and 73 percent of corn production. See "USA: Cultivations in 2007." Sweet potatoes and rice also come in

GM varieties. Other countries with significant GM food production are Argentina, Canada, Brazil, and China.

15. Material in this paragraph and the following comes from "Genetically Modified Food Controversies"; Schubert, "A Different Perspective on GM Food"; and Kimbrell, ed., *Fatal Harvest*, 211. An instructive discussion of the practices and hazards of genetic engineering may be found in Cummings, *Uncertain Peril.*

16. Kimbrell, ed., *Fatal Harvest*, 211.

17. Other environmental contaminants are under suspicion as well, including pesticides and heavy-metal factory emissions. See "Genetic Engineering," and "The 'Mysterious Bee Killer.'"

18. Until recently, an organic label certified absence of GM ingredients. But as demand for organic food increases, resistance to GM components has been eroding. New standards in the European Union (as of 2009) will allow 0.9 percent GM content under the organic label (Spevack, "Organic Foods Can Legally Contain GM").

19. Kingsolver, "Lily's Chickens," in *Small Wonder*, 123. A bountiful source of information on the complexities of getting food to the table is Pollan's *The Omnivore's Dilemma.*

20. The results of Dr. King's boycott are prominent in recent U.S. history. Comparably important for migrant farm workers (mostly Mexican), the actions of Cesar Chavez and his United Farm Workers led to an estimated seventeen million Americans refusing to buy grapes (White, "Cesar Chavez, Latino Champion of Civil Rights").

21. According to its sponsors, this boycott was actively supported by three million consumers and had cost the company $60 million in lost sales by 1990. See "The Dark Side of GE" and "The Pepsi Challenge."

22. "The Tuna Boycott."

23. Exxon Mobil was singled out particularly among oil companies because of its active role in opposing U.S. participation in the Kyoto protocol for climate change (Gregory, "Consumer Boycott to 'Stop Esso'").

24. "Esso Pays for Global Warming Sabotage."

25. For Coca-Cola, see "Campaign to Hold Coca-Cola Responsible"; for Monsanto, see Cummins and Lilliston, "U.S. Consumers & Farmers Battle Genetically-Engineered Soybeans."

26. A suitable model might be the Ethical Consumer's boycott list ("Current Consumer Boycotts").

27. Political neutrality is threatened from the start by the fact that the United States has control of all thirteen root name servers that direct Internet traffic to the right locations ("U.S. Keeps Hold on Internet"). Three of these servers have military connections (Karrenberg, "DNS Root Name Servers Explained for Non-Experts"). In 2005, Secretary General Kofi Annan initiated a

failed attempt to bring the Internet under U.N. control (Annan, "The U.N. Isn't a Threat to the Net"). Commercial neutrality is threatened by ongoing attempts of communication giants including AT&T, Verizon, Comcast, and Time Warner, which already dominate 98 percent of the broadband-access market ("Net Neutrality 101/FAQ's"), to charge for Internet use. By in effect taking over ownership of Internet services, these giant corporations would be able to put smaller competitors out of business. They also would be able to block nonprofit websites, like those proposed here, that might be viewed as antithetical to their interests.

28. According to a 2009 UN report, "millions of people in the U.S. . . . live in overcrowded and substandard conditions" ("UN Special Rapporteur on Adequate Housing Leaves the U.S."). According to a 2009 *New York Times* article, forty-nine million Americans lack consistent access to adequate food (DeParle, "Hunger in U.S. at a 14-Year High"). An online journalist reported in January 2010 that thirty-one million Americans cannot afford health insurance (Sanders, "Lack of Universal Health Care").

29. See Adler, "U.S. 2009 Foreclosures Shatter Record Despite Aid."

30. This often can be accomplished in two or three days. Whereas cooperative barn raising was practiced among secular communities as well during the nineteenth century, the practice is continued only among Amish (and certain Mennonite) communities today. A more broadly based communal group that currently constructs housing with unpaid labor is Habitat for Humanity, which has constructed nearly two hundred thousand houses worldwide ("Habitat for Humanity Builds 200,000th House").

31. These societies might follow the pattern of the original building societies in the United Kingdom that enabled working people to pool their savings in support of each other's home-building efforts. The essential thing about these societies is that they are owned by their members. As of March 2009, U.K. building societies included 2.9 million borrowing members and had over £395 billion of combined assets ("Becoming an Associate," 2). The U.S. counterpart is the savings and loan association, a thrift institution under the management (at least originally) of its depositors and borrowers.

32. DeMuth, "Defining Community Supported Agriculture."

33. Currently there are over twenty-five hundred local currencies operating worldwide, including about a hundred in the United States ("Local Currency"; "List of Community Currencies"). Among those in the United States, one of the better known is Ithaca Hours (http://www.ithacahours.com). The Ithaca Hours system (where "hour" signifies an hour's labor) began in 1991 and initially was associated with CSA farms around Ithaca, New York, along with a number of local stores (Glover, "Creating Community Economics"). It has since been expanded to include a local credit union and a health insurance

program, in which some local doctors participate (Stonington, "Local Currencies Aren't Small Change").

Another well-known local currency is issued by BerkShares, Inc., operating in Berkshire County, New York. Within less than two years after its founding in 2006, BerkShares was supported by five local banks, had issued shares of its own currency amounting to $1.5 million, and was aligned with over 350 local businesses ("What Are BerkShares?"). Issued in denominations ranging from one to fifty shares (all attractively printed bills), it exchanges with U.S. dollars at the ratio of 10:9 in participating banks. Inasmuch as the currency system itself operates without profit, no interest is charged on these bank transactions.

34. An often cited model is the Sustainable South Bronx project (http://www.ssbx.org), featuring community gardens and green jobs training.

35. See Skeffington, "Organic Fruit and Vegetable Growing."

36. The permaculture movement advocates a combination of ecologically modeled food-production systems and environmentally friendly lifestyles. Inaugurated during the 1970s, the movement now is represented in over a hundred countries, with sites as far south as Australia and as far north as Great Britain. For more information, see Diver, "Introduction to Permaculture"; and "The Basics."

37. Data in this paragraph come from "The World Health Organization's Ranking of the World's Health Systems" and The Commonwealth Fund, "Overall Ranking of 6 Countries."

38. Berestein, "Insurance Ills Put Squeeze on Consumers."

39. Goldstein, "On Top of Tax Breaks, Nonprofit Hospitals Reap Big Profits."

40. Families USA, "New Report Links High Prescription Drug Prices to Marketing Costs."

41. Data for this and the following paragraph are from "Off the Charts."

42. Not only beyond reach of the present study, but beyond reach of the body politic as well. However the current impasse over health care is resolved, it boils down to conflicting perceptions regarding the privilege of profit on the part of large businesses and the privilege of good health on the part of the indigent.

43. Like most local currencies, the Ithaca Hours program is not a branch of local government and certainly not a form of "socialized medicine." A generally more advantageous arrangement, nonetheless, would be for local government to sponsor health cooperatives open to the entire community. Health care might well join garbage disposal and police protection as services people come to expect from their local communities. The extra tax burden on individuals would be more than compensated by reductions in health insurance premiums.

44. Preventive procedures like colonoscopy and echocardiography are usually quite expensive. The suggestion here is only that preventative measures be emphasized, not that all such procedures be available free of charge.

45. Another cause of poor health in recent decades is a diet high in fat and carcinogenic additives (like benzene in processed food preservatives). Since diets of fast food (mostly fatty) and soda pop (high in calories) tend to be cheaper than healthy food, people with low incomes are particularly prone to obesity (Barbassa, "Working Poor Face Higher Obesity Rates"). In addition to sponsoring CSA programs to make fresh food readily available, communities might impose special taxes on food known to cause health problems (analogous to special taxes on cigarettes).

Bibliography

"Acid Rain—Are Ecosystems Recovering?" Science Encyclopedia, n.d. http://www.libraryindex.com/pages/1143/Acid-Rain-ARE-ECOSYSTEMS-RECOVERING.html (accessed January 2010).

Adam, David. "Suffocating Dead Zones Spread Across World's Oceans." *Guardian* (Manchester), August 14, 2008. http://www.guardian.co.uk/environment/2008/aug/14/pollution.endangeredhabitats.

"Adbusters—Culture Jamming." Enlightenment: The Experience Festival, http://www.experiencefestival.com/adbusters_-_culture_jamming (accessed January 2010).

Adler, Lynn. "U.S. 2009 Foreclosures Shatter Record Despite Aid." Reuters, January 14, 2010. http://www.reuters.com/article/idUSTRE60D0LZ20100114 (accessed January 2010).

"Al Gore Is Optimistic about Solar Energy, and Pretty Accurate, Too." *St. Petersburg Times,* Politifact, February 18, 2009. http://www.politifact.com/truth-o-meter/statements/2009/feb/18/al-gore/al-gore-optimistic-about-solar-energy-and-pretty-a/ (accessed May 2010).

"Ancient Roman Economy." UNRV History, n.d. http://www.unrv.com/economy.php (accessed December 2009).

Andrews, Edmund L., Michael J. de la Merced, and Mary Williams Walsh. "Fed's $85 Billion Loan Rescues Insurer." *New York Times,* September 16, 2008. http://www.nytimes.com/2008/09/17/business/17insure.html.

Annan, Kofi A. "The U.N. Isn't a Threat to the Net." *Washington Post,* November 5, 2005. http://www.washingtonpost.com/wp-dyn/content/article/2005/11/04/AR2005110401431.html.

Arabe, Katrina C. "Automotive Industry: The Big Picture." Industry Market Trends, October 8, 2003. http://news.thomasnet.com/IMT/archives/2003/10/automotive_indu.html (accessed May 2010).

Ariely, Dan. "Irrationality Is the Real Invisible Hand." Predictably Irrational, Technology Review, April 20, 2009. http://www.technologyreview.com/blog/post.aspx?bid=355&bpid=23416 (accessed December 2009).

Aubuchon, Vaughn. "World Population Growth History." Vaughn's One-Pagers, 2004. http://www.vaughns-1-pagers.com/history/world-population-growth.htm (accessed January 2010).

Baker, Roger. "Gunpowder." Industrial History of Cumbria, April 15, 2008. http://www.cumbria-industries.org.uk/gunpowder.htm (accessed December 2009).

Barbassa, Juliana. "Working Poor Face Higher Obesity Rates in U.S." Associated Press, March 3, 2004. http://ro-d.redorbit.com/news/science/51969/working_poor_face_higher_obesity_rates_in_us/index.html (accessed December 2009).

"The Basics." Permaculture Assocation, n.d. http://www.permaculture.org.uk/knowledge-base/basics (accessed January 2010).

Battles, Stephanie J. "Trends in Residential Air-Conditioning Usage from 1978 to 1997." Energy Information Administration, August 2, 2000. http://www.eia.doe.gov/emeu/consumptionbriefs/recs/actrends/recs_ac_trends.html (accessed December 2009).

Bayon, Ricardo. "Trading Futures in Dirty Air." Washington Post, August 5, 2001. http://www.newamerica.net/publications/articles/2001/trading_futures_in_dirty_air.

"Becoming an Associate of the Building Societies Association." Building Societies Association, London, March 2009. http://www.bsa.org.uk/docs/publications/associate.pdf (accessed December 2009).

Benbrook, C. M. "Genetically Engineered Crops and Pesticide Use in the United States: The First Nine Years." Technical Paper #7. Biotech Infonet, October 2004. http://www.biotech-info.net/Full_version_first_nine.pdf (accessed January 2010).

Benjamin, Alison. "Threatened Species Red List Shows Escalating Global Extinction Crisis." Guardian (Manchester), September 12, 2007. http://www.guardian.co.uk/environment/2007/sep/12/internationalnews.greenpolitics.

Bentham, Jeremy. An Introduction to the Principles of Morals and Legislation. 1789. Reprint, Oxford: Clarendon Press, 1876.

Berestein, Leslie. "Insurance Ills Put Squeeze on Consumers." Union-Tribune (San Diego), August 22, 2004. http://legacy.signonsandiego.com/news/health/20040822-9999-1n22health.html (accessed January 2010).

Berger, Peter L., and Thomas Luckmann. *The Social Construction of Reality: A Treatise in the Sociology of Knowledge.* Garden City, NY: Doubleday, 1966.

Best, Ben. "Say's Law and Economic Growth." Benbest.com, n.d. http://www.benbest.com/polecon/sayslaw.html (accessed December 2009).

"Bioplastic." Wikipedia, n.d. http://en.wikipedia.org/wiki/Bioplastic (accessed December 2009).

The Biosphere. San Francisco: W. H. Freeman, 1970.

Boulding, Kenneth. "The Economics of the Coming Spaceship Earth." In *Environmental Quality in a Growing Economy: Essays from the Sixth RFF Forum,* edited by Henry Jarrett, 3–14. Baltimore: Resources for the Future and Johns Hopkins Press, 1966.

Bowyer, Peter. "Weighing Up the Benefits Against the Environmental Disadvantages, Is It Feasible or Desirable for Hydroelectric Power to Be Developed on a Much Larger Scale in the Future?" May 2005. www.peter.mapledesign.co.uk/writings/physics/2005_Hydroelectric_Power_Feasible_or_Desirable.pdf (accessed January 2010).

Bradley, M. J., and Associates. "Comparison of Energy Use & CO_2 Emissions from Different Transportation Modes." American Bus Associations, May 2007. http://www.buses.org/research (accessed January 2010).

Bradsher, Keith. "As Asia Keeps Cool, Scientists Worry About the Ozone Layer." *New York Times,* February 23, 2007. http://www.nytimes.com/2007/02/23/business/23cool.html.

———. "Push to Fix Ozone Layer and Slow Global Warming." *New York Times,* March 15, 2007. http://www.nytimes.com/2007/03/15/business/worldbusiness/15warming.html.

Bridgman, P. W. *The Nature of Thermodynamics.* Cambridge, MA: Harvard University Press, 1941.

"Brief Questions and Answers on Ozone Depletion." U.S. Environmental Protection Agency, August 25, 2008. http://www.epa.gov/ozone/science/q_a.html (accessed January 2010).

Brillouin, Léon. *Science and Information Theory.* New York: Academic Press, 1956.

Broder, John M. "Climate Change Seen as Threat to U.S. Security." *New York Times,* August 8, 2009. http://www.nytimes.com/2009/08/09/science/earth/09climate.html.

Brown, Dan. *The Da Vinci Code: A Novel.* New York: Doubleday, 2003.

Brown, Lester R. *Eco-Economy: Building an Economy for the Earth.* New York: W. W. Norton, 2001.

Brown, Paul. "Ozone Layer Most Fragile on Record: Fears Over Increase in Skin Cancer as Scientists Report that Climate Change Continues to

Destroy the Earth's Protection." *Guardian* (Manchester), April 27, 2005. http://www.guardian.co.uk/science/2005/apr/27/environment.research.

Burkart, Karl. "Could Solar Roadways Power the U.S.?" Mother Nature Network, September 7, 2009. http://www.mnn.com/technology/research-innovations/blogs/could-solar-roadways-power-the-us (accessed January 2010).

"Campaign to Hold Coca-Cola Accountable." India Resource Center, n.d. http://www.indiaresource.org/campaigns/coke/ (accessed December 2009).

Carlson, Royce. "15 Reasons to Stop Eating Meat." *Zenzibar,* February 25, 2001. http://www.zenzibar.com/Articles/15_reasons.asp (accessed January 2010).

Carson, Rachel. *Silent Spring.* Boston: Houghton Mifflin, 1962.

"China Energy Data, Statistics and Analysis—Oil, Gas, Electricity, Coal." Country Analysis Briefs. Energy Information Administration, July 2009. http://www.eia.doe.gov/emeu/cabs/China/Full.html (accessed December 2009).

Chydenius, Anders. *The National Gain.* 1765. Trans. Georg Schauman. London: E. Benn, 1931.

The CIA World Factbook 2008. Washington, DC: Central Intelligence Agency, 2008.

Clark, Eric. *The Want Makers: The World of Advertising: How They Make You Buy.* New York: Viking, 1988.

Cleveland, Cutler J. "Biophysical Economics: From Physiocracy to Ecological Economics and Industrial Ecology." In *Bioeconomics and Sustainability: Essays in Honor of Nicholas Georgescu-Roegen,* edited by Kozo Mayumi and John Gowdy, 125–54. Northampton, MA: Elgar, 1999.

Clifford, Stephanie. "Cable Companies Target Commercials to Audience." *New York Times,* March 3, 2009. http://www.nytimes.com/2009/03/04/business/04cable.html.

"Coke versus Pepsi: It's All in the Head." *Neuron* 44, no. 2 (October 14, 2004): 379–87. http://www.eurekalert.org/pub_releases/2004-10/cp-cvp101204.php (accessed January 2010).

"Common Household Appliance Energy Use." Ames City Government, n.d. http://www.city.ames.ia.us/ElectricWeb/energyguy/appliances.htm (accessed December 2009).

Common, Mick, and Sigrid Stagl. *Ecological Economics: An Introduction.* New York: Cambridge University Press, 2005.

The Commonwealth Fund. "Overall Ranking of 6 Countries" (table). West Virginians for Affordable Health Care, 2007. http://www.wvahc.org/downloads/MMWchart.pdf (accessed December 2009).

"Considerations for a Vegetarian Diet." Dharma Realm Buddhist Assocation, n.d. http://www.drba.org/dharma/veggie/vegconsider.asp (accessed January 2010).

Container Recycling Institute. "Plastic Bottle Waste Tripled Since 1995." Press release. Mindfully.org, September 15, 2003. http://www.mindfully.org/Plastic/Polyethylene/PET-Bottle-Waste15sep03.htm (accessed December 2009).

Cook, Earl. "The Flow of Energy in an Industrial Society." *Scientific American* (September 1971): 142.

———. *Man, Energy, Society.* San Francisco: W. H. Freeman, 1976.

"Coral Reef Biology." NOAA's Coral Reef Information System (CORIS), n.d. http://www.coris.noaa.gov/about/biology/ (accessed December 2009).

Cosgrove-Mather, Bottie. "Ethanol Pollution Surprise: EPA Finds Worrisome Levels of Toxic Air Pollutants at Ethanol Plants." CBS News, May 3, 2002. http://www.cbsnews.com/stories/2002/05/03/tech/main508006.shtml (accessed December 2009).

Costanza, Robert, et al. *An Introduction to Ecological Economics.* Boca Raton, FL: St. Lucie Press, International Society for Ecological Economics, 1997.

"Costing the Earth: Environmental Economics." University of Cambridge News and Events, August 9, 2002. http://www.admin.cam.ac.uk/news/dp/2002080901 (accessed December 2009).

"Country Comparison GDP per Capita." *CIA World Factbook 2009.* https://www.cia.gov/library/publications/the-world-factbook/index.html (accessed January 2010).

Cross, Jay. "Neuromarketing." Internet Time Blog, October 31, 2003. http://internettime.com/blog/archives/000998.html (accessed December 2009).

"Culture Jamming (tm): Brought to You by *Adbusters." Stay Free!* n.d. http://stayfreemagazine.org/9/adbusters.htm (accessed December 2009).

Cummings, Claire Hope. *Uncertain Peril: Genetic Engineering and the Future of Seeds.* Boston: Beacon Press, 2008.

Cummins, Ronnie, and Ben Lilliston. "U.S. Consumers & Farmers Battle Genetically-Engineered Soybeans." *In Motion Magazine,* December 3, 1996. http://www.inmotionmagazine.com/monsanto3.html (accessed January 2010).

"Current Consumer Boycotts." Ethical Consumer, n.d. http://www.ethicalconsumer.org/Boycotts/currentUKboycotts.aspx (accessed December 2009).

Daintith, John, ed., *A Dictionary of Chemistry.* 5th ed. Oxford: Oxford University Press, 2004.

Daly, Herman E. *Beyond Growth: The Economics of Sustainable Development.* Boston: Beacon Press, 1996.

——. ed. *Economics, Ecology, Ethics: Essays toward a Steady-State Economy.* San Francisco: W. R. Freeman, 1973.

——. "Five Policy Recommendations for a Sustainable Economy." In *FEASTA Review* 1, edited by R. Douthwaite and J. Jopling. Dublin: FEASTA, 1999. http://www.feasta.org/documents/feastareview/daly2.htm (accessed December 2009).

——. *Steady-State Economics: The Economics of Biophysical Equilibrium and Moral Growth.* San Francisco: W. H. Freeman, 1977.

"The Dark Side of GE." Behind the Lines. *Multinational Monitor* 10, nos. 1–2 (January–February 1989). http://multinationalmonitor.org/hyper/issues/1989/01/behind-the-lines.html (accessed December 2009).

Darmstadter, Joel, Joy Dunkerley, and Jack Alterman. *How Industrial Societies Use Energy: A Comparative Analysis.* Baltimore: Resources for the Future, Johns Hopkins University Press, 1977.

Davis, Rick. "Technofix-ation." *New Internationalist* 231 (May 1992). http://www.newint.org/issue231/techno.htm (accessed December 2009).

"The Dead Zone." The 1 Mississippi Campaign, 2010. http://www.1mississippi.net/public-site/about-mississippi-river/dead-zone (accessed January 2010).

Dean, Cornelia. "Coral Reefs and What Ruins Them." *New York Times,* February 26, 2008. http://www.nytimes.com/2008/02/26/science/earth/26reef.html.

——. "Rising Acidity Is Threatening Food Web of Oceans, Science Panel Says." January 30, 2009. http://www.nytimes.com/2009/01/31/science/earth/31ocean.html?_r=1.

——. "Scientists Warn Fewer Kinds of Fish Are Swimming the Ocean." *New York Times,* July 29, 2005. http://www.nytimes.com/2005/07/29/science/29fishing.html.

de Graaf, John, David Wann, and Thomas H. Naylor. *Affluenza: The All-Consuming Epidemic.* San Francisco: Barrett-Koehler Publishers, 2001.

"Delamp Vending Machines." Utility Savings Initiative, February 2004. http://www.docstoc.com/docs/5317200/Vending-Machines-UTILITY-SAVINGS-INITIATIVE-USI-FACT-SHEET-Delamp (accessed January 2010).

DeLiberty, Tracy. "Energy Interactions with the Atmosphere and at the Surface." Department of Geography, University of Delaware, September 23, 1999. http://www.udel.edu/Geography/DeLiberty/Geog474/geog474_energy_interact.html (accessed January 2010).

Delwiche, C. C. "The Nitrogen Cycle." *Scientific American* (September 1970): 136–47.

DeMuth, Suzanne. "Defining Community Supported Agriculture." Alternative Farming Systems Information Center, U.S. Department of Agriculture, September 1993. http://nal.usda.gov/afsic/pubs/csa/csadef.shtml (accessed December 2009).

DeParle, Jason. "49 Million Americans Report a Lack of Food." *New York Times,* November 17, 2009. http://www.nytimes.com/2009/11/17/us/17hunger.html.

"Desalination: Background from Wikipedia." Mongabay.com, 2007. http://mongabay.com/reference/eco/Desalination.html (accessed January 2010).

de Vries, Lloyd. "Are Today's Students Too Self-Centered? Study Finds Rising Narcissism, Fears It Could Hurt Personal Relationships, Society." CBSNews.com, February 27, 2007. http://cbsnews.com/stories/2007/02/07/health/main2519593.shtml (accessed December 2009).

Diamond, Jared. *Collapse: How Societies Choose to Fail or Succeed.* New York: Viking, 2005.

Digshotep. "Daily Life: Ancient Egypt." Web Owls, May 23, 2006. http://web-owls.com/ 2006/05/23/daily-life-ancient-egypt/ (accessed January 2010).

Diver, Steve. "Introduction to Permaculture." Minnesotans for Sustainability, May 14, 1996. http://www.mnforsustain.org/diver_steve_permaculture_faq_intro.htm (accessed January 2010).

Domhoff, G. William. "Wealth, Income, and Power." Who Rules America? September 2005, October 2009. http://sociology.ucsc.edu/whorulesamerica/power/wealth.html (accessed December 2009).

Dooley, Roger. "Brain Scans Predict Buying Behavior." Neuromarketing: Where Brain Science and Marketing Meet, January 4, 2007. http://www.neurosciencemarketing.com/blog/articles/brain-scan-buying.htm (accessed December 2009).

Drengson, Alan, and Yuichi Inoue, eds. *The Deep Ecology Movement: An Introductory Anthology.* Berkeley: North Atlantic Books, 1995.

Dyson, Freeman. "Energy in the Universe." *Scientific American* (September 1971): 50–59.

Ebenstein, Lanny. *Milton Friedman: A Biography.* New York: Palgrave Macmillan, 2007.

"Economy Statistics: Welfare State: The Welfare State and Social Expenditure." NationMaster.com, n.d. http://www.nationmaster.com/graph/eco_wel_sta_the_wel_sta_and_soc_exp_of_gdp-welfare-state-social-expenditure-gdp (accessed January 2010).

Edwards, Paul N. "Global Comprehensive Models in Politics and Policymaking." *Climate Change* 32 (January 1996): 149–61.

Ehrlich, Paul R. *The Population Bomb.* New York: Ballantine Books, 1968.

Ekelund, Robert B., Jr., Rand W. Ressler, and Robert D. Tollison. *Macroeconomics: Private Markets and Public Choice.* 7th ed. Boston: Pearson/Addison-Wesley, 2006.

"Electricity Consumption (Most Recent) by Country." NationMaster.com, n.d. http://www.nationmaster.com/graph-T/ene_ele_con (accessed December 2009).

Elert, Glenn, ed. "Volume of US Soft Drink Consumption." *The Physics Fact-book*. http://hypertextbook.com/facts/2001/ChristiaanRule.shtml (accessed January 2010).

"Emissions of Greenhouse Gases in the United States in 1998." Energy Information Administration, October 1999. ftp://ftp.eia.doe.gov/pub/oiaf/1605/cdrom/pdf/ggrpt/057398.pdf (accessed January 2010).

Energy Information Administration. *Annual Energy Report 2008*. DOE/EIA-0384(2008). Washington, DC: Energy Information Administration, U.S. Department of Energy, 2009. http://www.eia.doe.gov/aer/pdf/aer.pdf (accessed June 2010).

"The Environmental Management of Industrialized States." The United Nations Industrial Development Organization, May 30, 2008. www.neisd.net/isa/MUNSA/docs/unido2.pdf (accessed January 2010).

"Esso Pays for Global Warming Sabotage as Consumers Turn Their Backs." Greenpeace UK, September 3, 2002. http://www.greenpeace.org.uk/media/press-releases/esso-pays-for-global-warming-sabotage-as-consumers-turn-their-backs (accessed December 2009).

European Commission. *Eutrophication and Health*. Local Authorities, Health and Environment Briefing Pamphlet Series #40. World Health Organization, 2002. http://ec.europa.eu/environment/water/water-nitrates/pdf/eutrophication.pdf (accessed December 2009).

Evans, Simon. "The Mathematics of Desire: Producers and Consumers." The Hypermedia Research Centre, University of Westminster, n.d. http://www.hrc.wmin.ac.uk/theory-mathematicsofdesire2.html (accessed December 2009).

"Evaporative Swamp Coolers: Basics." House-Energy.com, n.d. http://house-energy.com/Cooler/Basics-Evaporative.htm (accessed December 2009).

"Explore the Solar Resource." New Mexico Solar Energy Association, n.d. http://www.nmsea.org/Curriculum/7_12/The_Solar_Resource.htm (accessed January 2010).

Ezstreet. "Potholes Are Gone Forever with Permanent Cold Asphalt Products." EZ Street Press, October 24, 2007. http://ezstreet.wordpress.com/2007/10/24/potholes-are-gone-forever-with-permanent-cold-asphalt-products (accessed December 2009).

Fahrenthold, David A., and Steven Mufson. "Cost of Saving the Climate Meets Real-World Hurdles." *Washington Post,* August 16, 2007. http://www.washingtonpost.com/wp-dyn/content/article/2007/08/15/AR2007081502432.html.

Families USA. "New Report Links High Prescription Drug Prices to Marketing Costs, Profits and Enormous Executive Compensation." NAMI of Santa Cruz County, July 2001. http://www.namiscc.org/newsletters/July01/DrugPrices.htm (accessed January 2010).

Flannery, Tim. *The Eternal Frontier: An Ecological History of North America and Its Peoples.* New York: Atlantic Monthly Press, 2001.

Fonseca, Gonçalo L. "David Ricardo, 1772–1823." The History of Economic Thought Website, n.d. http://cepa.newschool.edu/het/profiles/ricardo.htm (accessed December 2009).

"Frequently Asked Questions." Kyoto Forestry Association, n.d. http://www .kfoa.co.nz/faqs.htm (accessed January 2010).

Freund, Peter, and George Martin. *The Ecology of the Automobile.* New York: Black Rose Books, 1993.

Friedman, Benjamin M. *The Moral Consequences of Economic Growth.* New York: Knopf, 2005.

Gache, Gabriel. "Global Warming Triggers Cooling Mechanism." Softpedia, November 5, 2007. http://news.softpedia.com/news/Global-Warming-Triggers-Cooling-Mechanism-70049.shtml (accessed December 2009).

Galbraith, Kate. "Up to $80 Billion to Transmit Wind Power from Midwest to Northeast." Green Inc.: Energy, the Environment and the Bottom Line. *New York Times,* February 10, 2009. http://greeninc.blogs.nytimes.com/ 2009/02/10/up-to-80-billion-to-transmit-wind-power-from-midwest-to-northeast.

"Gas Escaping from Ocean Floor May Drive Global Warming." Press release. University of California, Santa Barbara, July 19, 2006. http://www.ia.ucsb .edu/pa/display.aspx?pkey=1482 (accessed January 2010).

Gates, D. M. "Biosphere." In *The New Encyclopaedia Britannica,* vol. 2, 1037–44. 15th ed. Chicago: Encyclopaedia Britannica, 1974.

"GDP Growth in China 1952–2009." Chinability: Latest News and Statistics on China's Economy and Business Climate, October 10, 2009. http://www .chinability.com/GDP.htm (accessed January 2010).

"Genetically Engineered Crops and Pesticide Use (2004)." Union of Concerned Scientists, November 2, 2004. http://www.ucsusa.org/food_and_agriculture/ science_and_impacts/impacts_genetic_engineering/genetically-engineered-crops.html (accessed December 2009).

"Genetically Modified Crops in the United States." Factsheet. Pew Initiative on Food and Biotechnology, August 2004. http://www.pewtrusts.org/news_ room_detail.aspx?id=17950 (accessed December 2009).

"Genetically Modified Food Controversies." Wikipedia, n.d. http://en .wikipedia.org/wiki/GM_food_controversy (accessed December 2009).

"Genetic Engineering: References." Sierra Club, 2008. http://www.sierraclub .org/biotech/references.asp (accessed December 2009).

Georgescu-Roegen, Nicholas. *The Entropy Law and the Economic Process.* Cambridge, MA: Harvard University Press, 1971.

Gever, John, et al. *Beyond Oil: The Threat to Food and Fuel in the Coming Decades* Cambridge, MA: Ballinger, 1986.

Glasgow, Nathan, and Lena Hansen. "Setting the Record Straight on Ethanol: Focusing on the Nexus of the Agriculture and Energy Value Chains." Rocky Mountain Institute Publication No. 14, October 31, 2005. http://www.renewableenergyworld.com/rea/news/article/2005/11/setting-the-record-straight-on-ethanol-38601 (accessed December 2009).

"Global Warming and Hurricanes: An Overview of Current Research Results." Geophysical Fluid Dynamics Laboratory, October 17, 2008. http://www.gfdl.noaa.gov/~tk/glob_warm_hurr.html (accessed December 2009).

Glover, Paul. "Creating Community Economics with Local Currency." Ithaca Hours, December 11, 2006. http://www.ithacahours.com/intro.html (accessed December 2009).

Goldstein, Jacob. "On Top of Tax Breaks, Nonprofit Hospitals Reap Big Profits." Health Blog, *Wall Street Journal,* April 4, 2008. http://blogs.wsj.com/health/2008/04/04/on-top-of-tax-breaks-nonprofit-hospitals-reap-big-profits/?mod=WSJBlog (accessed December 2009).

Goudie, Andrew. *The Human Impact on the Natural Environment.* 4th ed. Cambridge, MA: MIT Press, 1994.

"Gree Electric Appliances Holding 2009: A Company Profile." July 2009. http://www.researchandmarkets.co.uk/reportinfo.asp?cat_id=0&report_id=1054979&q=Gree%20Electric%20Appliances&p=1 (accessed March 2010).

Gregory, Mark. "Consumer Boycott to 'Stop Esso.'" BBC News, May 8, 2001. http://news.bbc.co.uk/1/hi/business/1318360.stm.

Grenthe, Ingmar. "The Nobel Prize in Chemistry 1995: Presentation Speech." Nobelprize.org, 1995. http://nobelprize.org/nobel_prizes/chemistry/laureates/1995/presentation-speech.html (accessed December 2009).

"Habitat for Humanity Builds 200,000th House, with Much to Celebrate, More to Build." Habitat for Humanity, July 29, 2005. http://www.habitat.org/celebrate_build/press/200k_house_announcement.aspx (accessed December 2009).

Hamilton, Clive, and Richard Denniss. *Affluenza: When Too Much Is Never Enough.* Crows Nest, NSW, Australia: Allen & Unwin, 2006.

Hanson, N. R. Introduction to John Maynard Keynes, *A Treatise on Probability.* New York: Harper & Row, 1962.

Hardy, Ian. "Cutting Traffic with Driverless Cars." BBC News, September 10, 2009. http://news.bbc.co.uk/2/hi/programmes/click_online/8236921.stm (accessed January 2010).

Hawken, Paul, Amory Lovins, and L. Hunter Lovins. *Natural Capitalism: Creating the Next Industrial Revolution.* Boston: Little, Brown, 1999.

"Health-Related Quality of Life: Methods and Measures." National Center for Chronic Disease Prevention and Health Promotion, December 1, 2005. http://www.cdc.gov/hrqol/methods.htm (accessed December 2009).

Hecht, Jeff. "Global Warming Stretches Subtropical Boundaries." *New Scientist,* May 26, 2006. http://www.newscientist.com/article/dn9229-global-warming-stretches-subtropical-boundaries.html (accessed December 2009).

Heckeroth, Steven. "Solar Building Design." *Backwoods Home Magazine,* n.d. http://www.backwoodshome.com/articles/heckeroth63.html (accessed December 2009).

Herring, David. "Does the Earth Have an Iris Analog?" Earth Observatory, June 12, 2002. http://earthobservatory.nasa.gov/Features/Iris/ (accessed December 2009).

Hindley, Dave. "Orion-list Water Consumption." Orion List-serv, December 9, 2000. http://orion.huji.ac.il/orion/archives/2000b/msg00246.html (accessed December 2009).

"How Coal Works." Union of Concerned Scientists, 2009. http://www.ucsusa.org/clean_energy/coalvswind/brief_coal.html (accessed December 2009).

Hughes, J. Donald. *An Environmental History of the World: Humankind's Changing Role in the Community of Life.* New York: Routledge, 2001.

Hutchison, Fred H. "About Electricity." Clean-energy.us, March 27, 2009. http://www.clean-energy.us/facts/electricity.htm (accessed January 2010).

"Hydro Electric Power." UK Energy Saving, 2009. http://www.uk-energy-saving.com/hydro_electric_power.html (accessed December 2009).

"Internet Increases Global Inequality." BBC News, July 13, 1999. http://news.bbc.co.uk/2/hi/392171.stm (accessed January 2010).

"Iron Seeding (Fertilization) of the Ocean." Fact Sheet. Climate Change & Global Warming, n.d. http://climatechange.110mb.com/nations-iron-seeding-oceans.htm (accessed December 2009).

Irvine, Martha. "Wealth Top Priority for Today's Youth." SaukValley.com, January 23, 2007. http://www.saukvalley.com/articles/2007/01/23/news/state/300714976024485.txt (accessed December 2009).

Jackson, Janine. "Good News! The Rich Get Richer: Lack of Applause for Falling Wages Is Media Mystery." Fairness & Accuracy in Reporting (FAIR), March/April 2006. http://www.fair.org/index.php?page=2854 (accessed December 2009).

Jones, Adam. "Case Study: Female Infanticide." Gendercide Watch, 1999–2000. http://www.gendercide.org/case_infanticide.html (accessed December 2009).

"Junk Food Tactics Hit the Supermarket Shelves." The Parents Jury, August 14, 2006. www.parentsjury.org.au/downloads/junk_food_tactics_media_release.pdf (accessed January 2010).

Kahn, James R. *The Economic Approach to Environmental and Natural Resources.* Fort Worth, TX: Dryden Press, 1995.

Kahn, Joseph, and Jim Yardley. "As China Roars, Pollution Reaches Deadly Extremes." *New York Times,* August, 26, 2007. http://www.nytimes.com/2007/08/26/world/asia/26china.html.

Karrenberg, Daniel. "DNS Root Name Servers Explained for Non-Experts." Internet Society, September 2007. http://www.isoc.org/briefings/019 (accessed December 2009).

Kay, James. "Ecosystems, Science and Sustainability." In *Advances in Energy Studies: Exploring Supplies, Constraints and Strategies: Proceedings of the International Workshop,* edited by Sergio Ulgiati et al., 319–28. Padova, Italy: SGEditoriali, 2001.

Kay, Jane Holtz. *Asphalt Nation: How the Automobile Took Over America, and How We Can Take It Back.* New York: Crown Publishers, 1997.

Kayne, R. "What Is Neuromarketing?" Wisegeek.com, n.d. http://www.wisegeek.com/what-is-neuromarketing.htm (accessed December 2009).

Keynes, John Maynard. *Essays in Persuasion.* New York: W. W. Norton, 1931.

———. *The General Theory of Employment, Interest, and Money.* London: Macmillan, 1936.

———. *A Treatise on Money.* 2 vols. New York: Harcourt, Brace, 1930.

———. *A Treatise on Probability.* London: Macmillan, 1921.

Khatib, Hisham. *Financial and Economic Evaluation of Projects in the Electricity Supply Industry.* Institution of Electrical Engineers Power and Energy Series 44. London: Institution of Engineering and Technology, 1996.

Kimbrell, Andrew, ed. *Fatal Harvest: The Tragedy of Industrial Agriculture.* Washington, DC: Foundation for Deep Ecology, Island Press, 2002.

Kingsolver, Barbara. *Small Wonder.* Illustrations by Paul Mirocha. New York: HarperCollins, 2002.

Kluger, Jeffrey. "Hollywood's Smoke Alarm." *Time,* April 23, 2007.

Koci, Elina. "An Overview of the Regulatory Framework for Media in Albania." *Communications Law in Transition Newsletter* 2, no. 3 (February 18, 2001). http://pcmlp.socleg.ox.ac.uk/archive/transition/ (accessed January 2010).

Kolbert, Elizabeth. "The Darkening Sea." Annals of Science. *New Yorker,* November 20, 2006, 66.

Komin, Petr, and Kirsi Nikkola. "Ice Ages (Pleistocene Glaciation)." Arctic Studies Program, n.d. http://arcticstudies.pbworks.com/IceAges (accessed January 2010).

Krauss, Clifford. "Move Over, Oil, There's Money in Texas Wind." *New York Times,* February 23, 2008. http://www.nytimes.com/2008/02/23/business/23wind.html.

Kuhn, Thomas S. *The Structure of Scientific Revolutions.* 2nd ed., enl. Chicago: University of Chicago Press, 1970.

Lang, Susan. "Seven-Year Glitch: Cornell Warns that Chinese GM Cotton Farmers Are Losing Money Due to 'Secondary' Pests." *Cornell Univer-*

sity Chronicle Online, July 25, 2006. http://www.news.cornell.edu/stories/
July06/Bt.cotton.China.ssl.html (accessed December 2009).

Lapper, Richard, and Laurie Morse. "Market Makers in CO_2 Permits." *Financial Times* (London), March 1, 1995.

Leahy, Stephen. "New Studies Back Benefits of Organic Diet." IPSNews.net via *Tierramérica,* February 24, 2006. http://ipsnews.net/news.asp?idnews =32375 (accessed December 2009).

Leakey, Richard, and Roger Lewin. *The Sixth Extinction: Patterns of Life and the Future of Humankind.* New York: Doubleday, 1995.

Leonardi, Christine, ed. "Economic Growth versus Genuine Progress." *Gordon Institute of Business Science Review* 4 (April 2006). http://www.gibsreview .co.za/home.asp?pid=11&toolid=2&itemid=144&reviewed=143 (accessed January 2010).

Leopold, Aldo. *A Sand County Almanac.* New York: Ballantine, 1970 [1949].

Levine, Raphael D., and Myron Tribus, eds. *The Maximum Entropy Formalism: A Conference Held at the Massachusetts Institute of Technology on May 2-4, 1978.* Cambridge, MA: MIT Press, 1979.

LeVine, Steve. "ExxonMobil's Algae Exploration." *Business Week,* July 15, 2009. http://www.businessweek.com/bwdaily/dnflash/content/jul2009/ db20090715_064110.htm (accessed January 2010).

"List of Community Currencies in the United States." Wikipedia, n.d. http:// en.wikipedia.org/wiki/List_of_community_currencies_in_the_United_ States (accessed December 2009).

"Local Currency." Human Science Wikia, n.d. http:/humanscience.wikia.com/ wiki/Local_currency (accessed December 2009).

Lorenz, Konrad. *Evolution and Modification of Behavior.* Chicago: University of Chicago Press, 1965.

Lyon, Shanon. "CEO Salaries: What Is the Average Salary of a CEO?" PayScale.com, July 31, 2008. http://blogs.payscale.com/content/2008/07/ ceo-salaries--1.html (accessed January 2010).

MacArthur, R. H. "Fluctuations of Animal Populations and a Measure of Community Stability." *Ecology* 6 (1955): 533–36.

Malthus, Thomas. *An Essay on the Principle of Population.* London: J. Johnson, 1798.

Mandel, Michael. "What's So Good About Growth." *Business Week,* November 7, 2005. http://www.businessweek.com/magazine/content/05_45/ b3958122.htm (accessed December 2009).

Margalef, Ramón. *Perspectives in Ecological Theory.* Chicago: University of Chicago Press, 1968.

Marino, Vivian. "From Cereal to Juice, a Year of Innovation." Bulletin Board. *New York Times,* January 4, 2004. http://www.nytimes.com/2004/01/04/ business/bulletin-board-from-cereal-to-juice-a-year-of-innovation.html.

"Market Failure." Environmental Terminology Discovery Service, European Environment Agency, n.d. http://glossary.eea.europa.eu/EEAGlossary/M/market_failure (accessed December 2009).

Marx, Karl, and Friedrich Engels. *The Manifesto of the Communist Party*. 1848. Reprint, New York: International Publishers, 1932.

Maxwell, Ian. "Methane Menace from Siberia Threatens Global Climate." IT-Wire, September 10, 2006. http://www.itwire.com.au/content/view/5553/53/ (accessed December 2009).

McCain, Roger A. "The Neoclassical Model." Drexel University, n.d. http://faculty.lebow.drexel.edu/mccainr/top/prin/txt/gro/gro10.html (accessed December 2009).

The McIlvaine Company. "Air Pollution Market Soars, But Investment Opportunities Are Few." EnvironmentalExpert.com, August 16, 2006. http://www.environmental-expert.com/resulteachpressrelease.aspx?cid =5122&codi=7253 (accessed January 2010).

McKibben, Bill. *The End of Nature*. New York: Random House, 1989.

McNeill, J. R. *Something New Under the Sun: An Environmental History of the Twentieth-Century World*. New York: W. W. Norton, 2000.

Meadows, Donella H., et al. *The Limits to Growth: A Report for the Club of Rome's Project on the Predicament of Mankind*. New York: Universe Books, 1972.

Melosi, Martin V. "The Automobile and the Environment in American History." Automobile in American Life and Society, 2004. http://www.autolife.umd.umich.edu/Environment/E_Overview/E_Overview1.htm (accessed December 2009).

Meredith, Mike. "First GM Hybrid Transit Buses Go to Work: New Technology Offers Large Gains in Fuel Economy and Reduced Emissions." MSN Autos.MSN.com, 2009. http://autos.msn.com/advice/article.aspx?contentid =4022529 (accessed December 2009).

"Microbes Convert 'Styrofoam' into Biodegradable Plastic." Press release. American Chemical Society, February 23, 2006. http://www.eurekalert.org/pub_releases/2006-02/acs-mc022306.php (accessed December 2009).

Mill, John Stuart, *On Liberty*. 1859. Reprint, London; Watts & Co., 1929.

———. *Principles of Political Economy*. Boston: C. C. Little & J. Brown, 1848.

———. *Utilitarianism*. 1863. Reprint, Indianapolis: Bobbs-Merrill, 1957.

Milloy, Steven. "Carbon Offsets—Buyer Beware." JunkScience.com, July 19, 2007. http://www.junkscience.com/ByTheJunkman/20070719.html (accessed December 2009).

Moffatt, Mike. "Definition of Economic Good." About.com, 2009. http://economics.about.com/od/termsbeginningwithe/g/economic_good.htm (accessed December 2009).

"Montreal Protocol on Substances that Deplete the Ozone Layer." Alternative Fluorocarbons Environmental Acceptability Study (AFEAS), June 2, 2006. http://www.afeas.org/montreal_protocol.html (accessed December 2009).

Moore, Charles J., Shelly L. Moore, Molly K. Leecaster, and Stephen B. Weisberg. "A Comparison of Plastic and Plankton in the North Pacific Central Gyre." *Marine Pollution Bulletin* 42, no. 12 (December 2001). http://www.mindfully.org/Plastic/Moore-North-Pacific-Central-Gyre.htm (accessed December 2009).

"More Aluminum Cans Trashed Last Year Than Recycled." Mindfully.org (Summer/Fall 2003). http://www.mindfully.org/Sustainability/2003/Aluminum-Cans-RecycledSep03.htm (accessed January 2010).

"MTV Networks International Launches First Youth-Focused, Global, Multi-Platform Climate Change Campaign—MTV Switch." MTV Networks, June 13, 2007. http://www.mtvnetworks.co.uk/mtvswitch (accessed January 2010).

Murphy, Elaine. "Should There Be a 'Twinkie Tax'?" The Diet & Weight Loss Blog. NutritionData.com, January 13, 2008. http://blog.nutritiondata.com/dieting_weight_loss_blog/2008/01/should-there-be.html (accessed December 2009).

Myers, David. "The Secret to Happiness: Look Inside at What Actually Gives You Joy, and the Good Life May Be Closer than You Thought." *Yes!* June 18, 2004. http://www.yesmagazine.org/article.asp?id=866 (accessed December 2009).

"The 'Mysterious Bee Killer.'" *Idaho Observer,* May 2007. http://www.proliberty.com/observer/20070508.htm (accessed December 2009).

"Net Neutrality 101/FAQ's." Media Alliance, n.d. http://www.media-alliance.org/article.php?id=1600 (accessed January 2010).

Nietzsche, Frederick Wilhelm. *Human, All Too Human.* 1878. Trans. R. J. Hollingdale. Reprint, Cambridge: Cambridge University Press, 1996.

Norgaard, Richard B. "Ecological Economics: A Short Description." Introduction to Economics. Forum on Religion and Ecology, Yale University, 2000. http://fore.research.yale.edu/disciplines/economics/ (accessed December 2009).

"Nuclear Energy and Greenhouse Gas Emissions Avoidance in the European Union." *European Nuclear Society News,* July 2008. http://www.euronuclear.org/e-news/e-news-21/greenhouse-gas-emissions.htm (accessed January 2010).

Nutting, Rex. "Consumers Unable to Lead Economy Forward: GDP May Grow in Third Quarter, but It Won't Be Sustainable Growth." MarketWatch.com, July 24, 2009. http://www.marketwatch.com/story/consumers-unable-to-lead-economy-forward-2009-07-24 (accessed December 2009).

O'Brien, Daniel, and Mike Woolverton. "Trends in U.S. Energy Use and Supplies: How Biofuels Contribute." Agricultural Marketing Resource Center, November 2009. http://www.agmrc.org/renewable_energy/energy/trends_in_us_energy_use_and_supplies__how_biofuels_contribute.cfm (accessed January 2010).

Odum, Eugene. *Ecological Vignettes: Ecological Approaches to Dealing with Human Predicaments.* Amsterdam: Harwood Academic Publishers, 1998.

———. *Fundamentals of Ecology.* Philadelphia: Saunders, 1953.

Odum, Howard T. *Environment, Power, and Society.* New York: Wiley-Interscience, 1971.

"Off the Charts: Pay, Profits and Spending by Drug Companies." ACT UP New York, n.d. http://www.actupny.org/reports/drugcosts.html (accessed December 2009).

Oort, Abram H. "The Energy Cycle of the Earth." In *The Biosphere,* 14–23. San Francisco: W. H. Freeman, 1970.

"Ozone: Ozone Balance in the Stratosphere." Earth Observatory, n.d. http://earthobservatory.nasa.gov/Features/Ozone/ozone_2.php (accessed December 2009).

Pahl-Wostl, Claudia. *The Dynamic Nature of Ecosystems: Chaos and Order Entwined.* New York: Wiley, 1995.

"Penetrating Eastern Europe." International Development Research Centre, February 10, 2008. http://www.idrc.ca/en_foco_pobreza/ev-28838-201-1-DO_TOPIC.html (accessed January 2010).

Penman, H. L. "The Water Cycle." In *The Biosphere.* San Francisco: W. H. Freeman, 1970.

"The Pepsi Challenge." Behind the Lines. *Multinational Monitor* 11, no. 6 (June 1990). http://multinationalmonitor.org/hyper/issues/1990/06/lines.html (accessed December 2009).

Petersen, Karl. *Ergodic Theory.* Cambridge: Cambridge University Press, 1983.

Pfanner, Eric. "Gore to Bring Talk of Green to Ad Festival." *New York Times,* June 18, 2007. http://www.nytimes.com/2007/06/18/business/media/18green.html.

Pimm, Stuart. "The Complexity and Stability of Ecosystems." *Nature* 307, January 26, 1984, 321–26.

———. *The World According to Pimm: A Scientist Audits the Earth.* New York: McGraw-Hill, 2001.

"Plastics." Environmental Literacy Council, April 3, 2008. http://www.enviroliteracy.org/article.php/1188.html (accessed December 2009).

Pollan, Michael. *In Defense of Food: An Eater's Manifesto.* New York: Penguin, 2008.

———. *The Omnivore's Dilemma: A Natural History of Four Meals.* New York: Penguin, 2006.

Ponting, Clive. *A Green History of the World*. London: Sinclair-Stevenson, 1991.

"Population Growth Rate." *The World Factbook 2008*. http://bartleby.com/151/fields/27.html (accessed January 2010).

"Population in China." Iran Society of Travel Agents, 2007. http://www.irantour.org/china/populationchina.html (accessed January 2010).

Power, Thomas M. *The Economic Value of the Quality of Life*. Boulder: Westview Press, 1980.

Rabikowska, Marta. "Commercial Consumer TV Advertising in Poland." Faculty of Arts, University of Glasgow, n.d. http://www.arts.gla.ac.uk/Slavonic/Epicentre/Marta%20art.htm (accessed December 2009).

"Rainforest Facts." Raintree Nutrition, Inc., n.d. http://www.rain-tree.com/facts.htm (accessed December 2009).

Rawls, John. *A Theory of Justice*. Cambridge, MA: Belknap Press of Harvard University Press, 1971.

"Reference: Wind Energy Frequently Asked Questions." BWEA, n.d. http://www.bwea.com/ref/faq.html (accessed December 2009).

Reisman, George. *Capitalism: A Treatise on Economics*. Ottawa, IL: Jameson Books, 1996.

Rempel, Gerhard. "Mercantilism." History Today, n.d. http://mars.wnec.edu/~grempel/courses/wc2/lectures/mercantilism.html (accessed January 2010).

Robbins, Richard H. *Global Problems and the Culture of Capitalism*. Boston: Allyn and Bacon, 1999.

Ross, Julia. "Management of Municipal Solid Waste in Hong Kong and Taiwan: A China Environmental Health Project Research Brief." China Environment Forum, March 13, 2008. http://www.wilsoncenter.org/index.cfm?topic_id=1421&fuseaction=topics.item&news_id=397721 (accessed December 2009).

Ryan, Lindsay. "City Hopes New Bicycle Lanes Will Get More People Pedaling." *Washington Post*, September 1, 2005. http://www.washingtonpost.com/wp-dyn/content/article/2005/08/31/AR2005083101029.html.

Sachs, Jeffrey D. "The Social Welfare State: Beyond Ideology." *Scientific American* (November 2006). http://www.scientificamerican.com/article.cfm?id=the-social-welfare-state (accessed January 2010).

Sadowski, Dennis. "Nuncio Says by Living Simply Catholics Can Help Protect the Earth." Catholic News Service, May 7, 2007. http://www.catholicnews.com/data/stories/cns/0702263.htm (accessed January 2010).

Sagan, Carl. "Life." In *The New Encyclopaedia Britannica*, 10:894. 15th ed. Chicago: Encyclopaedia Britannica, 1974.

Sagrillo, Mick. "Putting Wind Power's Effect on Birds in Perspective." Wind Energy Technical Info. American Wind Energy Association, 2003. http://www.awea.org/faq/sagrillo/swbirds.html (accessed May 2010).

Sanders, Skeeter. "Lack of Universal Health Care Puts U.S. in Non-Compliance with International Law." *Seminal,* January 25, 2010. http://seminal .firedoglake.com/diary/26330 (accessed January 2010).

Sayre, Kenneth M. "Cybernetics." In *Philosophy of Science, Logic, and Mathematics in the Twentieth Century,* edited by Stuart Shanker, chap. 6. Vol. 9 of *Routledge History of Philosophy.* New York: Routledge, 1996.

———. *Cybernetics and the Philosophy of Mind.* London: Routledge & Kegan Paul, 1976.

———. "Information Theory." In *Routledge Encyclopedia of Philosophy,* edited by Edward Craig, vol. 4, 782–86. London: Routledge, 1998.

Schmitt, R., et al. "The Chinese Service Industry as a Challenge for European SME: A Systematic Approach for Market Entry." Proceedings of the 1st CIRP Industrial Product-Service Systems (IPS^2) Conference, Cranfield University, April 2009.

Schmitz, David. "Energy 'Consumption' and GDP." The Second Law of Life blog, May 17, 2007. http://secondlawoflife.wordpress.com/2007/05/17/ (accessed December 2009).

Schrödinger, Erwin. *What Is Life? And Other Scientific Essays.* Garden City, NY: Doubleday, 1956.

Schubert, David. "A Different Perspective on GM Food." *Nature Biotechnology* 20, no. 10 (October 2002): 969. http://www.biotech-info.net/different_ perspective.html (accessed January 2010).

Schumacher, E. F. *Small Is Beautiful: Economics As If People Mattered.* New York: Harper & Row, 1973.

Schumpeter, Joseph A. *History of Economic Analysis.* Edited by Elizabeth Boody Schumpeter. New York: Oxford University Press, 1954.

Schurman, Rachel, and Frances Moore Lappé. *The Missing Piece of the Population Puzzle.* Food First Development Report no. 4. Institute for Food and Development Policy, September 1988.

Schwartz, Nelson D. "A History of Public Aid During Crises." *New York Times,* September 6, 2008. http://nytimes.com/2008/09/07/business/07bailout .html.

Schwartz, Richard H. "Abolishing Intensive Livestock Agriculture: A Global Imperative." The Schwartz Collection on Judaism, Vegetarianism, and Animal Rights, n.d. http://www.jewishveg.com/schwartz/livstock.html (accessed January 2010).

"Sea Level Projections." National Research Flagships, November 18, 2008. http://www.cmar.csiro.au/sealevel/sl_proj_21st.html (accessed January 2010).

"Sea Level Rise." Greenpeace USA, n.d. http://www.greenpeace.org/usa/ campaigns/global-warming-and-energy/impacts/sea-level-rise (accessed January 2010).

Sevin, Josh. "Car Facts." *Grist Magazine,* n.d. http://www.journeytoforever.org/biofuel.html#carfacts (accessed January 2010).

Shanker, Stuart, ed. *Philosophy of Science, Logic, and Mathematics in the Twentieth Century.* Vol. 9 of *Routledge History of Philosophy.* New York: Routledge, 1996.

Shannon, Claude. "A Mathematical Theory of Communication." *Bell System Technical Journal* (July/October 1948): 379–423, 623–56.

Shekhar, Manisha. "Environment Does Not Allow Further Economic Growth in the World?" *Ecommerce Journal,* January 30, 2009. http://www.ecommerce-journal.com/articles/12807_environment_does_not_allow_further_economic_growth_in_the_world (accessed December 2009).

Shiva, Vandana. *Stolen Harvest: The Hijacking of the Global Food Supply.* Boston: South End Press, 2000.

"Shutting Down the Oceans." Institute of Science in Society, July 26, 2006. http://www.i-sis.org.uk/AcidOceans.php (accessed December 2009).

Simmons, I. G. *Changing the Face of the Earth: Culture, Environment, History.* Oxford: Basil Blackwell, 1989.

Singer, Peter. *Animal Liberation: A New Ethics for Our Treatment of Animals.* New York: Random House, 1975.

Skeffington, Micheline Sheehy. "Organic Fruit and Vegetable Growing as a National Policy: The Cuban Story." *Energy Bulletin,* February 21, 2006. http://energybulletin.net/node/13067 (accessed January 2010).

Smalls, YaVonda. "Redbox Becomes Red Hot." *South Bend Tribune,* March 4, 2007.

Smil, Vaclav. *Energy in World History.* Boulder: Westview Press, 1994.

———. *General Energetics: Energy in the Biosphere and Civilization.* New York: Wiley, 1991.

Smith, Adam. "History of Astronomy." In *Essays on Philosophical Subjects.* 1795. Reprint, Oxford: Oxford University Press, 1980.

———. *An Inquiry into the Nature and Cause of the Wealth of Nations.* 1776. Reprint, Oxford: Oxford University Press, 1979.

———. *The Theory of Moral Sentiments.* 1759. Reprint, Cambridge: Cambridge University Press, 2002.

Smith, Malcolm. "Environmental Implications of the Automobile." SOE Fact Sheet No. 93-1. Environment Canada, 1993. http://www.ec.gc.ca/soer-ree/English/products/factsheets/93-1.cfm (accessed December 2009).

Snoeks, Jos. "*Lates Niloticus* (Fish)." Global Invasive Species Database, April 13, 2005. http://www.issg.org/database/species/ecology.asp?si=89 (accessed December 2009).

"Snowmobile Accidents, on the Rise in Canada." Bio-Medicine.org, January 27, 2006. http://www.bio-medicine.org/medicine-news/Snowmobile-accidents--on-the-rise-in-Canada-7334-1/ (accessed January 2010).

"Social Origins of Democracy." International Center for Peace and Development, n.d. http://www.icpd.org/democracy/index.htm (accessed December 2009).

Soddy, Frederick. *Wealth, Virtual Wealth, and Debt: The Solution of the Economic Paradox.* 2nd ed. New York: E. P. Dutton, 1933.

Solow, R. M. "Technical Change and the Aggregate Production Function." *Review of Economics and Statistics* 39, no. 3 (1957): 312–20.

Sparling, Brien. "Basic Chemistry of Ozone Depletion." NASA Advanced Supercomputing (NAS) Division. May 1, 2003. http://www.nas.nasa.gov/About/Education/Ozone/chemistry.html (accessed December 2009).

Spevack, Ysanne. "Organic Foods Can Legally Contain GM." Organicfoodee.com, June 15, 2007. http://www.organicfoodee.com/news/2007/06/gmcontamination.html (accessed December 2009).

"Steaz Leadz the Way with Innovative Environmental Initiatives by Supporting NativeEnergy Carbon Offset Programs." Press release. Steaz, November 26, 2008. http://www.libertyrichter.com/steaz.pdf (accessed December 2009).

Stern, David I. "Energy and Economic Growth." *Rensselaer Working Papers in Economics,* April 2003. http://www.sterndavidi.com/Publications/Growth.pdf (accessed January 2010).

Stonington, Joel. "Local Currencies Aren't Small Change." Utne.com, June 29, 2004. http://www.smallisbeautiful.org/publications/essay_stonington.html (accessed December 2009).

Story, Louise, and Eric Dash. "Bankers Reaped Lavish Bonuses During Bailouts." *New York Times,* July 30, 2009. http://www.nytimes.com/2009/07/31/business/31pay.html.

"The Strange History of Economics." Post-Autistic Economics Network, n.d. http://www.paecon.net/StrangeHistory.htm (accessed December 2009).

Sweney, Mark. "Supermarkets Feel Junk Food Ad Ban Bite." *Guardian* (Manchester), November 27, 2006. http://www.guardian.co.uk/media/2006/nov/27/advertising.supermarkets.

Taylor, Jerry, and Peter Van Doren. "Energy." In *The Concise Encyclopedia of Economics* online, 2nd ed., 2007. http://www.econlib.org/library/Enc/Energy.html (accessed December 2009).

Tierney, John. "The Voices in My Head Say 'Buy It!' Why Argue?" *New York Times,* January 16, 2007. http://www.nytimes.com/2007/01/16/science/16tier.html.

Tilford, David. "Biodiversity to Go: The Hidden Costs of Beef Consumption." New American Dream, n.d. http://www.newdream.org/newsletter/beefcost.php (accessed January 2010).

Tocqueville, Alexis de. *Democracy in America.* Edited by J. P. Mayer and Max Lerner. Translated by George Lawrence. New York: Harper and Row, 1966.

"The Tuna Boycott Which Led to the 'Dolphin Safe' Tuna Label." European Cetacean Bycatch Campaign, n.d. http://www.eurocbc.org/page322.html (accessed December 2009).

Tursi, Frank, Susan E. White, and Steve McQuilkin. "Chapter 7: The Ax Falls." In *Lost Empire: The Fall of R. J. Reynolds Tobacco Company. Winston-Salem Journal,* n.d. http://extras.journalnow.com/lostempire/tob7a.htm (accessed December 2009).

Tverberg, Gail E. "Corn-Based Ethanol: Is This a Solution?" *Energy Bulletin,* June 6, 2007. http://www.energybulletin.net/node/30685 (accessed January 2010).

Ulanowicz, Robert E. *Ecology, The Ascendent Perspective.* Complexity in Ecological Systems series. New York: Columbia University Press, 1997.

———. *Growth and Development: Ecosystems Phenomenology.* New York: Springer-Verlag, 1986.

"United States Electricity: Consumption." indexMundi, September 17, 2009. http://www.indexmundi.com/united_states/electricity_consumption .html (accessed January 2010).

"UN Special Rapporteur on Adequate Housing Leaves the U.S., but Not Without Criticism." Press release. United Nations, November 8, 2009. http:// cangress.wordpress.com/2009/11/09/un-special-rapporteur-on-adequate-housing-leaves-the-us-but-not-without-criticism/ (accessed January 2010).

"USA: Cultivations in 2007." GMO Compass, July 11, 2007. http://www .gmo-compass.org/eng/agri_biotechnology/gmo_planting/283.usa_ cultivations_2007.html (accessed January 2010).

"U.S. Keeps Hold on Internet." Associated Press, July 1, 2005. http://www .wired.com/politics/law/news/2005/07/68065 (accessed January 2010).

"U.S. Population, 1790–2000." United States History, n.d. http://www.u-s-history .com/pages/h980.html (accessed January 2010).

Veblen, Thorstein. *The Theory of the Leisure Class.* New York: Macmillan, 1899.

———. "Why Is Economics Not an Evolutionary Science?" *Quarterly Journal of Economics* 12 (1898): 373–97.

Vernadsky, Vladimir I. *The Biosphere.* Translated by David B. Langmuir. Revised by Mark A. S. McMenamin. New York: Copernicus, 1998.

Vitousek, Peter M., et al. "Human Alteration of the Global Nitrogen Cycle: Causes and Consequences." The Ecological Society of America (ESA), n.d. http://www.esa.org/science_resources/issues/TextIssues/issue1.php (accessed January 2010).

Wald, Matthew L. "G.M. Buys Stake in Ethanol Made from Waste." *New York Times,* January 14, 2008. http://www.nytimes.com/2008/01/14/business/ 14gm.html.

Walker, Andrew. "Richest 2% Own 'Half the Wealth.'" BBC News, December 5, 2006. http://news.bbc.co.uk/2/hi/6211250.stm?lsm (accessed January 2010).

"Water Desalination Plants." Physics Forums. January 22, 2006 (accessed March 2010).

Weehuizen, R. "Mental Capital: A Preliminary Study into the Psychological Dimension of Economic Development." The Dutch Council for Health Research, 2006.

Weier, John. "A Delicate Balance: Signs of Change in the Tropics." Earth Observatory, June 19, 2003. http://earthobservatory.nasa.gov/Features/DelicateBalance/ (accessed December 2009).

Weinberg, Steven. *The First Three Minutes: A Modern View of the Origin of the Universe.* New York: Basic Books, 1988.

"What Are BerkShares?" BerkShares, Inc., n.d. http://www.berkshares.org/whatareberkshares.htm (accessed December 2009).

White, Deborah. "Cesar Chavez, Latino Champion of Civil Rights: Fought for Fairness and Human Dignity of Farm Workers." About.com: US Liberal Politics, n.d. http://usliberals.about.com/od/patriotactcivilrights/a/CesarChavez.htm (accessed December 2009).

Wicken, Jeffrey S. *Evolution, Thermodynamics, and Information: Extending the Darwinian Program.* New York: Oxford University Press, 1987.

Williams, Eric Eustace. *Capitalism and Slavery.* Chapel Hill: University of North Carolina Press, 1944.

Wilson, Edward O. *The Diversity of Life.* Questions of Science. Cambridge, MA: Belknap Press of Harvard University Press, 1992.

"Wind Energy." U.S. Hybrid, n.d. http://www.ushybrid.com/windpower.html (accessed January 2010).

"Wind Power." Energy Technology Fact Sheet. United Nations Environmental Programme, n.d. http://www.uneptie.org/energy/information/publications/factsheets/pdf/wind.pdf (accessed January 2010).

Wolf, Martin. "Energy Revolution Will Continue to Power Ahead." *Financial Times* (London), June 27, 2006.

Wolfers, Justin. "What Does the Human Development Index Measure?" *New York Times,* May 22, 2009. http://freakonomics.blogs.nytimes.com/2009/05/22/what-does-the-human-development-index-measure.

Wolff, Emily, and Michael L. Dansinger. "Soft Drinks and Weight Gain: How Strong Is the Link?" *Medscape Journal of Medicine,* August 12, 2008. http://www.ncbi.nlm.nih.gov/pmc/articles/PMC2562148/ (accessed February 2010).

Woodwell, G. M. "The Energy Cycle of the Biosphere." In *The Biosphere,* 26–36. San Francisco: W. H. Freeman, 1970.

"World Air Conditioner Unit (ACU) Market." Litvinchuk HVAC Marketing Agency, n.d. http://www.litvinchuk.ru/en/articles/folder/3.html (accessed January 2010).

World Commission on Environment and Development. *Our Common Future.* Oxford Paperbacks. New York: Oxford University Press, 1987.

"World Consumption of Soft Drinks on Rise." *Food and Drink Weekly,* April 17, 2006. http://findarticles.com/p/articles/mi_m0EUY/is_16_12/ai_ n16133094/ (accessed January 2010).

"The World Health Organization's Ranking of the World's Health Systems." Photius.com, February 29, 2007. http://photius.com/rankings/healthranks .html (accessed December 2009).

Wright, D. H. "Species-Energy Theory: An Extension of Species-Area Theory." *Oikos* 41 (1983): 498–506.

Wyer, Kathy. "Political Engagement among College Freshmen Hits 40-Year High: Annual Survey Also Shows Money Concerns Drive Students' College Choice." UCLA Newsroom, January 22, 2009. http://newsroom.ucla .edu/portal/ucla/political-engagement-of-college-78404.aspx   (accessed December 2009).

Xenophon. *The Oeconomicus of Xenophon.* Trans. Hubert A. Holden. London: Macmillan, 1889.

"Yearning for Balance." *YES! A Journal of Positive Futures* (Spring/Summer 1996).   http://www.sdearthtimes.com/et0996/et0996s1.html   (accessed January 2010).

Young, Louise B., ed. *The Mystery of Matter.* New York: Oxford University Press, 1965.

Index

Kenneth M. Sayre

is professor of philosophy and director of the Philosophic Institute at the University of Notre Dame. He is the author of numerous books ranging in topic from Plato to cybernetics to public values. His books include *Values in the Electric Power Industry* (1977), *Plato's Literary Garden* (1995), and *Parmenides' Lesson* (1997), all published by the University of Notre Dame Press.

CPSIA information can be obtained
at www.ICGtesting.com
Printed in the USA
LVHW081258151019
634258LV00013B/356/P